Light, Lasers, and Synchrotron Radiation

A Health Risk Assessment

NATO ASI Series

Advanced Science Institutes Series

A series presenting the results of activities sponsored by the NATO Science Committee, which aims at the dissemination of advanced scientific and technological knowledge, with a view to strengthening links between scientific communities.

The series is published by an international board of publishers in conjunction with the NATO Scientific Affairs Division

A	**Life Sciences**	Plenum Publishing Corporation
B	**Physics**	New York and London
C	**Mathematical and Physical Sciences**	Kluwer Academic Publishers Dordrecht, Boston, and London
D	**Behavioral and Social Sciences**	
E	**Applied Sciences**	
F	**Computer and Systems Sciences**	Springer-Verlag
G	**Ecological Sciences**	Berlin, Heidelberg, New York, London,
H	**Cell Biology**	Paris, and Tokyo

Recent Volumes in this Series

Volume 238—Physics, Geometry, and Topology
edited by H. C. Lee

Volume 239—Kinetics of Ordering and Growth at Surfaces
edited by Max G. Lagally

Volume 240—Global Climate and Ecosystem Change
edited by Gordon J. MacDonald and Luigi Sertorio

Volume 241—Applied Laser Spectroscopy
edited by Wolfgang Demtröder and Massimo Inguscio

Volume 242—Light, Lasers, and Synchrotron Radiation: A Health Risk Assessment
edited by M. Grandolfo, A. Rindi, and D. H. Sliney

Volume 243—Davydov's Soliton Revisited: Self-Trapping of Vibrational Energy in Protein
edited by Peter L. Christiansen and Alwyn C. Scott

Volume 244—Nonlinear Wave Processes in Excitable Media
edited by Arun V. Holden, Mario Markus, and Hans G. Othmer

Volume 245—Differential Geometric Methods in Theoretical Physics: Physics and Geometry
edited by Ling-Lie Chau and Werner Nahm

Series B: Physics

Light, Lasers, and Synchrotron Radiation

A Health Risk Assessment

Edited by

M. Grandolfo

National Institute of Health
Rome, Italy

A. Rindi

Sincrotrone Trieste
Trieste, Italy

and

D. H. Sliney

U.S. Army Environmental Hygiene Agency
Aberdeen Proving Ground, Maryland

Springer Science+Business Media, LLC

Proceedings of a NATO Advanced Study Institute/Ninth Course of the
International School of Radiation Damage and Protection on
Optical Sources, Lasers and Synchrotron Radiation:
Biological Effects and Hazard Potential,
held May 9-20, 1989,
in Erice, Sicily, Italy

Library of Congress Cataloging-in-Publication Data

NATO Advanced Study Institute/Ninth Course of the International School
of Radiation Damage and Protection on Optical Sources, Lasers, and
Synchrotron Radiation: Biological Effects and Hazard Potential (1989
: Erice, Italy)
Light, lasers, and synchrotron radiation : a health risk
assessment / edited by M. Grandolfo, A. Rindi, and D.H. Sliney.
p. cm. -- (NATO ASI series. Series B, Physics ; v. 242)
"Published in cooperation with NATO Scientific Affairs Division."
Includes bibliographical references and index.
ISBN 978-1-4899-0663-2

1. Light--Health aspects--Congresses. 2. Laser beams--Health
aspects--Congresses. 3. Synchrotron radiation--Health aspects-
-Congresses. 4. Health risk assessment--Congresses. I. Grandolfo,
M. II. Rindi, Alessandro. III. Sliney, David H. IV. Title.
V. Series.
RA569.3.N38 1991
612'.01448--dc20 90-23013
CIP

ISBN 978-1-4899-0663-2 ISBN 978-1-4899-0661-8 (eBook)
DOI 10.1007/978-1-4899-0661-8

FOREWORD

This volume contains the formal record of the lectures presented at the 9th Course of the International School of Radiation Damage and Protection held at the "E.Majorana" International Centre for Scientific Culture in Erice (Italy) from May 9 to May 20, 1989.

This course was the last of a series of 4 courses, started in 1981, that were dedicated to the assessment of risk hazard from non-ionizing radiation. The proceedings of these courses were all published by Plenum Press with the following headings:

1) M.Grandolfo, S.M.Michaelson and A.Rindi, Eds.: "Biological Effects and Dosimetry of Nonionizing Radiation; Radiofrequency and Microwave Energy", Plenum Press, New York, NATO ASI Series A Life Sciences, Vol.49 (1983);
2) M.Grandolfo, S.M.Michaelson and A.Rindi, Eds.: "Biological Effects and Dosimetry of Static and ELF Electromagnetic Fields", Plenum Press, New York, E.Majorana International Science Series, Life Sciences, Vol.19 (1985);
3) M.H.Repacholi, M.Grandolfo and A.Rindi, Eds.: "Ultrasound; medical applications, biological effects and hazard potential", Plenum Press, New York (1987).

We hope that all these volumes together may represent a complete textbook and a reference for the students and scientists interested in the physics, biology, measurement and dosimetry, health effects and standard setting, in short, the risk assessment of that wide field of radiation presently classified as non-ionizing radiation.

We are indebted to the Associazione Italiana Protezione dalle Radiazioni (AIRP), The International Non Ionizing Radiation Committee of the International Radiation Protection Association (INIRC/IRPA), the Istituto Superiore di Sanita', the Italian Ministry of Scientific and Technological Research, the Sincrotrone Trieste, and the Sicilian Regional Government for sponsoring and supporting the Course.

We acknowledge with great appreciation the cooperation and skilled assistance of Mrs.Franca Grisanti and Mr.Giacomo Monteleone of the Physics Laboratory of the Italian National Institute of Health, Rome, Italy.

We take the opportunity to thank all the lecturers and students that, by participating to the courses from all around the world, made possible this endeavor.

M.Grandolfo
A.Rindi
D.H.Sliney

CONTENTS

THE SOURCES AND THE INTERACTION WITH MATTER

Fundamentals of Optical Physics 1
M. Grandolfo

Characteristics of Optical Noncoherent Sources 15
J. Bernhardt

Lamps and Lighting Systems 25
A.F. McKinlay

Characteristics of Synchrotron Radiation 61
E. Burattini

Characteristics of Coherent Optical Sources 77
F. Hillenkamp and W.J. Marshall

Radiation Interactions with Matter: Some General Considerations on the Ionizing/Non-Ionizing Boundary Region 91
M. Fabretto and A. Rindi

THE DETECTION

Radiometric Measurements for Broadband Optical Sources 121
J.H. Bernhardt

Laser Beam Diagnostics 133
W.J. Marshall

Laser Radiation Detectors 141
J. Franks

Quantitative Determination of Ultraviolet Radiation 147
W.J. Marshall

Measurements of Welding Arcs and Plasma Arcs 157
P. Eriksen

Ultraviolet Exposure in the Outdoor Environment: Measurements of Ambient Ultraviolet Exposure Levels at Large Zenith Angles 169
D.H. Sliney, R.L. Wood, Jr., P.M. Moscato, W.J. Marshall, and P. Eriksen

THE BIOLOGICAL EFFECTS

Anatomy and Physiology of the Eye.............................. 181
A. Checcucci

Anatomy and Physiology of the Skin.............................. 185
A. Checcucci

Biological Effects of Coherent and Noncoherent IR Radiation.. 191
L.A. Court, and D. Courant

Infrared Radiation and the Eye.............................. 219
P.G. Soderberg

Infrared Energy and Skin: Biological Effects and Health Hazards.. 229
A. Checcucci

Potential Ocular and Skin Hazards from Lamps and Projector Systems.. 233
J. Franks

Ultraviolet Radiation and the Eye.............................. 237
D.H. Sliney

Ultraviolet Mutagenesis.............................. 247
I. Quinto, M. Mallardo, M.R. Ruocco, A. Arcucci, and G. Scala

RISK ASSESSMENT, PROTECTION AND STANDARDS

Laser Output Parameters Necessary for Hazard Assessment.......... 259
J. Franks

Control of Outdoor Laser Hazards.............................. 263
J. Franks

Protective Screens and Filters for IR Protection............... 269
P. Eriksen

Welding Safety Measures and Welding Filters.................... 277
P.G. Soderberg

Laser Eye Protectors.............................. 287
P.K. Galoff

Visible and Infra-Red Radiation (Non Laser) Protection Standards.. 293
A.F. McKinlay

Ultraviolet Standards.............................. 299
B.F.M. Bosnjakovic

IRPA/INIRC Guidelines on Ultraviolet.............................. 319
J.H. Bernhardt

Laser Safety Standards: Historical Development and Rationale.. 329
D.H. Sliney

The IRPA/INIRC Guidelines on Limits of Exposure to Laser Radiation 341
D.H. Sliney

Medical Surveillance in Laser Working Sites 347
P.G. Soderberg and D.H. Sliney

APPLICATIONS AND RELATED TOPICS

Medical Lasers and Biological Criteria and Limits of Their Therapeutic Effects 353
L.A. Court and D. Courant

Current and Future Applications of Lasers in Medicine 373
M.H. Repacholi, G. McLennan, A. Pugatschew, and R. Hancock

Towards a Solution of the Ozone Layer Depletion Problem 395
B.F.M. Bosnjakovic

Participants 416

Index 419

FUNDAMENTALS OF OPTICAL PHYSICS

Martino Grandolfo

National Institute of Health, Physics Laboratory
and INFN-Sezione Sanita', Rome, Italy

INTRODUCTION

In the electromagnetic spectrum the wavelengths covered by optical radiation range from 1 nm to 1 mm. This wavelength region includes not only the visible part of the electromagnetic spectrum, but also the ultraviolet (UV) - up to the soft ionizing X-ray domain - and the infrared (IR) - up to the microwave domain. The region from 1 nm to 190 nm (vacuum UV) is completely absorbed in air and consequently has no direct biological effect.

There are quite a few reasons for treating this wavelength region as a separate entity, though any exact boundary is to a certain degree arbitrary. Optical radiation is produced by several radiation sources, such as conventional incandescent, fluorescent and phosphorescent lamps, electric arcs, and the laser. It is this last source that has engendered the greatest concern from the point of view of radiation protection.

As indicated, the boundaries of the optical radiation region cannot be precisely defined. Spectral designation schemes differ; the broad spectral regions defined by physicists, such as "near" or "far" IR are useful in discussing sources, whereas the CIE bands, such as UV-A or IR-B, relate to biological effects only. Both schemes are used in this book.

In contrast to X-rays, optical radiation is essentially nonionizing. Its action is either photochemical (as in the UV) or thermal (as in the IR); the visible region is a transition region characterized by both effects.

In contrast to radiofrequency and microwave radiation, optical radiation usually acts at the surface. Penetration in the skin is mostly restricted to a few millimetres or less. The eye is an exception in that it admits visible energy into the body. Even in this case the penetration rarely goes beyond the retinal pigment epithelium.

Until this century, the principal source of optical radiation was the sun, but solar radiation could not be considered very dangerous, primarily because of protective avoidance reactions naturally built into the organism, and the development of adaptive pigmentation. The development of artificial radiation sources, however, has made the protection problem more urgent. In the early 1920s the first health protection standards were laid down to protect against overexposure to the UV radiation and visible light produced by welding arcs, and against

Light, Lasers, and Synchrotron Radiation
Edited by M. Grandolfo *et al.*, Plenum Press, New York, 1990

the near-IR radiation related to the so-called "glassblower's cataract". When Meyer-Schwickerath in 1954 produced retinal lesions with a carbon arc light source, it became impossible to avoid the conclusion that even commonly used equipment can be hazardous. Although hazards and protective procedures relating to lasers will be emphasized in subsequent parts of this book, most of the material is also applicable to non-laser optical sources.

The material in this paper is provided as an introduction to those aspects of optics which play an important role in the subject of optical radiation safety. It may serve as a useful introductory presentation of the subject for the individual who has not recently worked in optics or optical engineering, but has responsibilities for safety of laser equipment and high intensity light sources.

HISTORICAL DEVELOPMENT

In the development of electromagnetic theory James Clerk Maxwell accepted the Faraday's idea that the electric and magnetic fields were the fundamental quantities and was able to devise a set of partial differential equations obeyed by these fields. He immediately noted that the equations predicted that an electromagnetic disturbance originating at one charged object would travel out as a wave with a speed that could be deduced from electrical and magnetic measurements. The predicted speed, within the small experimental error of measurement, agreed with the speed of light. Maxwell's contribution to physics was very significant. He had provided convincing evidence of the superiority of a field theory to an action-at-a-distance theory and also established the link between electromagnetism and optics. Neither the revolution of quantum mechanics nor the advent of special theory of relativity has in any way reduced the significance of the electromagnetic field equations Maxwell proposed over one hundred years ago.

During the early period of the development of optics there were two diametrically opposed theories of light. Isaac Newton on the one hand believed in a corpuscular theory in which light consisted of a stream of infinitesimal particles; Christian Huygens held the view that light was a type of wave motion. Newton objected to the wave theory on the basis that it could not explain the sharp shadows cast by objects illuminated with visible light. He failed to realize that, if light waves were of sufficiently short wavelength, the observation of sharp shadows presented no contradiction.

Early in the nineteenth century the work of Thomas Young and Augustin Fresnel provided the explanation of the interference and diffraction of light in terms of a wave theory. Fresnel also deduced laws of reflection and refraction which were subsequently shown to predict correctly the fraction of incident light in the reflected and refracted beams as well as the directions of these beams.

Physicists of the nineteenth century tried to understand a wave model of light in terms of the transverse vibrations of a hypothetical elastic medium which they named the ether. However, the properties required of the ether were unrealistic. This was the situation when Maxwell formulated his equations. In terms of these equations the electric and magnetic fields are the fundamental entities concerned with optics and electromagnetism and there is no need to introduce the ether at all. Furthermore, Maxwell's equations yield only transverse vibrations whereas an elastic-solid medium transmits longitudinal as well as transverse waves.

A few years later Heinrich Hertz produced electromagnetic waves of several centimeters wavelength in the laboratory. From this time Maxwell's equations have received universal acceptance, and optics has been treated as a branch of electromagnetism.

Early in the twentieth century Max Planck introduced the quantum concept and shortly afterward Albert Einstein predicted the existence of the photon and used Planck's quantum hypothesis to deduce the law of the photoelectric effect. Whereas a wave theory predicts a continuous distribution of energy in a wave, the existence of photons requires that the energy in an electromagnetic wave be quantized in discrete packets as for a corpuscular theory. For some time this apparent contradiction provided a serious impediment to the development of quantum theory. It is now recognized that both the corpuscular and wave points of view simultaneously have their merits - the wave picture correctly describes the average behavior of the photons but the energy is carried by the photons in discrete amounts. For long-wavelength radiation, such as radiofrequency radiation, the photon energy is small and ordinary sources provide such enormous numbers of photons that their discreteness is not of practical importance. On the other hand, for short wavelength radiation, such as gamma-ray radiation, the photon energy is large and the discreteness is of striking significance. The incorporation of wave-particle duality into a single theory of electromagnetism falls into the realm of quantumelectrodynamics, invented by Paul Dirac. Later significant developments of the theory were pioneered by Richard Feynman, Julian Schwinger, and Shin-Ichiro Tomonaga.

A second important development of electromagnetic theory in the twentieth century has been its role in the quantum theory of atomic and molecular structure. Niels Bohr proposed that the hydrogen atom consists of a proton and an electron which interact with each other through the electrostatic interaction but whose motion is governed by quantum mechanics. Louis de Broglie suggested that a wave phenomenon was associated with the motion of electrons. Erwin Schroedinger described this wave by a wave equation. The solutions of his equation provide accurate descriptions of the structure of atoms, molecules, and bulk matter. The forces that enter Schroedinger's theory are well described by classical electromagnetic theory; it is in the laws of mechanics that this theory differs from the classical theory. Other important contributions to quantum mechanics and the atomic theory have been made by Werner Heisenberg, Paul Dirac, Wolfgang Pauli, Max Born, and Eugene Wigner.

The development of optical masers or lasers in the 1960's gave new life to the field of optics. The invention of the maser and the formulation of the theory of coherent atomic radiation was carried through by Charles Townes, Nikolai Basov, and Aleksandr Prokharov.

LIGHT AS A WAVE PHENOMENON

The theory of wave motion forms a descriptive system appropriate, with minor modifications, to a wide range of observations as sounds, light and radio signal. The same system of equations may be used because the different groups of phenomena have many properties in common and are these common properties which are described by the wave equation.

It is possible to show that the expression for a transverse sinusoidal traveling wave is the solution of the wave equation:

$$\nabla^2 s = \partial^2 s/\partial x^2 + \partial^2 s/\partial y^2 + \partial^2 s/\partial z^2 = 1/v^2 \, \partial^2 s/\partial t^2 \qquad (1)$$

where $s = s(x,y,z,t)$ is the displacement at position (x,y,z) at time t.

Applying vector algebra to Maxwell's equations it is possible to show that, for the electric field:

$$\nabla^2 E = \mu_0 \varepsilon_0 \partial^2 E/\partial t^2 \tag{2}$$

and a similar calculation leads to the equation:

$$\nabla^2 B = \mu_0 \varepsilon_0 \partial^2 B/\partial t^2 \tag{3}$$

for the magnetic induction.

These equations for E and B both have the form of the wave equation (1) and we can conclude immediately that there exist plane, transverse electromagnetic waves of speed:

$$v = 1/\sqrt{(\mu_0 \varepsilon_0)} \tag{4}$$

Since the frequency or wavelength of the wave does not appear in the wave equation, we conclude that all electromagnetic waves regardless of frequency have this same speed. Using the well-known values of ε_0 (8.85×10^{-12} F/m^2) and μ_0 ($4\pi \times 10^{-7}$ H/m) we obtain $v = 2.998 \times 10^8$ m/s. The speed of light in vacuum is 2.9979×10^8 m/s which is the same as the predicted speed of electromagnetic waves within the accuracy of this calculation. A similar calculation prompted Maxwell to declare that light is a special case of propagation of electromagnetic waves.

Obvious solutions to the wave equations for E and B are the following:

$$E = E_0 \sin 2\pi(\nu t - x/\lambda) \tag{5a}$$
$$B = B_0 \sin 2\pi(\nu t - x/\lambda) \tag{5b}$$

where ν is the frequency of oscillation, and λ is the wavelength. The speed of the waves is:

$$v = \lambda \nu \tag{6}$$

and B_0 and E_0 are constant vectors.

THE PARTICLE PICTURE OF RADIATION

In the early years of this century it became apparent that some of the interactions of light with matter could be explained only by assuming that the energy carried by a beam of light of frequency ν is concentrated into discrete packages, called photons, of energy $h\nu$. We now accept the view that the energy carried by any electromagnetic radiation is localized in photons whose paths through space are governed by a wave equation. In many of the interactions of electromagnetic radiation with matter we can ignore the quantum character of the radiation and treat the radiation as a classical electromagnetic wave. On the other hand, there are many interactions in which the quantum nature of the radiation must be taken into account. This "wave-particle duality" causes no difficulty so long as we realize that the photon theory merely adds quantum properties to the already familiar wave properties of electromagnetic radiation. According to the "correspondance principle" we can always apply the classical theory in situations involving large numbers of photons. No confusion arises so long as we do not attempt to apply wave and quantum concepts simultaneously.

As an example, a photon from a ruby laser which has a wavelength of approximately 0.7 µm, and hence from Eq.(6) a frequency of 4.3×10^{14} Hz, would have an energy of 2.9×10^{-19} J, being the Planck's constant $h = 6.6 \times 10^{-34}$ Js. This also means that a ruby laser which emits a pulse of energy of 0.29 J actually has 10^{18} photons in that one burst of radiant energy. A single photon is obviously a very small quantity of energy, yet several studies have shown that the human eye is capable of detecting a few photons of light.

Although quantum field theory has unified those conflicting pictures of the nature of radiation to some degree, the elementary level still presents the complementary wave-particle picture. Our choice of an appropriate mode of description will depend upon the particular experimental situation. In general, the wave picture is most useful for low-energy, long-wavelength phenomena. Both quantum and wave concepts are used in the optical region of the spectrum.

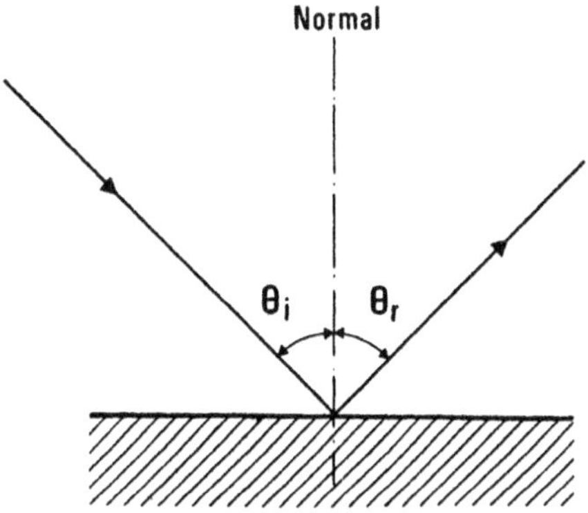

Fig. 1 In the specular reflection the angle of reflectance equals the angle of incidence

INTERACTION OF OPTICAL RADIATION AT AN INTERFACE

From a macroscopic point of view, the interaction of electromagnetic radiation with matter takes the form of absorption, transmission, reflection, refraction, and diffraction. In most instances one of these effects dominates almost to the exclusion of others. However, all effects are always present to some degree.

Reflection

Reflection takes place at an interface. There are two basic types of reflections that are of interest to us - specular and diffuse. When light undergoes specular reflection the angle of incidence is observed to be equal to the angle of reflection, as indicated in Fig. 1

Specular reflection can occur when the size of surface irregularities or roughness is less than the wavelength of the incident radiation. This latter description of specular reflection is important to keep in mind.

Diffuse reflection occurs when the surface irregularities are randomly oriented and are much greater than the wavelength of the incident radiation; for example, when light is reflected from chalk, a black iron skillet, or a rough granite surface. Most naturally-occurring surfaces are diffusely reflecting in the visible light region of the spectrum. Diffuse reflection is shown in Fig. 2.

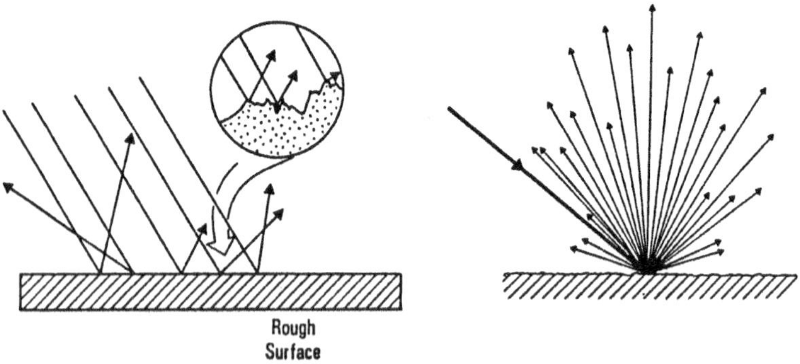

Fig. 2 Light reflected from a diffuse surface is redirected in a random fashion as a result of surface roughness

It is important to remember that diffuse and specular reflections are highly wavelength dependent. A given surface may produce a reflection which is specular at on wavelength but it may or may not be specular at a greatly different wavelength. As an example, microwave radiation is reflected from a black cast iron skillet as a mirror-like reflection but light is completely diffused by such a surface. Obviously the size of the irregularities in the reflecting surface must be considerably greater than the wavelength of the incident radiation. If the size of the reflecting surface irregularities is of the order of the wavelength then the phenomena of diffraction and scattering take place.

The reflection, R, and transmission, T, coefficients for oblique incidence of plane polarized light (the Fresnel formulas) follow directly from the wave equation. For light with the electric vector E vibrating in the plane of incidence (the plane formed by the propagation axis of an incident wave and a normal to the interface) the reflection coefficient is:

$$R_{\|} = \tan^2(\theta-\theta')/\tan^2(\theta+\theta') \qquad (7)$$

with $T_{\|} = 1 - R_{\|}$. For plane waves with the electric vector perpendicular to the plane of incidence, where $\theta' = \arcsin(\theta/n)$:

$$R_{\perp} = \sin^2(\Theta-\Theta')/\sin^2(\Theta+\Theta') \quad (8)$$

and $T_{\perp}$ and $R_{\perp}$ are equal to 1. For unpolarized light, an averaging of these results leads to:

$$R = 0.5R_{\perp} + 0.5R_{\parallel} \quad (9)$$

while T is (1-R). The Fresnel formulas are very useful in the analysis of specular reflection hazards.

At Brewster's angle:

$$\Theta_B = \tan^{-1}(n/n_0) \quad (10)$$

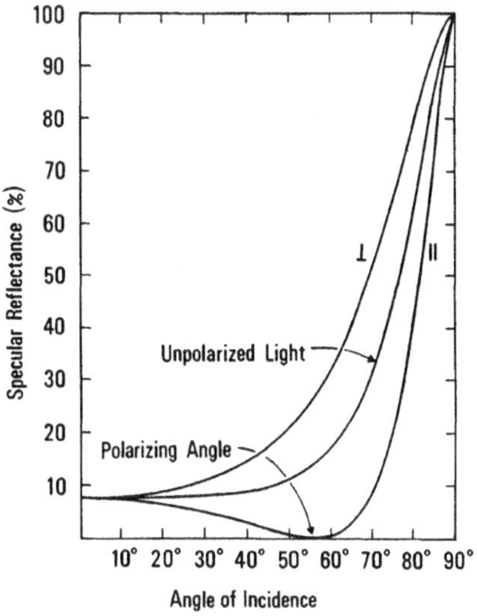

Fig. 3 The specular reflectance for plate glass depends upon the polarization of the incident light and the angle of incidence

No light which is plane-polarized in the plane of incidence can be reflected. This effect is shown in Fig. 3, where it is seen that the specular reflectance of glass vanishes at approximately 56° when n is 1.5. For water, with an index of refraction n of 1.33, Θ_B is 53°.

Refraction

Refraction also takes place at an interface. Refraction occurs whenever a beam of light passes from one transmitting medium to another having a different refractive index n. It is responsible for a bending of light near air-water and air-glass interfaces. The law of refraction, which is also known as Snell's Law states that the angle of incidence (Θ_1) and the angle of refraction (Θ_2) are related by the equation:

$$\sin\Theta_1/\sin\Theta_2 = n_2/n_1 \qquad (11)$$

where n_1 and n_2 are the indices of refraction of the first medium and second medium respectively.

INTERACTION OF OPTICAL RADIATION WITH A MEDIUM

Transmission

The transmitted light which emerges from a medium is dependent upon the phenomena of absorption and scattering and also upon the reflection of some of the light at the interfaces between media. The transmittance of a medium is usually represented by T and is for a certain wavelength and for a certain path length. The transmittance of most materials varies markedly across the optical spectrum. By the Law of Conservation of Energy the transmitted beam energy E_T is the incident beam energy E_I less that lost due to absorption E_A and reflection E_R:

$$E_T = E_I - E_R - E_A \qquad (12)$$

Absorption

Photons are removed from a beam of light by collisions that produce either excited states or free electrons and holes. If a beam of initial intensity I_0 is incident upon some absorbing medium, the intensity I of the beam after traversing a thickness x of the medium is observed to satisfy the relation:

$$I = I_0\exp(-\mu x) \qquad (13)$$

where μ, the optical absorption coefficient, is frequency dependent. This expression may be derived as follows (see Fig. 4). Light of initial intensity I_0 and frequency ν is absorbed in passing through a medium. The intensity I of light incident on a slab of thickness dx and perpendicular to the incident light beam is reduced by an amount dI on passing through the slab. The number of photons n per unit volume in the beam is proportional to the intensity of the beam. Assuming that for a homogeneous medium the probability of absorption of a specific photon in the slab is independent of the number of photons present, it follows that the fractional change dI/I in the intensity is independent of the position x in the medium and proportional only to the thickness dx causing the absorption. Therefore;

$$dI/I = -\mu dx \qquad (14)$$

where, for a given frequency, μ is a constant characteristic of the medium.

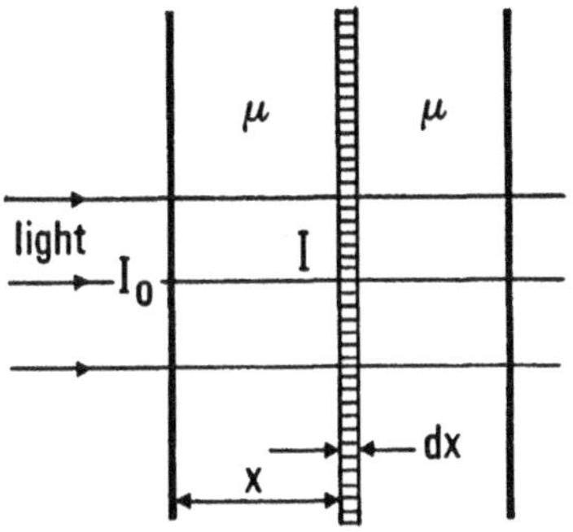

Fig. 4 The intensity I of light incident on a slab of thickness dx is reduced by an amount dI on passing through the slab

The minus sign is used since the intensity is reduced by the absorption. The intensity I of the beam after it has penetrated a thickness x of the medium is given by:

$$\int_{I_0}^{I} dI/I = -\int_{0}^{x} \mu dx \tag{15}$$

or:

$$\ln(I/I_0) = -\mu x \tag{16}$$

or:

$$I = I_0 \exp(-\mu x) \tag{17}$$

The absorption coefficient μ is dependent upon the frequency (or energy) of the photons in the beam. There are usually several processes that contribute to absorption at a given frequency so that the absorption coefficient is really a sum of several coefficients, each referring to a specific absorption process.

Absorption is a maximum at a certain natural frequencies of the atoms (or molecules). These natural frequencies correspond to the energy differences between certain of the allowed states. If a photon has an energy equal to the difference in energy between two appropriate quantum states, the photon may be absorbed, leaving the atom (or molecule) in an excited state. Not all possible excitations are allowed; only those satisfying the selection rules of quantum mechanics are permitted.

The electrons act as forced electric dipoles under the influence of the electric field in the wave. An electron so excited may immediately return to the ground state by reemitting the energy just absorbed. Alternatively, the energy may be reemitted at different frequencies if the excited electron passes through intermediate energy states during its return to the ground state. In either case, the direction of emission is random so that the intensity of the transmitted wave is reduced by this scattering process (see Fig. 5).

Scattering is most pronounced at those frequencies corresponding to the emission frequencies of the atom or molecule as just described. However, scattering may still be significant at frequencies other than those of the emission spectrum. This occurs because any system will respond to an external driving force even if the frequency of the external force does not correspond to a natural frequency of the system. The response to the forced oscillations, however, is very small far from a natural resonance, but it increases as the frequency of the external force approaches a resonance frequency. The blue color of the sky is due to such an effect. Molecules that make up the atmosphere have emission frequencies in the ultraviolet region. Therefore, their response to an electromagnetic wave decreases as the frequency of the wave decreases. The response to waves with frequencies in the blue region is greater than the response to waves with lower frequencies and the scattered radiation is predominantly blue. It can be shown that the scattered intensity is proportional to the fourth power of the driving frequency (Rayleigh's law).

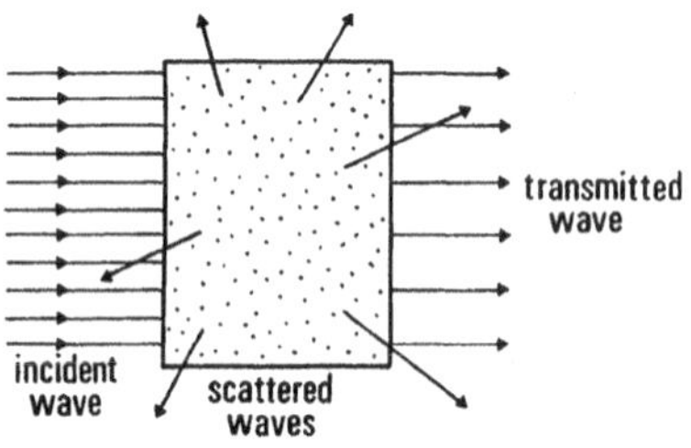

Fig. 5 In the scattering process the direction of emission israndom and the intensity of the transmitted wave is reduced

Small particles whose size approaches that of a wavelenght of light, λ, scatter light as well as atoms and molecules and, if the particles are much smaller than λ (e.g., gas molecules) Rayleigh scattering takes place.

If light goes through clouds of dust or condensed water vapor where the particle or droplet size is of the order of the wavelength of light or greater this strong wavelength preference is not seen in the scattered light. Clouds usually appear white or grey. This type of large-particle scattering is termed Mie scattering. Unlike Rayleigh scattering, Mie scattering is strongly directional. Normally, the forward component of Mie scattered radiation is much greater than backscatter.

The phenomena of scattering must be considered not only in terms of the atmospheric propagation of laser beams, but also in reference to the light scattered in the ocular media of the human eye. The scattering of light in a homogeneous medium from a light beam can also be expressed in terms of an exponential function:

$$I = I_0 \exp(-\sigma x) \tag{18}$$

where σ is the scattering coefficient ($\sigma_m + \sigma_r$) and is the sum of the Mie scattering coefficient σ_m and the Rayleigh scattering coefficient σ_r.

INTERFERENCE AND DIFFRACTION EFFECTS

Interference

If light does consist of waves, we should be able to observe interference phenomena, such as standing waves, beats, and the cancellation created by two waves of equal intensity but apposite phase.

Standing light waves were actually observed by Wiener in 1890. The earliest experimental observation that gave clear support to a wave theory of light involved interference effects and was made by Thomas Young in 1801. The experiment is sketched in Fig. 6. Light from the sun illuminated a single pinhole, which produced a single source of light waves in a darkened room. Two subsequent pinholes produced two sets of light waves that were guaranteed to have well-defined phases with respect to each other, since they were merely "samples" from a common wave front. When these light waves then overlapped, they were in phase in some regions and out of phase in other regions, thus producing light and dark zones on the screen. This effect would not have been observed if light were a stream of small particles.

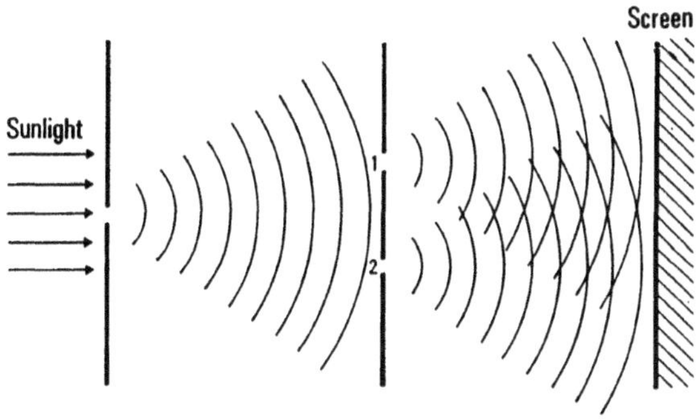

Fig. 6 Young's experiment demonstrated the interference of light waves from pinhole 1 with those from pinhole 2 by letting them combine on a screen

Observation of these interference effects by Young signaled the failure of Newton's corpuscolar theory of light and the acceptance of Huygens' wave theory. From these observations alone there was no way of deducing the type of waves constituting light, but it was clear that waves were involved.

One of the simplest experiment illustrating interference involves the use of two parallel slits with widths on the order of a few λ at a distance d apart. If θ is the angular distance on the screen measured in a plane normal to the slit axes, it is easy to show that:

$$d \sin\theta = n\lambda \qquad (19)$$

where n being a discrete integer (i.e., n = 0,1,2,3,...) describes the location of interference maxima.

The way that light waves propagate is most easily predicted by application of Huygens' principle.

When Huygens, in the seventeenth century, proposed that light is a wave phenomenon, he was able to explain the then known behavior of light by the use of a simple guess as to how these waves propagate. His guess is now called Huygens' principle: Every point of an advancing wave front can be considered as a point source of spherical secondary waves. Later positions of this wave front are given by the envelope of these secondary waves.

Huygens' principle seems a bit arbitrary - especially since we are asked to ignore the backward portion of spherical secondary waves. Indeed, today we know that Huygens' principle is not a fundamental principle of physics; instead, it was an intuitive guess by Huygens that proved to be extremely useful. In general, the motion of a wave front can be rigorously computed in any situation by solving the three-dimensional wave equation that describes the waves. Though this direct method of attack is quite difficult, Kirchhoff was able to show that application of Huygens' principle resulted from a reasonably accurate solution of the wave equation. This proof is known as Kirchhoff's theore and allows us to use simple Huygens' constructions to solve many problems in wave motion.

Diffraction

When a wave impinges on a very small opening in a screen, it spreads in all directions beyond the screen. The spreading of waves beyond an obstacle is called diffraction. If light traveled only in straight lines, all obstacles would cast sharp-edged shadows. The fact that light spreads beyond an obstacle can lead to interference effects in these "shadow" regions.

In general, whenever a wave front passes a sharp-edged obstacle, interference effects are visible beyond the obstacle. We are normally not aware of these diffraction effects because the incident light rays must be parallel to reveal these features. Light from normal, extended sources produces a penumbra region that obscures these interference features. The diffraction of light places a definite upper limit on the useful magnification of a microscope or telescope.

Huygens' principle can be applied to predict the locations of the regions of constructive and destructive interference in single-slit diffraction. In Fig. 7, secondary waves are shown emanating from two particular points separated by w/2, half the width of the slit. At an angle θ such that:

$$w/2 \sin\theta = \Delta l = \lambda/2 \qquad (20)$$

the distance Δl shown in Fig.8 will be one-half wavelenght. These two secondary waves will thus combine out of phase on a distant screen. Every point in the upper half of the slit opening can be paired with another point in the lower half of the slit in a similar fashion; thus at this particular angle, every such pair cancels. We conclude, therefore, that perfect cancellation results and the screen is dark at this angle. In fact, dividing the slit into any even number of segments will allow us similarly to pair off points of the slit for which the light waves will be one-half wavelength out of phase and thus lead to darkness on the screen. The angles at which darkness occurs are then given by:

$$\sin\Theta = n\lambda/w \qquad (n = 1,2,3,....) \qquad (21)$$

However, if the slit is subdivided into an odd number of segments, the above argument leading to cancellation can be made for all but the last segment. Points in this last odd segment cannot be cancelled off with other points, since all the points in the remaining even number of segments have already been paired. We expect, then, that at angles given by:

$$\sin\Theta = m\lambda/2w \qquad (m = 3,5,7.....) \qquad (22)$$

we will find light on the distant screen, but each successive maximum is less bright, since the uncancelled odd segment of the screen contains a smaller fraction of the total light as m increases. This argument breaks down for m = 1 since in that case there are no remaining segments of the screen to be paired. Thus the central maximum in a single-slit interference pattern is twice the width of all others. Furthermore, since all secondary wavelets are exactly in phase at $\Theta = 0^o$, this central maximum is far brighter than the others. The bulk of the light passing through the slit falls into the central maximum.

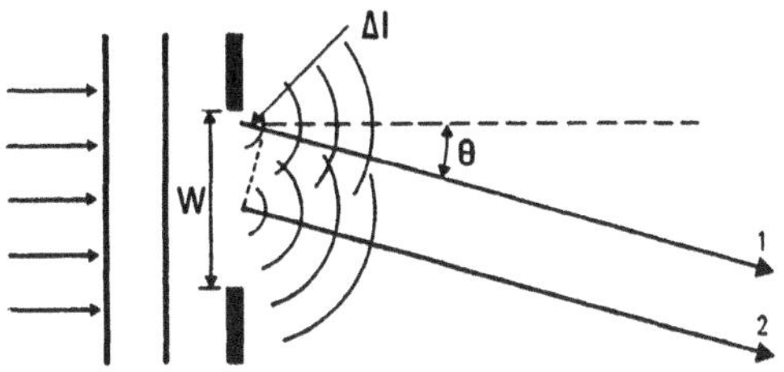

Fig. 7 When $\Delta l = \lambda/2$, secondary waves from all points separated by w/2 will be out of phase when they reach a distantscreen

The single-slit pattern shown above is actually produced by each of the two slits in a double-slit interference experiment. If the slits have a width that is not negligible compared to their separation, both the double-slit pattern and the single-slit pattern will be superimposed.

REFERENCES

1. D.H. Sliney, and M. Wolbarsht, "Safety with Lasers and Other Optical Sources," Plenum Press, New York and London (1980).
2. C. Zafiratos, "Physics," John Wiley & Sons, Inc., New York (1976).
3. R.P. Feynman, R.B. Leighton, and M. Sands, "Lectures on Physics," Addison Wesley, Reading (1964).
4. D. Landau, and E.M. Lifshitz, "The Classical Theory of Fields," Pergamon Press, Oxford (1962).
5. R.L. Armstrong, and J.D. King, "The electromagnetic Interaction," Prentice Hall Inc., London (1973).

CHARACTERISTICS OF OPTICAL NONCOHERENT SOURCES

Jürgen H. Bernhardt

Institute for Radiation Hygiene
of the Federal Health Office
D 8042 München-Neuherberg

INTRODUCTION

The sources of electromagnetic radiation can be categorized by several different methods. Sources can be grouped according to the type of material or the type of equipment which produces the radiation. The manner in which the radiation originated can be described in terms of nuclear, electronic or molecular transitions between energy states. Slight modifications of vibrational, rotational, or translational energy states in molecular bonds can give rise to infrared or microwave photons. When the temperature of a body is elevated, a variety of energy transitions take place and photons are emitted. The higher the temperature the greater percentage of high energy photons are released.

In order to explain the expression optical "noncoherent" sources we have to use the wave picture of radiation. The simplest way to understand the properties of coherent or noncoherent sources is to make a distinction between sources and point sources.

Light emanating from a point source has a characteristic known as spatial coherence; that is, light waves all along the sphere of radius from the point source have the same phase and are more or less in step. Light from an extended source close by is, for this reason, incoherent, since the rays are not all in step at a given radius from the source.

A further degree of coherence is present when a light source emits only one wavelength. Monochromatic light is thought of as temporally coherent because all of the waves can retain their same phase relationships (e.g., to be in step) only if the wavelengths of all the rays in the bundle are the same. The photons from an incandescent bulb is quite incoherent in both respects, because the lamp emits different wavelengths, therefore lacks temporal coherence, and emits in different directions in a random fashion, and thereby lacks spatial coherence until a sample is taken at a great distance from the light bulb.

BLACKBODY RADIATION

The chief source of electromagnetic optical radiation is the sun (described in more detail in the following section). Radiation from an

Light, Lasers, and Synchrotron Radiation
Edited by M. Grandolfo *et al.*, Plenum Press, New York, 1990

incandescent source like the sun is due to molecular and atomic vibration and collision and appears as a continuous spectrum. The spectral emission is characterized by basis physical laws.

The Stefan-Boltzmann Law

The Stefan-Boltzmann law expresses the total radiation emitted from a perfect temperature radiator as a function of the body's absolute temperature. The power radiated per unit area, W_B ($W \cdot m^{-2}$), is proportional to the fourth power of the absolute temperature, i.e.,

$$W_B = \sigma \cdot T^4 \tag{1}$$

where T is the body's temperature in kelvins (K) and σ is the Stefan-Boltzmann constant, given by

$$\sigma = \frac{2\ \pi^5\ K^4}{15\ h^3\ c^3} = 5.6697 \cdot 10^{-8}\ W \cdot m^{-2} \cdot K^{-4}, \tag{2}$$

where k = Boltzmann's constant,
h = Planck's constant,
c = speed of light.

A perfect temperature radiator is called a blackbody. A real, non-blackbody temperature radiator in equilibrium at the same temperature emits less radiant flux than the black body does. Whereas the radiant flux of a nonblackbody temperature radiator is determined by the material character of its surface as well as by its temperature, the blackbody's radiant flux is determined only by its temperature. The radiance of a blackbody is the same in all outward directions. A real source radiates less according to its emissivity, ε, defined by the equation

$$\varepsilon = W/W_B, \tag{3}$$

where W is the actual power emitted per unit area.

The Planck Radiation Law and Wien's Displacement Law

The spectral distribution of blackbody radiation is described by the Planck radiation law. Stated in terms of spectral radiant existance, the law is

$$W_{\lambda,\ B}\ (\lambda,T) = \frac{c_1}{\lambda^5 \left(e^{c_2/\lambda T} - 1 \right)}\ W \cdot m^{-3}, \tag{4}$$

where the first radiation constant $c_1 = 2\ \pi\ h\ c^2 = 3.7415 \times 10^{-6}\ W \cdot m^3$, and the second radiation constant $c_2 = hc/K = 1.43879 \times 10^{-2}\ m \cdot K$.

The spectral radiance of a black body is given by

$$L_{\lambda,B} = \frac{W_{\lambda,B}}{\pi}\quad W \cdot m^{-3} \cdot sr^{-1}. \tag{5}$$

If the formulas (4) and (5) are graphed for various values of the absolute temperature T, the curves of Fig. 1 are obtained. The figure shows the absolute values of spectral radiance, the power per unit area of the source per unit wavelength and per unit solid angle into which radiation is emitted, as a function of the wavelength. Wien's displacement law gives the wavelength of maximum emission, λ_{max} (μm), namely

$$\lambda_{max} = 2898/T \tag{6}$$

This expressed the familiar fact that the spectral distribution of a blackbody radiator and the wavelength λ_{max} of maximal spectral radiance shifts to shorter wavelengths with increasing temperature. As the substance surrounding a cavity is heated up, the cavity appears deep red, red hot, yellow, then "white hot", and finally blue or violet.

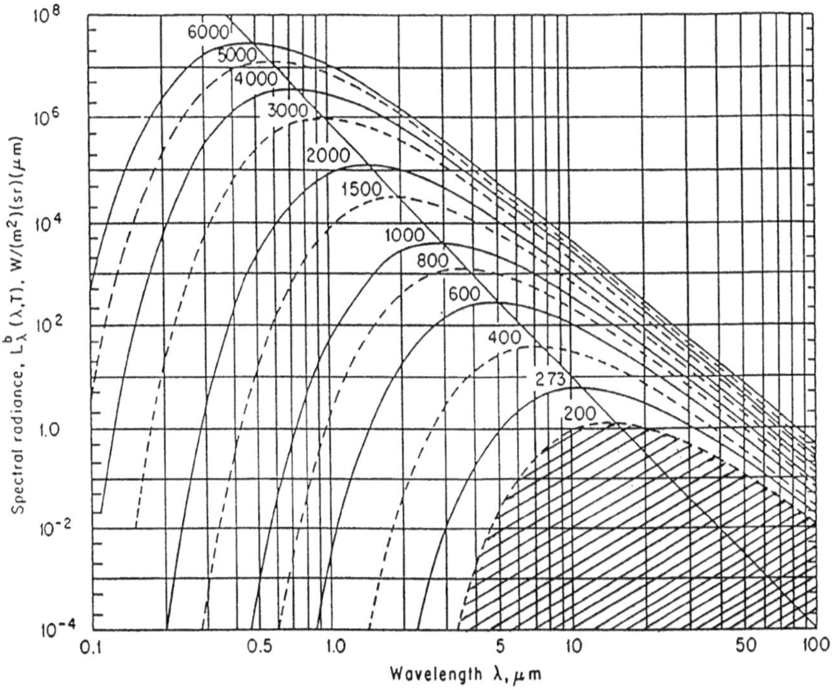

Fig. 1 Spectral radiance L_λ of a blackbody at the temperature in kelvins shown on each curve. The diagonal line intersecting each curve at its maximum shows Wien's displacement law. The hatched area gives an example for the total radiation emitted from a blackbody for, in this case, 200 K, as given by the Stefan-Boltzmann law.

Temperature as a parameter of radiation: Colour temperature

Limited aspects of the radiation from nonblack sources, e.g., incandescent sources, can often be approximated by chose of blackbody radiation. A blackbody temperature that results in a best approximation is said to be the equivalent temperature of the nonblack source. The convenience of such an approximation lies in specifying the nonblack radiation by a single parameter. A commonly used concept of equivalent temperature is the colour temperature. The colour temperature of a light source is the temperature at which a blackbody must be operated to have the same relative visible spectral distribution as the light source.

The extraterrestrial source of greatest importance, the sun, has been measured accurately at high altitudes, yielding results which are relatively uninfluenced by atmospheric effects. The solar constant has been determined from measurements of sunlight by different observers and represents the value of the total irradiance in the whole spectrum from the sun at a point outside the influence of the atmosphere.

The value for the solar constant for an average earth-sun distance is 1.39 kW $\cdot$ m^{-2}. The sun is usually considered to be represented nearly by a blackbody at a temperature of 6000 K.

After passage of the sunlight through the atmosphere, the spectrum barely resembles that of a blackbody. Fig. 2 shows three spectra, a 5900 k blackbody, the extraterrestrial solar spectrum and the sea-level spectrum. Implicit in the last of these spectra is the effect of the amount of absorbing constitutents from the atmosphere. The absorption in mainly due to water vapor, to atmospheric gases (ozone) and to atmospheric pollution.

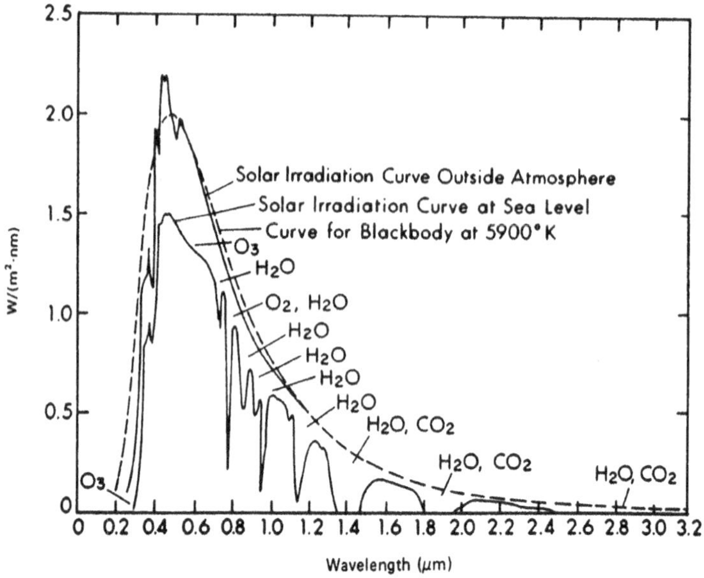

Fig. 2 Spectral distribution curves related to the sun: extraterrestrial solar irradiation and irradiation curve at sea level (adapted from Sliney and Wolbarsht, 1980).

Ozone layer

At normal temperatures and pressures, the ozone in the earth's atmosphere would comprise a layer only about 3 mm thick. In fact it is distributed throughout the stratosphere (a region between 10 and 50 km above sealevel) with maximum concentration at about 25 km, the exact location and thickness being subject to seasonal and geographical variations. The normal biatomic oxygen molecule is split by absorption of UVR of wavelengths less than 242 nm. The single atoms (free radicals) then combine with an oxygen molecule to form ozone, which absorbs UV of wavelengths less than 320 nm. The naturally occurring ozone layer is currently under attack from artifical sources, in particular halogenated hydrocarbons such as chlorofluoromethanes and chlorofluorocarbons. Current estimates of the reduction in ozone due to continued production of chlorofluorocarbons are about 10 %. A reduction in stratospheric ozone will produce a shift in the UV-spectrum at the surface of the earth towards shorter wavelengths.

Rayleigh scattering

Attenuation of solar radiation by gaseous molecules, particularly oxygen and nitrogen, is referred to as Rayleigh scattering. The particle diameter is small compared with the wavelength ($D \ll \lambda$). This is not an absorption process but a scattering of radiation from the direct beam. Attenuation by this means is inversely proportional to the fourth power of the wavelength. Thus, when the direct beam from the sun enters the earth's atmosphere, there will be greater scattering of blue light than red. It is because of this that the sky appears blue in colour. Moreover, when the sun is very low in the sky, the direct beam must traverse a considerable thickness of atmosphere and, therefore, light is depleted of blue and, in such circumstances, the sun appears red.

Aerosol scattering

An aerosol is a colloidal system in which the dispersing medium is a gas. Aerosols comprising small suspended particles in air are mostly to be found within one or two kilometres of the eath's surface. Their density varies considerably with humidity and their occurrence in the atmosphere contributes to the attenuation of solar radiation. An atmosphere containing aerosols is turbid or hazy. The term Mie scattering is often used to include attenuation by small particles such as aerosol constituents and dust; the particle diameter is of the same order as the wavelength.

Altitude

Altitude affects total solar irradiance and also the relative contributions of different wavelengths. On first entering the earth's atmosphere sunlight is attenuated principally by Rayleigh scattering. Then comes the region of ozone concentration at approximately 20 km altitude. To the combined effects of Rayleigh scattering and ozone absorption can be added attenuation due to aerosol and dust particles in the last few kilometres before the radiation reaches the earths surface. The extent of the altitude effects depends on latitude; in the Alps, total UV (combined direct sun plus scattered from sky) increases by some 15 % per kilometre).

Table 1 gives a brief account of the effect of the atmosphere on solar radiation showing the consituents responsible and the regimes of effectiveness.

The following tables include data for the global spectral irradiance. The total or global irradiance is the sum of the direct component of the solar radiation and of the diffuse component of the sky radiation.

Tab. 1. Attenuation of Solar Radiation by the Atmosphere

Wavelength regions, μm									Altitude, km
0.12–0.20	0.20–0.29	0.29–0.32	0.32–0.35	0.35–0.55	0.55–0.9	0.9–2.5	2.5–7	7–20	
O_2 absorbs almost completely	Solar irradiation intensity approximates extra atmospheric attenuation by scattering increases markedly toward shorter wavelengths								Above 60
	(0.20–0.21 μm; absorption by O_2); absorption by O_3 appreciable	O_3 absorption not important					Energy small	Energy very small	60–33
	No radiation penetrates below about 11 km	O_3 absorption attenuates more than loss by scattering	O_3 absorption significantly attenutates radiation	Irradiation diminished mostly by scattering by permanent gases in atmosphere	H_2O responsible for major absorption; CO_2 absorbs slightly at 2 μm; water vapor (or ice crystals) is found up to about 70,000 ft			Strong O_3 absorption at 9.6 μm; strong CO_2 absorption 12–17 μm	33–11
			Highly variable dust, haze (H_2O) (and smoke) responsible for attenuation in regions 0.32 to 0.7 μm		Energy transmitted with small loss down to 2 km	Energy penetrates to sea level only through windows at approximately 1.2, 1.6, and 2.2 μm	No significant penetration below 2 km, except in windows at approximately 3.8 and 4.9 μm	Energy transmitted with moderate loss; many absorption bands due to atmospheric gases	11–2
		Appreciable penetration through "clear" atmosphere to sea level About 7% / About 30%	Penetration through "clear" atmosphere to sea level about 40%	Dust may rise to more than 4 km					2–sea level

Table 2 shows the comparism between the spectral distribution of the extraterrestrial irradiance and the global irradiance for a solar zenith angle of 0°.

Tab. 2 Spectral distribution of the extraterrestrial irradiance and the global irradiance for a zenith angle of 0°

Wavelength μm	Extraterrestr. Irradiance W/m^2	Global Irradiance W/m^2
< 0,28	8	0
0,28 – 0,32	22	5
0,32 – 0,36	42	27
0,36 – 0,40	46	36
0,40 – 0,44	67	56
0,44 – 0,48	80	73
0,48 – 0,52	77	71
0,52 – 0,56	71	65
0,56 – 0,60	68	60
0,60 – 0,64	64	61
0,64 – 0,68	59	55
0,68 – 0,72	55	52
0,72 – 0,76	50	46
0,76 – 0,80	47	41
0,80 – 1,0	182	156
1,0 – 1,2	120	108
1,2 – 1,4	80	65
1,4 – 1,6	58	44
1,6 – 1,8	40	29
1,8 – 2,0	26	20
2,0 – 2,5	38	35
2,5 – 3,0	21	15
> 3,0	29	–

Table 3 presents characteristic values for global (direct plus diffuse) UV spectral irradiance.

The ratio of diffuse to direct solar radiation is strongly dependent on the elevation angle of the sun, especially in the UV-region. The diffuse ultraviolet irradiance in the UV-B-region is comparable to the direct irradiance in contrast to visible and IR-A wavelengths. Thus, by using the diffuse component of ultraviolet radiation one can actually shield oneself from the direct sun while still receiving sufficient ultraviolet radiation to produce suntanning.

Celestial bodies of less importance as sources of radiation are the moon, planets, and stars. The moon and planets, producing illumination as a result of the reflection of sunlight, are nearly replicas of sunlight, yielding radiation of an effective temperature of 5900 K at a very much reduced level from that of sunlight (full moon: E_λ (peak = 0.235 $\mu W \cdot cm^{-2} \cdot \mu m^{-1}$; Venus: E_λ (peak) $\approx 10^{-2}$ $\mu W \cdot cm^{-2} \cdot \mu m^{-1}$). Details on stellar radiation can be found in Wolfe (1965).

Tab. 3 Approximate Spectral Irradiance for Global Ultraviolett for cloudless days at sea level and 3.2 mm atmospheric ozone based upon the measurements of Bener (1972) (adapted from Sliney and Wolbarsht, 1980)

Wavelength (nm)	*Global (Direct + Diffuse) Spectral Irradiance in $\mu W/(cm^2 \cdot nm)$ for Solar Zenith Angle in Degrees*					
	0°	*30°*	*60°*	*70°*	*80°*	*85°*
297.5	0.218	0.107	0.0116	0.00	0.00	0.00
300	0.720	0.404	0.0766	0.0023	0.00	0.00
302.5	2.08	1.29	0.363	0.0116	0.0018	0.00072
305	4.97	3.43	1.17	0.0661	0.0079	0.00253
307.5	8.22	6.04	2.48	0.220	0.0293	0.00934
310	13.0	9.66	3.99	0.523	0.0706	0.0208
312.5	18.6	14.6	7.46	1.36	0.242	0.0613
315	24.1	18.9	10.2	2.51	0.492	0.131
317.5	29.3	23.7	13.2	3.56	1.01	0.267
320	33.9	27.7	16.6	5.06	1.44	0.449
325	46.5	37.9	22.0	7.71	2.71	1.03
330	53.5	44.1	28.1	11.5	4.62	1.92
340	57.8	47.5	30.9	12.5	5.32	2.32
360	62.4	51.5	33.1	13.5	6.34	3.03
380	72.7	60.0	37.4	14.2	6.65	3.35

ARTIFICIAL NON.COHERENT RADIATION SOURCES

Since artificial Non coherent radiation sources are treated more thoroughly in the chapter "Lamps and Lighthing Systems" by Dr. A.F. McKinley, this section is restricted to a short summary of the most important facts.

General properties

Prior condition for radiation emission is the existence of a source for the excitation energy of photons. The most important kinds of excitation are

- thermal excitation
- electric excitation
- radiation excitation
- chemical excitation

Most of the artifical radiation sources are temperature radiators, i.e., radiation is produced when a body is heated (incandescence). The most important laws of temperature radiation were treated in a previous section of this chapter (blackbody radiation). Actually, no material emits a perfect blackbody spectrum. The ratio of the theoretically possible spectral emittance to the actual emittance of a gray body is the emissivity ε. For instance the emissivity of tungsten throughout the visible is approximately 0.4.

The electric excitation plays a mayor part in gas discharge lamps. In these cases a stream of electrons flowing through a gas creates a release of photons characteristic of that particular gas. If gas has a low pressure and the current is not great, a line spectrum is emitted. As the gas pressure and the current density are increased, a continuous spectrum due to the temperature of the gas begins to appear. At very high current densities and gas pressures, this portion of the emission predominates. Electric excitation is also important for semiconductor-sources (LED).

Excitation by radiation is important for fluorescent tubes (see below). Finally, chemical excitation (i.e., in biological process) is practically of no importance for radiation sources.

Quartz halogen source

Quartz halogen lamps incorporate a tungsten filament and a halogen gas. They differ from ordinary tungsten lights in several aspects. The quartz envelope permits transmission of shorter wavelengths than is possible with glass. The tungsten is evaporated from the filament, combines with the halogen gas inside the lamp at high temperature and is redeposited onto the filament (the halogen cycle). There is a shift in spectral emission towards shorter wavelengths as a result of the higher temperature. A quartz halogen lamp may be usually employed as a standard light source.

Gas discharge source

Another method of producing radiation is to pass an electric current through a gas. Atoms of the gas may be ionised if sufficient energy is transferred from a moving electron. Alternatively, the travelling electron may not impart sufficient energy for ionisation but, instead, it may raise the electrons of the gas to an excited energy level. When they return to a lower energy level, or their ground state, radiation of one or more characteristic wavelengths is emitted.
The wavelengths are determined by the type of gas present in the lamp and appear as spectral lines. The width of the lines and the amount of radiation in the interval between them (the continuum) depends on the pressure in the lamp. At low pressures, fine lines with little or no continuum are produced. As pressure is increased the lines broaden and their relative amounts alter. The magnitude of the continuum also increases.

Arc sources

These lamps operate at high pressures (20 to 100 atmospheres) and are very intense sources of UVR. The two electrodes in an arc lamp are positioned only a few millimetres apart and the arc between them supports a high current density. High temperatures are creates and broadened spectral lines are superimposed on a continuum arising from the glowing plasma. The output spectrum from a high-pressure mercury lamp exhibits a marked self-absorption at 254 nm. Commonly available lamps contain xenon, mercury, or a mixture of the two elements.
A high voltage is required for initial ionisation of the gas between the electrodes, but once the arc is established it can be maintained by a low-voltage supply.

Fluorescent lamps

The primary source of radiation in a fluorescent lamp arises from a low-pressure mercury discharge. This produces a strong emission at 254 nm. A phosphor coating is applied to the inner wall of the tube. The 254 nm photons produced by the mercury discharge are absorbed in the phosphor coating, were they excite the molecules causing a fluorescent emission to take place. By altering the composition and thickness of the phosphor and the glass envelope a wide variety of emission spectral characteristics may be obtained. The output is thus chiefly the fluorescent emission spectrum from the coating, with a certain amount of breakthrough of the mercury lines.

Metal halide lamps

The addition of other metallic elements to a mercury discharge lamps allows for the addition of extra lines to the mercury spectrum. Unfortunately other metals are not sufficiently volatile and also attack the silica envelope. The solution is to introduce the extra metal in halide form. During the warm-up phase the mercury spectrum dominates, but the metal spectrum takes over as the halide evaporates. A variety of metal halides has been used. Advantage has been taken of the strong lead emission lines at 364, 368 and 406 nm in the lead iodine lamp were there is a 50 % increase in output in the region between 355 and 380 nm compared to a conventional mercury lamp. Antimony and magnesium halide lamps provide spectral lines in the UVB and UVC regions.

REFERENCES

Driscoll, W.G., Vaughan, W. (Eds.) 1983, Handbook of Optics.
Kiefer, J., (Ed.) 1977, Ultraviolette Strahlen, Verlag de Gruyter, Berlin.
Koller, L.R., 1965, Ultraviolet Radiation, 2nd ed., Wiley, New York.
Moseley, H., 1988, Non-Ionising Radiation, Medical Physics Handbook 18, Adam Hilger, London.
Naumann, H., Schröder, G., 1987, Bauelemente der Optik, Hanser-Verlag, München.
Sliney, D., Wolbarsht, M., 1980, Safety with Lasers and Other Optical Sources, Plenum Press, New York and London.
Schulze, R., 1970, Strahlenklima der Erde, Dr. Dietrich Steinkopf Verlag, Darmstadt.
Valley, S.L. (Ed.) 1965, Handbook of Geophysics and Space Environments, Air Force Cambridge Research Laboratories, Bedford, Mass..
Wolfe, W.L. (Ed.), 1985, Handbook of Military Infrared Technology, GPO, Washington.

LAMPS AND LIGHTING SYSTEMS

A.F. McKinlay

National Radiological Protection Board
Chilton, Didcot, Oxon, UK

INTRODUCTION

The subject of this Chapter is primarily the physical characteristics of lamps and in particular their optical radiation emission characteristics. The uses of lamps in general and specialized lighting and in other specific non-lighting applications such as printing, germicidal and tanning applications are also considered. Data on comparative hazard assessment measurements are also presented. A summary of the lighting applications of different types of lamps is presented in table 1.

Apart from lighting applications, lamps are found in many forms, in a wide range of environments and used by many trades and professions for a diverse range of applications that is ever increasing. A summary of some such applications together with the types of lamps used ranging over industry, medicine, commerce, agriculture and the home etc is provided in tables 2 to 4.

Lamps can be conveniently grouped under 2 broad headings; those that produce radiation by incandescence and those that produce it by a gas discharge. The latter group can be sub-divided according to the pressure of the filling gas into those lamps that emit primarily within narrow wave-bands (so-called line-emissions) and those that emit broad-band (so-called continuum) radiation. The band-width of the emission from a gaseous discharge is dependent on the pressure of the gas and because of this and the possible presence of specific additives (eg, metal halides) the emissions from some lamps can be considered as a combination of both line and continuum.

Optical radiation may be emitted from a lamp either deliberately by the nature of the lamp and its intended purpose; as for example the emission of light for the purpose of illumination or the emission of ultra-violet radiation (UVR) for the purpose of sterilization (germicide) or adventitiously by the nature of the source and its inadequate or inefficient design; as for example the emission of hazardous levels of UVR from a general lighting double envelope high intensity discharge lamp whose outer envelope has broken.

It is important therefore in hazard evaluation terms to consider not only the type of lamp being used and particularly its spectral emission characteristics but also the spectral emission requirements of its intended purpose and the design and construction of its luminaire or filter containment system.

Light, Lasers, and Synchrotron Radiation
Edited by M. Grandolfo *et al.*, Plenum Press, New York, 1990

Table 1 Physical characteristics and applications of lamps (adapted from McKinlay et al 1988)

Lamp type	Sub-group	Construction and Operation	Efficacy lm W^{-1}	Typical Applications
Incandescent	Tungsten filament	A tungsten filament heated to incandescence in a glass envelope usually filled with an inert gas; does not require any control gear.	8-18	Homes, hotels, restaurants, for general lighting and for display lighting; emergency lighting.
	Tungsten halogen filament	A tungsten filament heated to incandescence in a small envelope containing halogens; does not require any control gear, but may require low voltages.	18-24	Display, area floodlighting, shops and for homes.
Gaseous Discharge	High pressure mercury tungsten discharge (blended)	An electric discharge in a high pressure mercury atmosphere contained in an arc tube in series with a tungsten filament heated to incandescence, the whole contained within a glass envelope with a fluorescent coating. Does not need control gear.	10-26	As a replacement for tungsten filament lamps where lamp life is important, eg, because access is difficult.
	High pressure mercury discharge (fluorescent)	An electric discharge in a high pressure mercury atmosphere contained in an arc tube within a glass envelope with a fluorescent coating; needs control gear.	36-54	General lighting in factories, area floodlighting.
	High pressure mercury discharge (metal halide)	An electric discharge in a high pressure mercury atmosphere with metal halide additives in an arc tube sometimes contained within a glass envelope; the outer envelope may have a fluorescent coating; needs control gear.	66-84	Industrial and commercial lighting, eg, general lighting in high bay factories, shops and offices; area floodlighting.
	Very high pressure and mercury/xenon	An electric discharge in a very high pressure mercury or mercury/xenon atmosphere: needs control gear.	40-50	Projection systems photochemical reactors.
Gaseous Discharge	Low pressure mercury discharge	An electric discharge in a low pressure mercury atmosphere contained in a glass tube internally coated with a fluorescent material.		Homes, factories, offices, shops, public buildings etc.
	Low pressure sodium discharge	An electric discharge in a low pressure sodium atmosphere in a glass arc tube contained in a glass envelope; needs control gear.	101-175	Industrial area lighting where distortion of most colours is acceptable; road lighting.
	High pressure sodium discharge	An electric discharge in a high pressure sodium atmosphere in an arc tube contained in a diffusing or clear outer envelope, needs control gear.	67-121	General lighting in factories, warehouses, commercial buildings; area floodlighting.
	Xenon discharge	An electric discharge in a xenon atmosphere in an arc tube.	20-50	Projection systems, area floodlighting and photography.

Table 2 Some examples of the agricultural application of lamps (adapted from CIE 1988)

Application	Uses	Species	Lamps
Plants - cultivation	Photosynthesis	Tomatoes, lettuce, tobacco, cucumbers etc	High pressure mercury lamp Fluorescent lamp Metal halide lamp Incandescent lamp
	Photomorphogenesis	Barley, oats, rice, potatoes etc	Incandescent lamp
	Photoperiodism	Chrysanthenum carnation, tomatoes, strawberry etc	Fluorescent lamp Incandescent lamp High pressure sodium lamp
- Insect protection	Insect lures Insect invasion	Rice, orange etc Orange, peach, pear, grape, etc	Fluorescent lamp Fluorescent lamp
Stock raising	Germicide Health Heating Photoperiodism	Chicken, cattle, pig Pig, chicken Pig, cattle, chicken Chicken, silkworm	Low pressure mercury lamp Fluorescent lamp Incandescent lamp Incandescent lamp
Fishery - Fishing	Fish luring	Sardine, mackerel, squid	Incandescent lamp High pressure mercury lamp
- Farming	Photoperiodism	Trout	Incandescent lamp High pressure mercury lamp
	Breeding control Photosynthesis	Sea bream, trout, shellfish seaweed	Fluorescent lamp
Plants in domestic and office environments	Photoperiodism	Decorative plants	High pressure mercury lamp Fluorescent lamp

Table 3 Some examples of the medical, cosmetic and germicidal application of lamps (adapted from CIE 1988)

Application	Uses	Lamps	Useful spectral emissions
Medical treatment	Skin diseases, eg, psoriasis, vitiligo	Fluorescent lamp	UVA, UVB
		High pressure mercury lamp	UVA, UVB
		Metal halide lamp	UVA, UVB
	Muscle injury	Incandescent lamp	IR
		Tungsten halogen lamp	IR
	Hyperbilirubinaemia (neo-natal jaundice)	Fluorescent lamp	blue
		Metal halide lamp	blue
Cosmetic	Skin tanning	Fluorescent lamp	UVA
		Metal halide lamp	UVA
Germicidal/ sterilizing	Water, food, medical areas, containers, instruments, etc.	Low pressure mercury lamp	UVC
		High pressure mercury lamp	UVC
		Metal halide lamp	UVC, UVB
	Insect lures	Fluorescent lamp	UVA

Table 4 Some examples of the industrial application of lamps (adapted from CIE 1988)

Industrial field	Application	Lamps	Useful spectral emissions
Printing	Ink curing	High pressure mercury lamp Metal halide lamp	UVA, B, C UVA
	Ink drying	Incandescent lamp Tungsten halogen lamp	IR IR
	Plate making	High pressure xenon lamp Metal halide lamp High pressure mercury lamp Fluorescent lamp Tungsten halogen lamp	Visible, UVA UVA UVA UVA, visible Visible
Document copying - diazo system	Exposure	Fluorescent lamp High pressure mercury lamp Metal halide lamp	UVA, blue UVA UVA
- plain paper system/ZnO system	Exposure	Fluorescent lamp Tungsten halogen lamp	Blue-green Visible
	Fixing	Tungsten halogen lamp	Infra-red
Painting	Paint curing Paint drying	High pressure mercury lamp Incandescent lamp Tungsten halogen lamp	UVA, B, C IR IR
Semiconductors	Exposure	High pressure mercury lamp	UVA
Printed circuits	Exposure	High pressure mercury lamp Fluorescent lamp	UVA UVA
Chemical processes	Photochemical reactions	High pressure mercury lamp Metal halide lamp	UVA UVA, blue, green etc
General processes	Drying, baking, shrinking, softening	Incandescent lamp Tungsten halogen lamp	IR IR

Incandescent (Heated Filament) Lamps

The incandescent lamp is the oldest type of lamp still in common use, figure 1. Its durability has been due to its simplicity of construction enabling cheap mass-production methods. However, in the face of the requirement to save energy and the potential competition of 'compact' gas discharge fluorescent lamps research and manufacturing efforts are still being made to increase both the comparatively short life and the low luminous efficacy of these lamps. The latter is a major drawback of incandescent lamps as about 90% of the power supplied to the lamp is dissipated as infra-red radiation. The optical radiation from incandescent lamps results from the heating of tungsten filaments which, with the exception of lamps for some photographic and graphic arts applications, are not heated to temperatures in excess of about 3000 K. The maximum luminous efficacy of a tungsten filament lamp is lower than that of the equivalent temperature 'black-body' radiator because the emissivity of tungsten varies with wavelength and at all wavelengths is less than 1.00. As the temperature of a tungsten filament is raised the level of radiation emitted at shorter wavelengths rises proportionally faster than that at longer wavelengths, ie, the visible radiation emissions rise faster than those of the infra-red, hence the luminous efficacy increases. The luminous efficacy of tungsten at its melting point 3655 K is approximately 53 lm W^{-1}, however to obtain a reasonable life it is necessary to operate tungsten filament lamps at temperatures well below the melting point. The usual range of luminous efficacies of such lamps is 8 - 18 lm W^{-1}, table 1. Typical gas filled incandescent filament lamps operate between 2700 and 3000 K with powers up to 500 W and the peaks of their spectral emissions are in the infra-red (IRA) region.

Figure 1. Photograph of a 60 W tungsten filament lamp (left) and a 100 W tungsten halogen lamp (right) [Photograph by courtesy of the National Radiological Protection Board, UK].

In applications where much more power (up to 5 kW) and/or higher luminous efficacies (up to 24 lm W^{-1}) are required (but at extra cost) tungsten (quartz) halogen lamps are often used, figure 1. These are tungsten filament lamps to which a halogen vapour (usually iodine) is added. The presence of the vapour results in reversible cyclic chemical reactions between tungsten metal lost by evaporation from the filament and the halogen vapour. This prevents deposition of tungsten metal on the inside of the wall of the bulb. In addition, because of their small size they operate at an increased gas pressure compared with 'conventional' incandescent lamps and evaporation from the filament is minimized. These

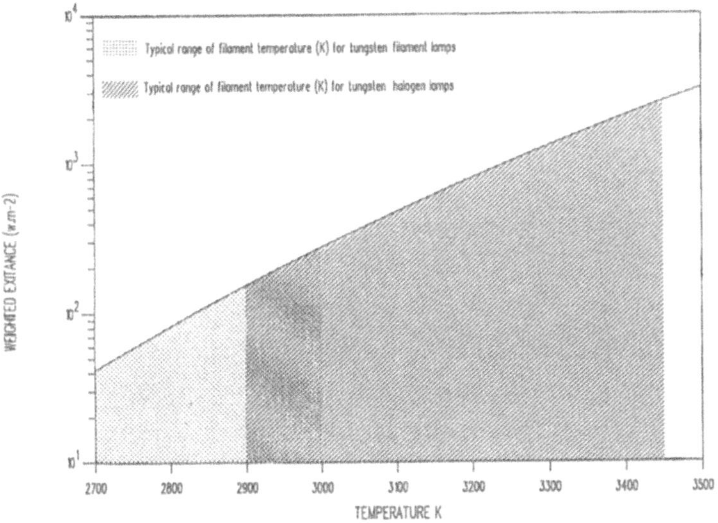

Figure 2. ACGIH (1988) weighted exitance as a function of temperature of a 'Black Body Radiator', showing typical temperature ranges for tungsten filament and tungsten halogen lamps [McKinlay et al 1989].

features result in improved luminous efficacy and in longer life. In order to maintain the efficiency of the regenerative chemical reactions the temperature of the wall of the bulb must be maintained at not less than 260°C. The majority of the bulbs of tungsten halogen lamps are made from silica (quartz) whose thermal properties are most suitable. The combination of filament temperatures which are likely to be in the range 2900 to 3450 K and quartz bulbs results in a significantly higher level of emission of UVR compared with 'ordinary' tungsten filament lamps. In addition, the blue-light emission, which is a measure of the potential retinal photochemical hazard of a lamp (ACGIH 1988) is higher. These points are illustrated in figures 2 to 4 for incandescent 'black-body' sources over the range of temperature 2900 to 3450 K.

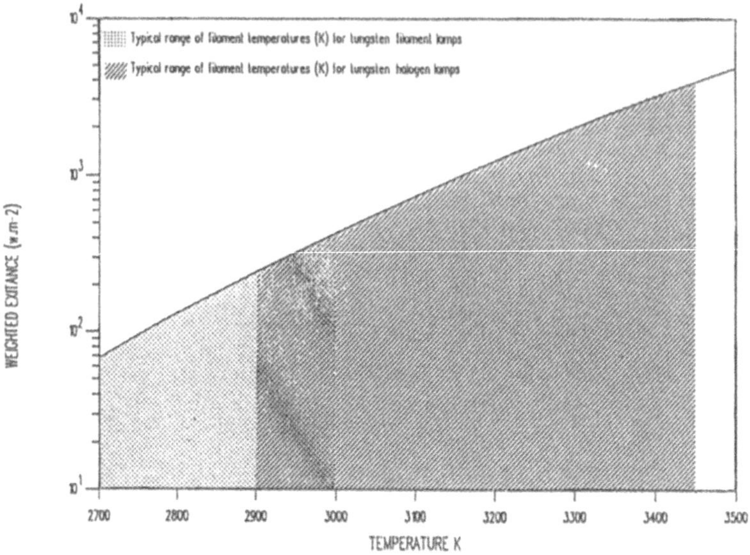

Figure 3. Erythemally (McKinlay and Diffey 1987) weighted exitance as a function of temperature of a 'Black Body Radiator', showing typical temperature ranges for tungsten filament and tungsten halogen lamps [McKinlay et al 1989].

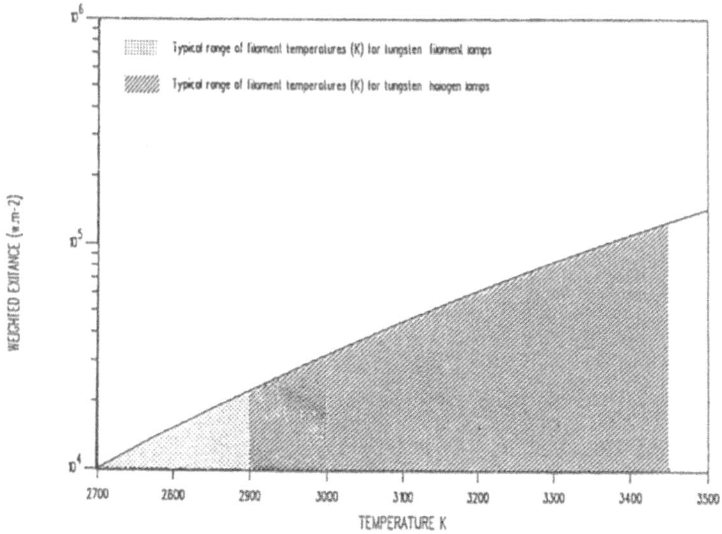

Figure 4. Blue-light (ACGIH) weighted exitance as a function of temperature of a 'Black Body Radiator', showing typical temperature ranges for tungsten filament and tungsten halogen lamps [McKinlay et al 1989].

The incorporation of suitable secondary filtration to reduce UVR emissions to an acceptable level is an important feature of the design of any illumination system using tungsten halogen lamps. Such lamps are used safely in filtered luminaires for a diverse range of applications. However, some tungsten halogen lamps and particularly some desk-top lamps have no protective secondary filtration (McKinlay et al 1989). Values of occupational hazard weighted (ACGIH 1988) and biologically (erythemally) weighted (McKinlay and Diffey 1987) UVR irradiances from some such lamps are presented in table 5. Graphical representations of the spectral emissions of two such lamps are illustrated in figures 5 and 6.

Table 5. Ultra-violet radiation emissions from tungsten halogen lamps

Lamp Number	Effective irradiance[ϕ] ($mW\ m^{-2}$ effective)		Estimated MEDs per annum[≠]		Illuminance[ϕ] (lux)
	ACGIH*	Erythema**	Domestic	Occupational	
1	6.1	10.1	61	182	1400
2 (H)	2.2	3.8	23	68	1600
2 (L)	0.7	1.2	7	21	800
3	1.0	2.0	12	36	900
4 (1)	2.0	3.4	20	61	3000
5	25.0	41.0	245	735	14000
6	2.1	3.4	20	60	1200
7	2.2	3.3	20	60	7600
8	3.3	5.7	34	102	1400
9	1.2	1.9	11	33	700
10	6.3	10.0	60	180	1900
11	35.1	56.2	337	1012	20500
12	23.0	38.0	230	690	15900

ϕ All measurements at 30 cm from lamps

* ACGIH (1988) occupational hazard weighted irradiance. 1 maximum permissible exposure for 8 h working day is equivalent to 1 $mW\ m^{-2}$ effective

** International Lighting Commission (CIE) erythema reference action spectrum (McKinlay and Diffey 1987)

≠ Domestic exposure estimated as a maximum of 500 $h\ a^{-1}$; occupational exposure estimated as a maximum of 1500 $h\ a^{-1}$, 1 MED = 300 $J\ m^{-2}$ effective

H High power switch position   L Low power switch position

Data on the optical radiation emissions from a variety of high power stage, studio and photocopier tungsten halogen lamps are presented in table 6.

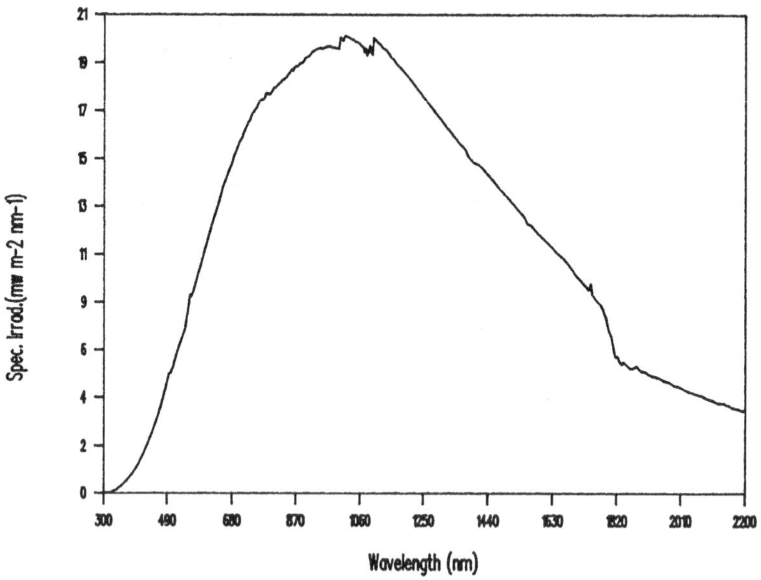

Figure 5. Typical spectral emission from a 50 W tungsten halogen lamp fitted in an aluminium reflector as a spot-light [McKinlay et al 1989].

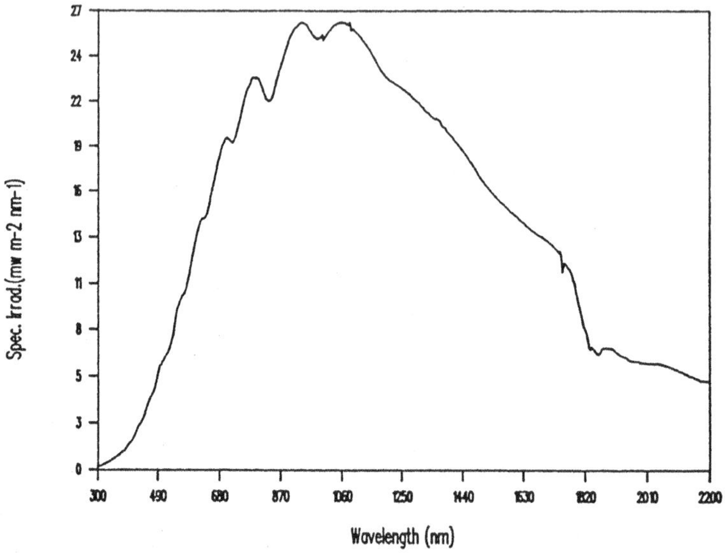

Figure 6. Typical spectral emission from a 50 W tungsten halogen lamp fitted in a dichroic reflector as a spot light [McKinlay et al 1989].

Table 6 Optical radiation emissions from a variety of tungsten (quartz) halogen lamps [data from Mohan et al 1980]

Lamp type	Power W	Ultra-violet radiation UVA irradiance $W\ m^{-2}$	Ultra-violet radiation ACGIH weighted * $W\ m^{-2}$ effective	Visible radiation illuminance lux	Visible radiation radiance $W\ m^{-2}\ sr^{-1}$	Infra-red radiation $W\ m^{-2}$	Type of lamp and measuring distance
FCV 1000 W	1000	1.1	3×10^{-2}	9.7×10^{3}	9.2×10^{4}	170	Stage & studio lamp 50 cm
DSF 1500 W	1500	0.35	2.2×10^{-4}	3.5×10^{3}	1.7×10^{4}	62	Stage & studio lamp 100 cm
EHC/EHB 500 W	500	0.48	1.9×10^{-2}	4.1×10^{3}	8.6×10^{4}	75	Stage & studio lamp 50 cm
EGR 750 W	750	1.5	5.8×10^{-2}	1.1×10^{4}	10^{5}	185	Stage & studio lamp 50 cm
EER 2000	2000	0.75	8×10^{-4}	6.6×10^{3}	1.3×10^{5}	113	Stage & studio lamp 100 cm
EFX 500 W	500	0.4	1.2×10^{-2}	4.1×10^{3}	7.2×10^{4}	86	Stage & studio lamp 50 cm
FAE MOVIE LITE	550	1.26	4.3×10^{-4}	1.4×10^{4}	3.9×10^{4}	222	Projection lamp with reflector 100 cm
DXT 1000 W	1000	0.26	10^{-5}	1.8×10^{3}	6×10^{3}	24	Flood light lamp 100 cm
FFW PAR 64	1000	1.3	7.3×10^{-5}	1.4×10^{4}	7.7×10^{4}	229	Spot light lamp 450 cm
Q1000T3/CL	1000	0.24	7×10^{-3}	2.4×10^{3}	5.9×10^{4}	41	Photocopier lamp 100 cm
BRH 1000 W	1000	1.97	8.2×10^{-2}	1.3×10^{4}	1.6×10^{5}	183	Photocopier lamp 50 cm
BCK 500 W	500	1.78	7×10^{-2}	1.26×10^{4}	1.94×10^{5}	188	Photocopier lamp 50 cm
QH650T3/CL	650	3.7×10^{-3}	9×10^{-6}	204	4.2×10^{5}	17	Photocopier lamp 100 cm
Q500T2.5/2CL	500	0.70	2.2×10^{-2}	6.3×10^{3}	1.1×10^{5}	100	Photocopier lamp 50 cm

* ACGIH (1988)

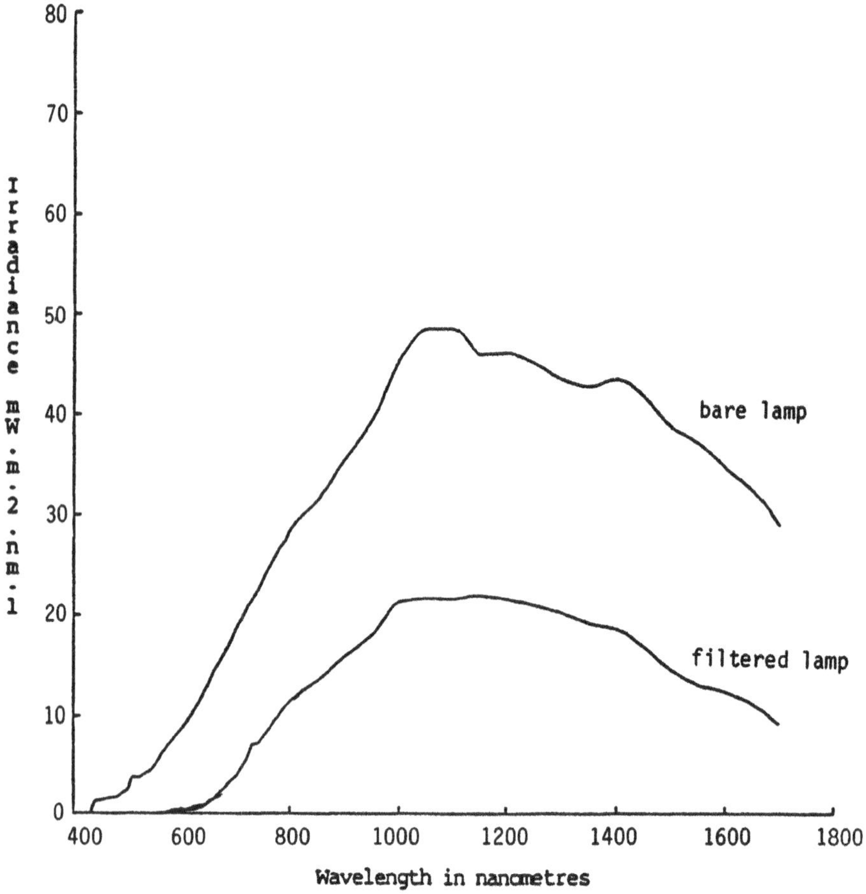

Figure 7. Spectral emissions from a 450 W tungsten halogen cooking hob fitted with two different filters. The large infra-red radiation emissions of incandescent lamps and particularly of tungsten halogen lamps, makes them useful as heating tools and more recently as cooking appliances (halogen hobs). The spectral emission associated with a halogen cooker is illustrated.

Low Pressure Gas Discharge Lamps

In low pressure gas discharge lamps the filling gas is usually an inert gas. The commonest type of low pressure (non phosphor) discharge lamp is the 'neon' lamp. The colours of the emissions of these lamps are characteristic of the filling gas. Neon, argon and krypton are commonly used. Because of their characteristic multiplicity of spectral emission lines they are often used in laboratories for instrumentation wavelength alignment and calibration purposes. The use of 'soft glass' envelopes effectively attenuates any UVR emissions. A notable exception is the low pressure mercury discharge lamp which is often used for the purpose of germicide and disinfection. Such lamps are very efficient emitters of UVR. Approximately 50% of the electrical power is converted to UVR of which up to 95% is emitted at a wavelength of 253.7 nm. Germicidal type lamps are available in a range of sizes, shapes and powers. Small low-wattage (5-10 W) germicidal lamps are often used as fluorescence-inducing lamps for the purposes of, for example, chromatographic analysis, the fluorescence identification and authentication of documents and the location of finger-prints. The quartz envelopes of some lamps in this category transmit 185 nm wavelength radiation characteristic of mercury vapour. The interaction of this radiation with oxygen in the surrounding air results in the production of ozone which is a powerful oxidizing and potentially toxic agent.

Fluorescent Lamps. The most common application of the low pressure discharge is in fluorescent lamps. The fluorescent lamp operates by means of a discharge between two electrodes through a mixture of mercury vapour and a rare gas, usually argon. The mercury vapour is present at a pressure between 0.8 and 1.4 Pa. This corresponds to a liquid mercury temperature of 40°C, the optimum temperature at which most fluorescent lamps are designed to operate. The argon, or other rare gases is intended to aid the initiation (striking) of the discharge and is present at a pressure of between 100 and 400 Pa. Light is produced by conversion of 253.7 nm mercury emission radiation to longer wavelength radiations by means of a phosphor coating on the inside of the wall of the lamp. Lamps are available with many different phosphors and envelopes to produce a wide range of spectral emissions covering the visible (light), UVA and UVB regions. While the continuum emissions of fluorescent lamps are characteristic of the phosphors the narrow peak spectral emissions are dominated by the characteristic line emission spectrum of the low pressure mercury vapour discharge. These lines are extremely narrow as illustrated in figure 8 ($\sim 10^{-3}$ nm) (McKinlay et al 1988) a characteristic which is often hidden in the presentation of emission spectra because of the usually comparatively broad measurement bandwidth (usually a few nanometres).

General Lighting Fluorescent Lamps. These lamps are available in a range of physical sizes, powers and phosphors. The range of phosphors includes a large selection of 'near white' and 'special colour' lamps.

In relation to other light sources the fluorescent lamp is particularly efficient, with some 20% or so of the input energy resulting in useful light with an overall luminous efficacy of about 80 lm W^{-1} (typical cool white).

Detailed spectral measurement data of general lighting lamps commonly used in the USA and occasionally in Europe are presented in table 7 (Mohan et al 1980). Data from a study of the amount of UVR emitted by white fluorescent lamps used in the UK for general lighting purposes are presented in table 8 (Whillock et al 1988).

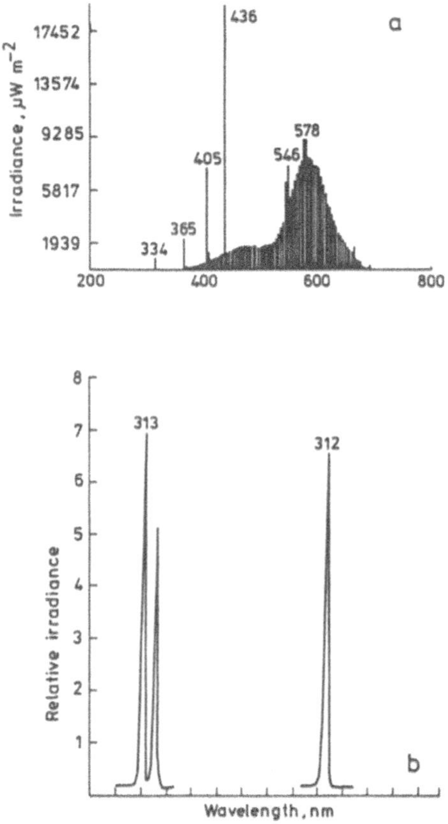

Figure 8. Spectral emission from a white fluorescent lamp measured using a monochromator with (a) 3 nm and (b) 0.02 nm resolution [McKinlay et al 1988].

These data differ somewhat from those of Cole et al (1986) obtained from measurements on general lighting lamps being used in the USA where generally the levels of UVR emissions were higher, table 9. In particular, the emissions from some of the so-called super high output (SHO) tubes displayed the presence of 253.7 nm radiation. It has been suggested that this might have been due to a low iron content in the glass envelope.

During the past few years the further development and improved design of general lighting fluorescent lamps has been evident principally in two areas; compact fluorescent lamps and energy efficient lamps. The former lamps are essentially low wattage small diameter fluorescent tubes folded in a compact form, for example, the so-called '2D' or 'U' or 'dual U' types, figure 9. They are most readily available commercially with cool white phosphors but other phosphors are also available.

Illuminance and UVR data from a measurement study on some such lamps are presented in table 10 (Whillock et al 1989). Compact fluorescent lamps have typical luminous efficacies of up to 80 lm W^{-1} (CIE 1988).

Development work on energy efficient lamps has produced improved high efficiency phosphors resulting in lamps with luminous efficacies as high as 100 lm W^{-1}; a green light emitting silicate phosphor has been reported providing a luminous efficacy of over 130 lm W^{-1}. Other approaches to the task of producing energy efficient lamps have related to the design of filament switches and improved ballast systems.

Table 7 Measurement data for some American general lighting fluorescent lamps (after Mohan et al 1980)
All data have been normalised for an illuminance of 500 lux

Type of lamp	Length (mm)	Diameter (mm)	Colour type	UVA ($W m^{-2}$)	UVB* ACGIH ($W m^{-2}$)	IRA ($W m^{-2}$)
F48T12-CW-HO	1200	38	Cool white	5.2×10^{-2}	4.3×10^{-4}	2.1×10^{-2}
F84T12-CW-HO	2100	38	Cool white	4.5×10^{-2}	1.8×10^{-4}	1.9×10^{-2}
F96T12-CW-1500	2400	38	Cool white	5.2×10^{-2}	3.7×10^{-4}	2.4×10^{-4}
F96PG17-CW (Power groove)	2400	38	Cool white	5.3×10^{-2}	2.8×10^{-5}	2.2×10^{-2}
F96T12.UHO Durotest	2400	38	?	9.2×10^{-2}	3.6×10^{-4}	9.9×10^{-2}
F40.T12	?	38	?	1.9×10^{-2}	$\nless 2 \times 10^{-4}$	1.7×10^{-1}
F40-CW-RS-WM	?	?	Cool white	1.6×10^{-1}	5.9×10^{-3}	8.8×10^{-2}

* ACGIH (1988)

Table 8 Emissions from fluorescent lighting lamps- white (Whillock et al 1988)

	Total time that tubes have been in 'on' state before measurement (h)	Irradiance totals in each waveband (mW m^{-2})		
		Manufacturer		
		A	B	C
UV-A	0	34.98	40.18	31.82
	100	28.53	34.90	26.11*
	2000	30.06	31.83	26.11
UV-B	0	4.16	5.02	4.38
	100	3.15	4.05	4.05*
	2000	3.17	3.58	3.22
UV-B (ACGIH)	0	69 10^{-3}	79 10^{-3}	79 10^{-3}
	100	49 10^{-3}	63 10^{-3}	68 10^{-3}*
	2000	47 10^{-3}	51 10^{-3}	50 10^{-3}
UV-B/UVR (%)	0	10.6	11.1	12.1
	100	9.9	10.4	11.8*
	2000	9.5	10.1	11.1
UV-B (ACGIH)/ UVR (ACGIH) (%)	0	0.20	0.20	0.25
	100	0.17	0.18	0.22*
	2000	0.16	0.16	0.19
Irradiance weighted for erythema+	0	0.27	0.31	0.29
	100	0.19	0.25	0.26*
	2000	0.19	0.21	0.20

Notes:
Data normalised for an illuminance of 500 lux.
Lamp dimensions = 150 cm x 38 mm.
UVR range – 200–400 nm.
UV-A range – 315–400 nm.
UV-B range – 280–315 nm.
UV-B (ACGIH) – UV-B weighted according to the standard recommended by ACGIH (1988).
*These values are after 122 h.
+Weighted according to McKinlay & Diffey (1987) action spectrum.

Table 9 Measurement data for some American lighting fluorescent lamps (after Cole et al 1986) All data have been normalised for an illuminance of 500 lux

Type of lamp	Length (mm)	Diam (mm)	Colour Type	UVA ($W.m^{-2}$)	UVB (ACGIH) ($W.m^{-2}$)*
F96 T12 WW/HO - 3 - lamps +F96 T12 CW - 3 - lamps	2400	38	WARM WHITE and COOL WHITE	1.9×10^{-2}	1.6×10^{-4}
F40 T10 CW/99	1000	38	COOL WHITE	3.5×10^{-2}	3.0×10^{-4}
F96 T12 CW/HO - 2 - lamps	2400	38	COOL WHITE	2.9×10^{-2}	4.1×10^{-4}
F48 T12 CW/HO	1200	38	COOL WHITE	1.5×10^{-2}	1.0×10^{-4}
F96 T12 CW/HO - 2 - lamps	2400	38	COOL WHITE	3.7×10^{-2}	4.0×10^{-4}
F40 WW RS WM	1000	38	WARM WHITE	0.1×10^{-2}	0
F40 WW RS WM	1000	38	WARM WHITE	1.7×10^{-2}	1.1×10^{-4}
F40 CW RS SS	1000	38	COOL WHITE	2.5×10^{-2}	2.5×10^{-4}
F40 CW RS SS	1000	38	COOL WHITE	2.6×10^{-2}	2.2×10^{-4}
F15 T8 CW	380	?	COOL WHITE	3.8×10^{-2}	2.7×10^{-4}
F96 T12 CW HO - 2 - lamps	2400	38	COOL WHITE	5.6×10^{-2}	7.1×10^{-4}
F96 T12 CW SHO - 2 - lamps	2400	38	COOL WHITE	5.9×10^{-2}	12.7×10^{-4}

*One MPE = 10^{-3} W m^{-2} ACGIH (1988)

Low Pressure Sodium Lamps. Like mercury, sodium metal can be vapourized and used together with a rare gas in a discharge and both low and high pressure discharges are used in practical lamp design. Sodium lamps are particularly used for street and other large area light applications where colour rendering is relatively unimportant. In low pressure sodium lamps the sodium vapour is held at a pressure of typically 0.4 Pa and the rare gas at a pressure of a few hundred pascals.

Low pressure sodium lamps have the highest luminous efficacies, typically greater than 150 lm W^{-1}, of commercial light sources. This is due to; optimizing the vapour pressure to reduce thermal losses; to the low excitation energy required to excite the sodium vapour and; to the high photopic response (V_λ) of the human eye at the principal sodium spectral emission wavelengths, 589 and 589.6 nm (yellow), figure 10. Due to the high chemical reactivity of sodium a special 'borate glass' is required for the discharge tube.

'Special' Applications Fluorescent Lamps. Apart from a number of special colour rendering fluorescent lamps which are essentially variations of general lighting fluorescent lamps a number of special applications fluorescent lamps have been developed and are commercially available. An example is the so-called 'blacklight' lamp which uses a nickel/cobalt oxide (Woods' glass) envelope. The phosphor chosen for these lamps emits around 370 nm in the UVA. As the glass is almost entirely opaque to light the lamps are used for a number of commercial, scientific and industrial fluorescence purposes as well as for display and entertainment. An example of the spectral emission from a typical 'Blacklight' fluorescent lamp is illustrated in figure 11.

Table 10. Spectral emissions from compact fluorescent lamps (Whillock et al 1989)

Lamp type	Erythemally* Weighted irradiance (μW m^{-2})	ACGIH≠ Weighted irradiance (μW m^{-2})		Illuminance at 65 cm (lux)
		UVA	UVB	
Luma LC7 - 7 W (A)	18.87	47 x 10^3	0	131
Luma LC7 - 7 W (B)	18.71	41 x 10^3	0	121
Luma LC7 - 7 W (B) with diffuser	42 x 10^{-3}	197	0	106
Osram Dulux EL - 11 W (A)	14.60	36 x 10^3	0.1	101
Osram Dulux EL - 11 W (B)	15.15	38 x 10^3	0.1	101
Philips SL9 - 9 W	12.71	37 x 10^3	0	81
Sylvania Lynx CFD - 13 W	158.18	114 x 10^3	30.81	121
Thorn 2D - 16 W	35.87	54 x 10^3	2.48	258
Tungsram Globolux - 16 W	9 x 10^{-2}	663	0	111

Codes A and B distinguish between 2 different lamps of the same type
* McKinlay and Diffey (1987) ≠ ACGIH (1988)

A medical treatment for the condition hyperbilirubinaemia in neonates (neonatal jaundice) consists of irradiating the newborn child with 'blue light' in the wavelength range approximately 400 to 470 nm. Lamps often used for this purpose have emission spectra as illustrated in figure 12.

Three applications of fluorescent lamps in printing and copying merit mention viz; lamps suitable for diazo printing with a principal emission around 360 nm, figure 13; those intended for modern 'fast' diazo printing with a main emission around 420 nm and; those used for 'photocopying' with predominantly green emissions, figure 14.

A number of lamps have been designed for horticultural use including the 'Grolux' lamp whose spectral emission characteristics are illustrated in figure 15. The emission of this lamp corresponds with the spectral requirements of chlorophyll photosynthesis.

The development of a range of phosphors with enhanced UVA emissions has led to the widespread use of fluorescent lamps in sunbeds, solaria and PUVA (Psoriasis UVA) treatment cabinets. A summary of the spectral emission characteristics of different UVA fluorescent lamps is presented in table 11 (Diffey and McKinlay 1983) and measurement data on the most recently available types of such lamps are presented in table 12 (Whillock et al 1988). The spectral emissions of UVA lamps used for tanning purposes are illustrated in figures 16 and 17.

Fluorescent lamps intended for UVB irradiation also are available. An example of these is the Westinghouse FS type lamp whose spectral emission is illustrated in figure 18.

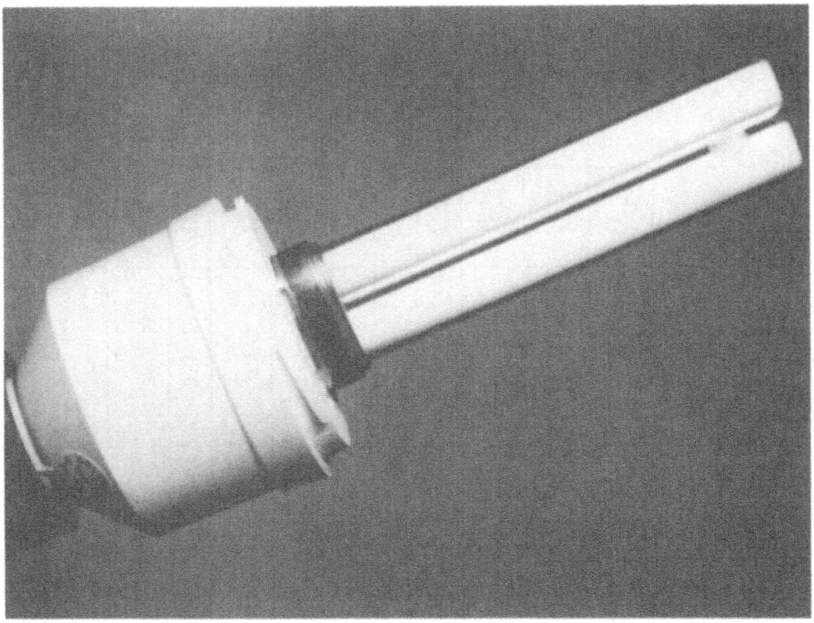

Figure 9. Photograph of a typical 'U' type 16 W compact fluorescent lamp [photograph courtesy of the National Radiological Protection Board, UK].

Table 11 Summary of spectral irradiance data for 10 types of UVA fluorescent lamps. All measurements on bare tubes at 1 m on mid-point axis with no reflectors (Diffey & McKinlay 1983)

Lamp code	Fluorescent peak (nm)	UVA $W\ m^{-2}$	UVB $W\ m^{-2} \times 10^2$	$\frac{UVB}{UVA}$%	$\frac{UVB(EE)}{UVA}$% Summed to 325 nm	$\frac{UVB(EE)}{UVA}$% Summed to 315 nm	$\frac{UVB(297\ nm)}{total\ UVB}$%	$\frac{UVB(EE)(297\ nm)}{total\ UVB\ (EE)}$%
Wolff Belarium SA-1-10 (100W)	348 - 354	1.18	1.82	1.55	0.116	0.093	0.48	6.0
GEC (100W) T.12.BL51	366	1.35	1.06	0.79	0.181	0.180	8.0	32
Osram (100W) Liteguard	351	1.55	0.86	0.56	0.039	0.030	0.25	3.5
Muster 75/85W	350	1.84	2.03	1.10	0.096	0.084	1.5	16.7
Harpo F75/85W	351/2	1.51	1.71	1.14	0.099	0.086	1.8	19.2
Thorn 75/85W	351/2	1.86	0.81	0.44	0.028	0.020	0.23	3.5
Philips (100W) TL85-09	351/2	1.83	1.27	0.69	0.045	0.035	0.36	5.2
Wotan (100W) L100/79	351/2	1.68	0.96	0.57	0.036	0.027	0.30	4.6
FL85 UVA (100W)	351/2	1.54	2.19	1.42	0.149	0.135	2.2	19.7
Wolff Relarium RA1-10 (100W)	350 - 357	1.17	1.29	1.10	0.088	0.072	0.7	8.1
Mean	-	1.55	1.40	0.94	0.088	0.076	**0.87	**9.6
Coefficient of variation %	-	16	34	37	56	64	82	67

** GEC Lamp omitted because of very different UVA emission.

Table 12 UVA fluorescent lamps (180 cm x 38 mm - Unless stated otherwise) (data from Whillock et al 1988) All lamps measured at a point 135 cm on the normal through the midpoint of the lamp Irradiance totals in each waveband ($mW.m^{-2}$)

	UVA		UVB		UVB (ACGIH)		UVB/UVR (%)		UVB(ACGIH)/UVR(ACGIH) (%)	
T =	0 hrs	100 hrs	0 hrs	100 hrs	0 hrs	100 hrs	0 hrs	100 hrs	0 hrs	100 hrs
MANUFACTURER										
A (150 cm x 38 mm)	968	751	7.20	5.22	120×10^{-3}	84×10^{-3}	0.74	0.69	0.01	0.01
B1 (with reflector)	2055		1.40		19×10^{-3}		0.07		9×10^{-4}	
B2 (150 cm x 38 mm)	1017	890	7.25	6.08	106×10^{-3}	90×10^{-3}	0.71	0.68	0.01	0.01
C1	1096	1033	34.64	32.28	1134×10^{-3}	1044×10^{-3}	3.06	3.03	0.10	0.10
C2	691	613	6.92	5.86	221×10^{-3}	176×10^{-3}	0.99	0.95	0.03	0.03
C3 (150 cm x 38 mm)	1014	990	9.78	9.65	216×10^{-3}	215×10^{-3}	0.96	0.97	0.02	0.02
C4	1035	990	63.49	61.52	2842×10^{-3}	2817×10^{-3}	5.78	5.85	0.27	0.28

UVR - 200 - 400 nm
UVA - 315 - 400 nm
UVB - 280 - 315 nm
UVB (ACGIH) - UVB weighted according to the standard recommended by the American Conference of Governmental Industrial Hygienists (1988). One MPE is 10^{-3} W m^{-2}.
T = total time (in hours) that tubes have been in the 'on' state before measurement

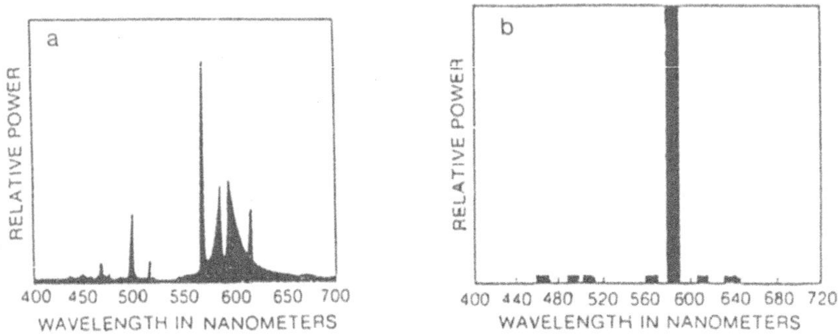

Figure 10. Spectral emissions from (a) low pressure sodium lamp and (b) high pressure sodium lamp [after IES 1981].

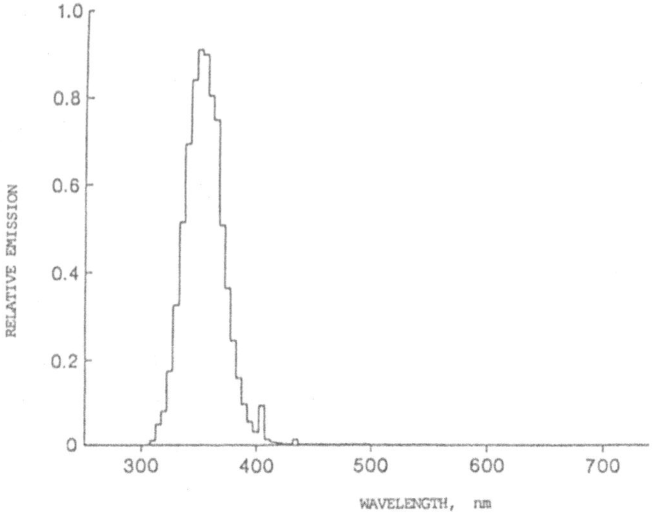

Figure 11. Spectral emission from a 40 W blacklight fluorescent lamp [data from CIE 1988].

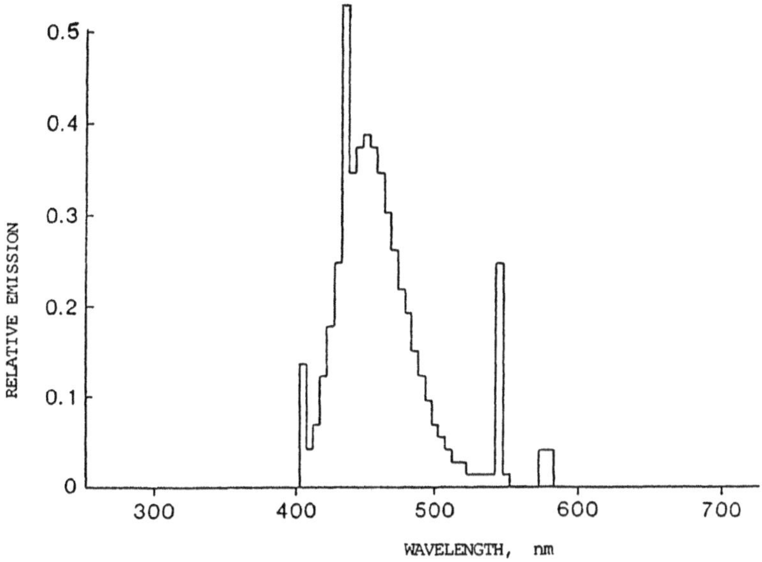

Figure 12. Spectral emissions from a 35 W fluorescent lamp with a blue emitting phosphor used for the treatment of hyperbilirubinaemia in neonates [data from CIE 1988].

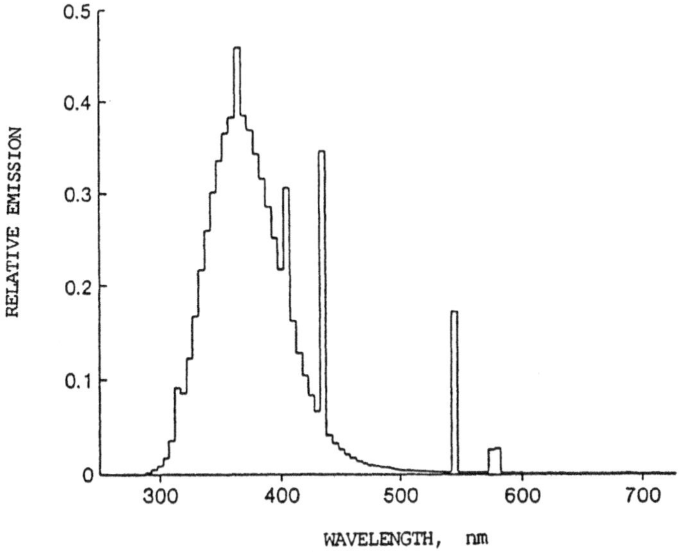

Figure 13. Spectral emission from a 40 W, UVA emitting fluorescent lamp used for diazo photo-printing [data from CIE 1988].

High Intensity Discharge (HID) Lamps

The emission characteristics of a discharge alter significantly when the pressure of the filling vapour is increased. In general the line emissions that are characteristic of the vapour broaden, often being subsumed by a continuum and there is a shift in the spectral emissions to longer wavelengths.

The designation 'high intensity discharge' (HID) lamps includes the families of lamps often called high pressure mercury vapour, metal halide and high pressure sodium vapour lamps.

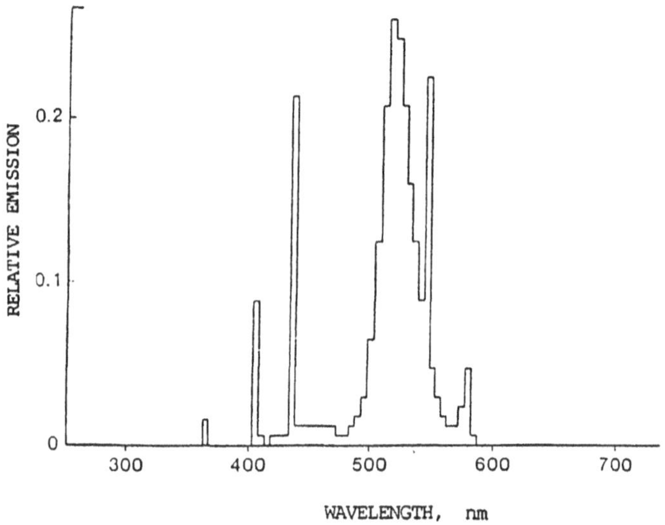

Figure 14. Spectral emission from a 25 W fluorescent lamp with a green emitting phosphor used for photocopying [data from CIE 1988].

Mercury Vapour Lamps. High pressure (200-400 kPa) mercury vapour lamps are widely used for industrial and commercial lighting, street lighting, display lighting, floodlighting and a large number of printing, curing and other industrial applications. The light spectral emissions of the discharge are in the blue, green and yellow regions of the spectrum with major emission lines at 405, 436, 546, 577 and 579 nm, and a large amount of UVR is also generated, figure 19. Mercury discharges have characteristically no red emissions. Typical spectral efficacies range from 50 - 60 lm W^{-1}. The general construction of high pressure mercury lamps is a fused silica (quartz) discharge tube containing the mercury/

argon vapour discharge mounted inside an outer envelope of soda-lime or borosilicate glass, figure 20. The colour range of the lamp can be made more acceptably white by the addition of a red emitting phosphor on the inside wall of the outer envelope. Another similar type of lamp has an outer envelope that is additionally coated internally with a reflective layer designed to direct most of the light downwards.

The outer glass envelope effectively absorbs most residual UVR; consequently the quantity of potentially harmful UVR emitted by such lamps depends critically on the integrity of this envelope.

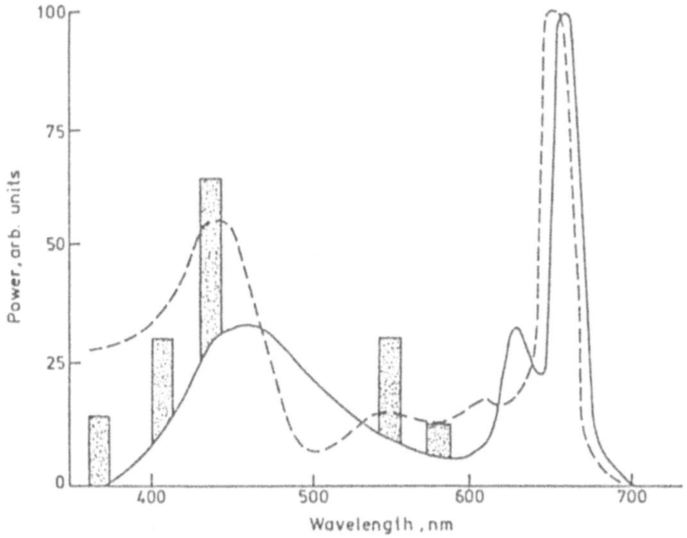

Figure 15. Spectral emission from a 'Grolux' lamp (full curve) and the spectral activity for chlorophyll photosynthesis (dotted curve) [McKinlay et al 1988, original data from Henderson and Marsden 1972].

In the USA, but apparently not in Europe, it is a legally enforceable manufacturing requirement that either breakage of the outer envelope must cause the lamp to fail to operate, in which case the lamps are described as "self-extinguishing" and are marked with the letter "T", or if not "self-extinguishing" they should be marked with the letter "R". In the latter case a warning notice must be included with the packaging of the lamp (FDA 1988). Data from measurements made at 2 m from a mercury HID lamp with the outer envelope removed illustrate the importance of this aspect of safety design, table 13.

Table 13 Ultra-violet radiation emissions from 400 W high pressure mercury vapor discharge lamps used in the USA for general lighting purposes. Effective Irradiances at 2 m in W m^{-2} x 10^{-3} (MPEs) (ACGIH weighted) [data from Piltingsrud et al 1978]

Test condition	A		B		C		D	
Lamp type	With outer bulb	Without outer bulb	With outer bulb	Without outer bulb	With outer bulb	Without outer bulb	With outer bulb	Without outer bulb
General Electric H400 A33-1 Clear	2.5	110*	<0.1	160	12.5	3640	0.2	0.2
Westinghouse H33 GL 400/DX White	<0.1	1040	<0.1	190	0.2	3680	<0.1	0.3
General Electric H400D x 33-1 White	<0.2	75*	<0.1	180	0.5	3900*	<0.1	<0.1*
General Electric MV400/BUH [Metal halide]	<0.3	122*	<0.1	6*	1	510*	<0.1	<0.1*

Test Conditions:

A: Lamp mounted vertically - measurements at 2 m on mid-line axis of lamp.

B: Lamp mounted horizontally - measurements at 2 m on central axis of lamp.

C: Lamp mounted horizontally in reflector shield with no face plate - measurements at 2 m on central axis of lamp.

D: Lamp mounted horizontally in reflector shield with glass face plate fitted - measurements at 2 m on central axis of lamp.

*Lamps did not operate at normal intensity.

Table 14 Ultra-violet irradiances from typical graphic arts metal halide mercury lamps (data from Philips 1988)

Lamp type*	Power (W)	UVC‡ Irradiance ($W\ m^{-2}$)	UVB‡ Irradiance ($W\ m^{-2}$)	UVA‡ Irradiance ($W\ m^{-2}$)
HPA 400	400	0.5	3.2	9.0
HPA 1000	930	2.3	9.0	23.0
HPA 2000	1750	4.5	19.0	48.0

* Philips

‡ Measured at 1 m

Table 15 Visible light emissions from compact (XE) and linear (XB) xenon lamps (after Henderson and Marsden 1972)

Lamp type	Power (W)	Light output (lumens)	Arc luminance* ($cd\ m^{-2}$)	Arc length (mm)
XE	150	3,000	9×10^7	2.5
XE	250	5,000	10^8	3
XE	500	12,000	2×10^8	5
XB	1,000	20,000	10^6	85
XB	1,500	31,500	1.6×10^6	155
XE	2,000	70,000	1.2×10^9	6/7
XB	10,000	250,000	1.4×10^6	750

*Luminance of the sun 1.65×10^9 $cd\ m^{-2}$

Metal Halide Lamps. The family of 'metal halide' lamps encompasses a number of different types of high intensity mercury lamps whose discharges all contain additives. The additives are most typically metal halides chosen to produce either a strongly coloured emission (usually a single halide), to produce a more broadly spectrally uniform emission (multi-halide) or to enhance the UVR (most often UVA) emission.

Examples of metal halides used either singly or in combination are lead, iron, cobalt, sodium and thallium, indium, scandium, dysprosium, gallium and other rare earths. The spectral emissions from a metal halide lamp used for illumination purposes is illustrated in figure 21.

Compared with 'ordinary' mercury high intensity discharge lamps the spectral efficacies of metal halide lamps are high, typically 75–125 lm W^{-1}. The spectral emission of a typical graphic arts metal (iron and cobalt) halide lamp designed to have a high UVA emission is illustrated in figure 22 and its UVR emission data are presented in table 14 (Philips 1988).

High Pressure Sodium Lamps. High pressure sodium lamps are similar in construction to but generally have a smaller diameter than high pressure mercury lamps, figure 23. The inner tube is made from polycrystalline alumina and the outer envelope from borosilicate glass. The emissions are characteristic of the sodium vapour and as the vapour pressure is raised the emissions broaden across the visible spectrum. At very high vapour pressures (greater than about 27 kPa) the characteristic line emissions of sodium vapour at 589 and 589.6 nm disappear due to 'self absorption' by the gas, figure 10. The luminous efficacies of high pressure sodium lamps are typically high, 60 - 140 lm W^{-1} depending on tube size. No starting electrodes are used and the arc is initiated by means of a high voltage, high frequency electrical pulse.

Very High (Super) Pressure Mercury, Mercury Xenon and Xenon Lamps (Compact and Linear Arc Lamps). Where an optical source of very high radiance/luminance is required and of small size a very high pressure arc lamp may be used. These have a filling gas of mercury vapour, mercury vapour plus xenon gas or xenon gas. Metal halide types are also available. Two physical types are commonly used; the compact (or short) arc and the linear arc.

Typical applications of compact arcs include projectors, searchlights and solar radiation simulators. Mercury compact arcs operate at pressures of between 2 and 4 MPa and and at very high operating currents (usually dc for increased arc stability and for longer life). These arcs require several minutes warm-up period to achieve full operating pressure and radiation (luminous) output. The warm-up period is reduced by half when xenon gas is added to the discharge. They are available in a range of powers from 30 W to 5 kW with luminous efficacies from 40 to 55 lm W^{-1} depending on wattage.

Linear mercury and mercury-xenon arcs are often used in photochemical reactors with water cooling often used to maintain a stable temperature. The discharge pressure is typically of the order of 5 to 20 MPa.

The spectral emission of xenon lamps, which at wavelengths shorter than infra-red, closely matches that of a black-body radiator at about 6000 K enables their use as solar radiation simulators. Their emission spectrum is continuous from the UV through to the IR regions, figure 24. The emission peaks in the infra-red principally between 800 and 1000 nm may be effectively removed by filtration. As with the mercury and mercury-

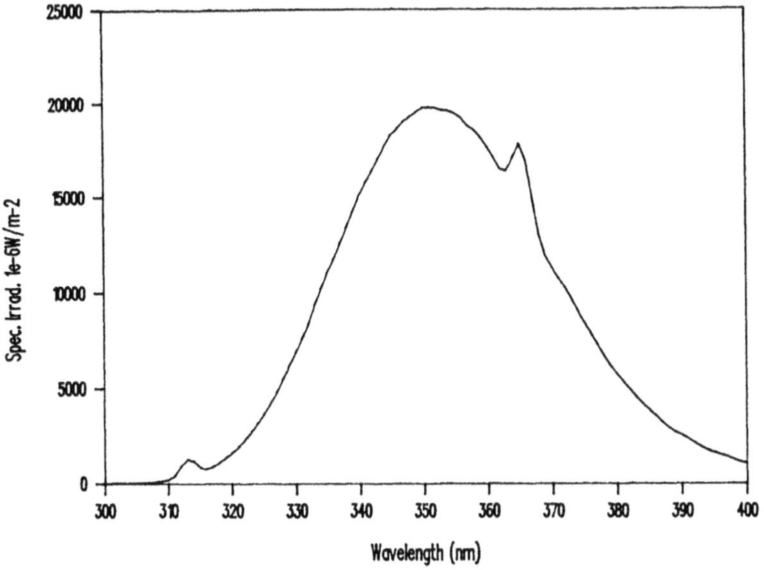

Figure 16. Spectral emission from a 100 W Philips TL09 UVA fluorescent lamp used for tanning (sunbeds) [Whillock et al 1988].

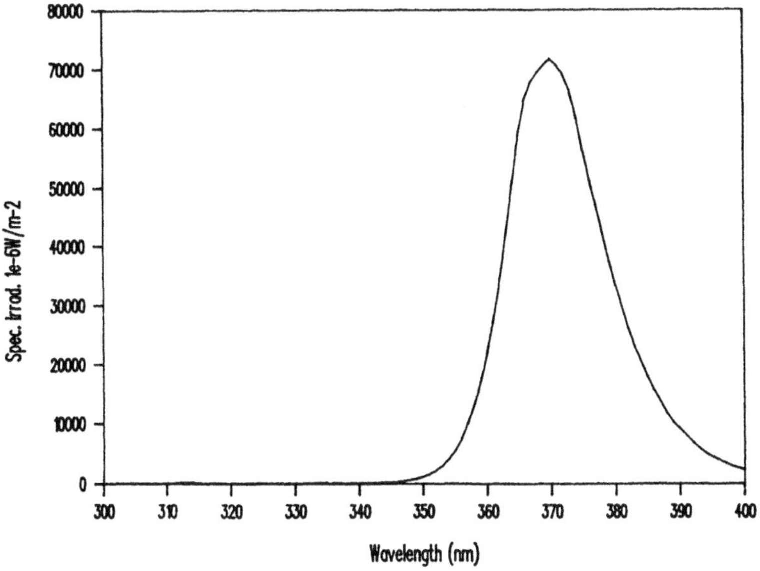

Figure 17. Spectral emission from a 100 W Philips TL10R UVA fluorescent lamp used for tanning (sunbeds) [Whillock et al 1988].

xenon compact arcs, xenon compact arcs are operated at high pressures (typically several hundred thousand pascals) and at powers of between a few watts and over 30 kW. Their luminous efficacies range from between 20 and 50 lm W^{-1} depending on their operating power.

The luminance of compact xenon arcs may approach that of the sun (1.65 x 10^9 cd m^{-2}) as shown in table 15 and in some lamps of greater than 10 kW rating the luminance of the cathode spot may exceed 10^{10} cd m^{-2}. All high pressure xenon lamps have the advantage over mercury lamps of instant starting.

Miscellaneous Lamps

Flash Lamps and Flash Tubes. Flash lamps and flash tubes are designed to produce brief duration but high peak luminous intensity flashes. They are used for various purposes including photographic illumination, laser stimulation and stroboscopic illumination. A typical flashtube contains xenon (and sometimes additionally argon, hydrogen and krypton) gas at a pressure of between approximately 25 - 80 kPa in a borosilicate glass or (more usually) quartz envelope. A trigger electrical pulse, usually applied to an external electrode, causes ionization in the gas causing a luminous discharge to take place. The typical flash tubes used by professional and amateur photographers have a pulse duration of between 2 and 4 ms with a correlated colour temperature of about 6000 K. Their luminous efficacies range from 30 to 60 lm s J^{-1}. Using a range of associated circuitry, flashtubes can be designed to operate with flash energies of a fraction of a joule to several kJ and with pulse durations between 1 μs and several ms. Pulse repetition frequencies of up to several kHz can be obtained.

Pulsed Xenon Arc Lamps. These are non-polarized xenon arc lamps, often used for printing and other graphic arts applications where instant start and high stability are important. They operate on each ½ cycle (100 pulses per second (50 Hz)) of an ac power supply. The emission spectrum is close to that of solar radiation (600 K). They are available with a range of mean powers from a few hundred watts to approximately 10 kW with luminous efficacies of about 30 to 40 lm W^{-1}. They are available in either helical or linear forms.

Glow Lamps. Glow lamps are low power gas discharge lamps, most typically (but not exclusively) containing neon gas (argon is also used). They are available in a number of geometries with powers from around 100 mW to a few watts for applications such as pilot (indicator) lamps, instrument panel lamps, 'night lights' etc. Their luminous efficacies are very low, around 0.3 lm W^{-1} (neon type).

Solid State Lamps. This category of 'lamps' includes light emitting diodes, electroluminescent lamps, and tritiated phosphor 'lamps'. The light emitting diode (LED) is a semi-conductor (p-n junction) in which the recombination of electrons and holes result in the emission of photons of infra-red or visible radiation. The wavelength of the emission of the LED depends on the 'base' semiconductor material and the added impurities. LEDs incorporating phosphors are also available. LEDs usually operate with a dc voltage of a few volts and currents of up to 100 mA or so. Typical luminous intensities for continuously operating red-emitting LEDs range from 0.5 to 2.0 mcd and for green- and yellow-emitting LEDs from 1 to 16 mcd (IES 1981).

Electroluminescent lamps are essentially thin parallel plate capacitors with a phosphor material incorporated in the dielectric and with

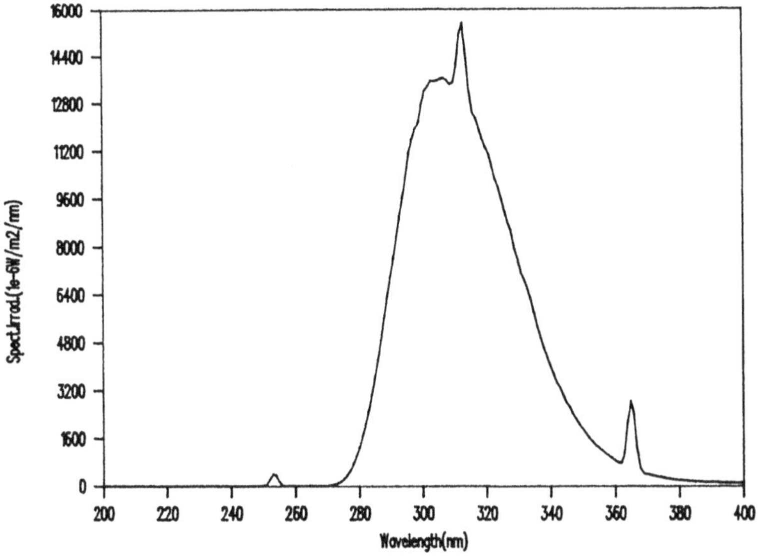

Figure 18. Spectral emission from a 80 W Westinghouse FS UVB fluorescent lamp [Whillock et al 1988].

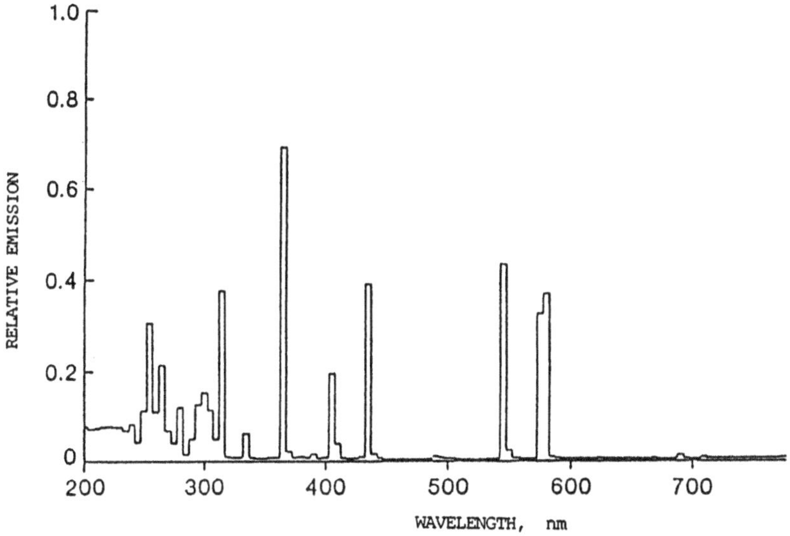

Figure 19. Typical spectral emission from a 2 kW high pressure mercury vapour (HID) lamp [data from CIE 1988].

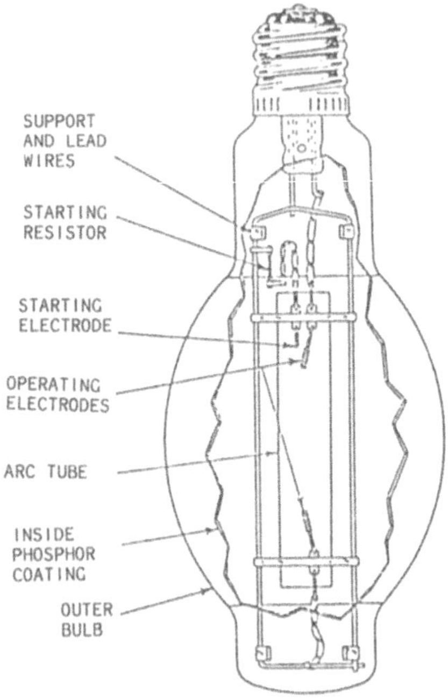

Figure 20. Basic construction of a 400 W high pressure mercury (fluorescent) (HID) lamp [IES 1981].

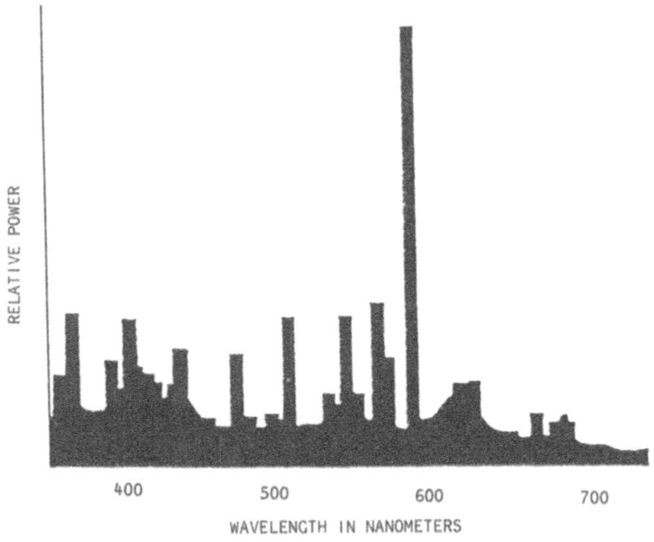

Figure 21. Spectral emission from a metal halide (sodium and scandium lamp used for illumination (IES 1981).

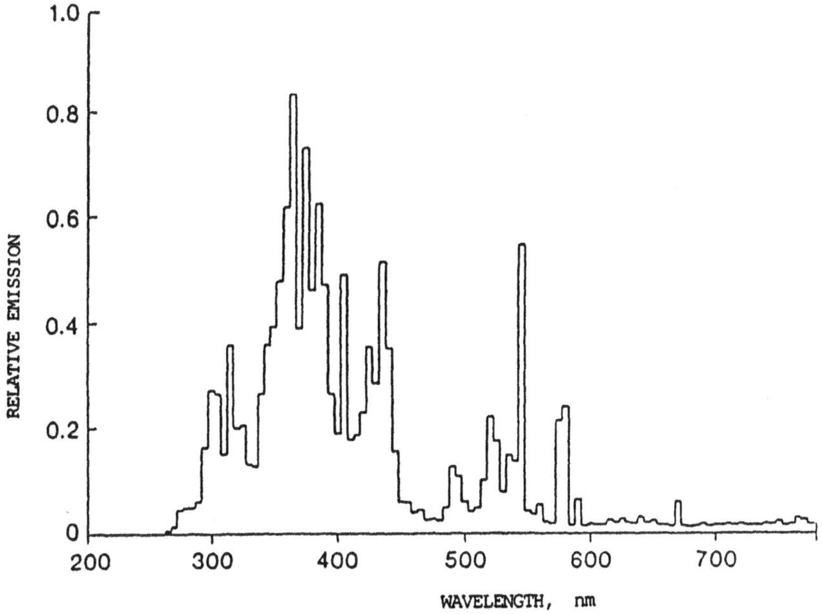

Figure 22. Spectral emission from a metal halide (iron and cobalt) lamp used for graphic arts (UVA) applications [data from CIE 1988].

either or both of the conducting 'plates' transparent. Because they can be constructed with a stiff or flexible backing and easily cut into a variety of different and complicated shapes they have a number of applications as illuminated signs, instrument panels and dials etc. They are available if required with different coloured phosphors but have a low luminance, up to 100 cd m^{-2}.

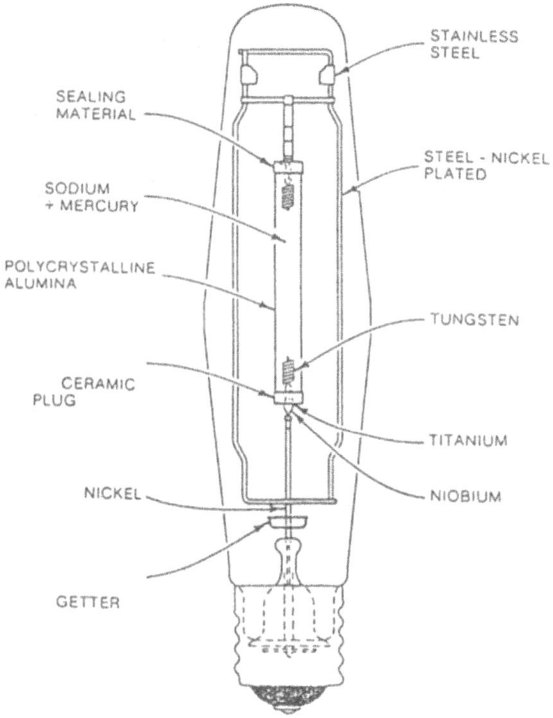

Figure 23. Basic construction of a high pressure sodium lamp [IES 1981].

Tritiated phosphor lamps consist of a sealed glass ampoule or tube containing a light emitting phosphor and the radioactive gas tritium. Tritium, which is an isotope (^3H) of hydrogen emits very low energy β radiation. The β radiation is absorbed by the phosphor which re-emits the energy in the form of visible radiation. The most commonly used phosphors emit green light and luminances up to about 10 cd m^{-2} can be achieved although around 2 cd m^{-2} is the norm. The glass envelope of the lamp effectively absorbs the residual β radiation and prevents the escape of tritium gas.

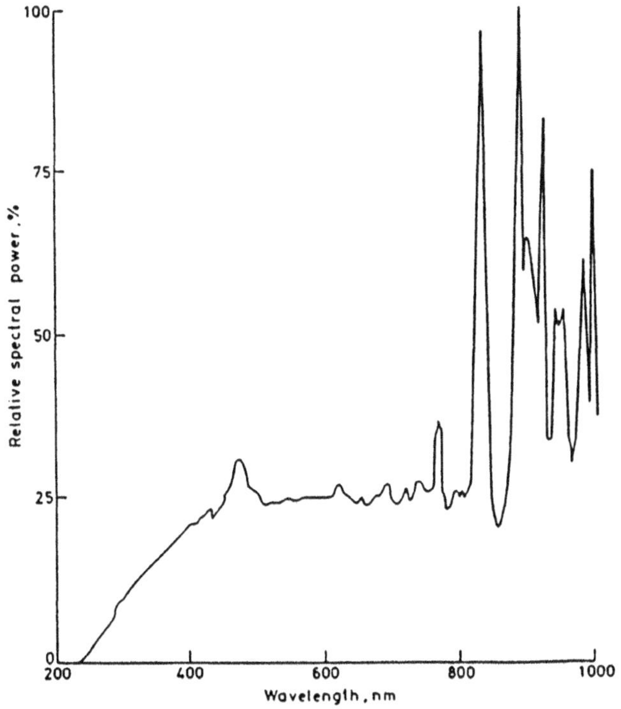

Figure 24. Spectral emission from a compact xenon lamp.

REFERENCES

ACGIH, (1988), "Threshold Limit Values for Chemical Substances and Physical Agents in the Workroom Environment". The American Conference of Governmental Industrial Hygienists, Cincinnati USA: ACGIH.

CIE, (1988), Electric Light Sources: State of the Art - 1987, Publication No 77, Vienna: CIE, (1988).

Cole, C. A., Forbes, P. D., Davies, R. E. and F. Urbach, "Effect of Indoor Lighting on Normal Skin". Report published by the New York Academy of Sciences, (1986).

Diffey, B. L. and McKinlay, A. F. (1983), "The UVB Content of 'UVA' Fluorescent Lamps and its Erythemal Effectiveness in Human Skin", Physics in Medicine and Biology, 28 (4), 351-358.

FDA (1988) "Performance standards for light-emitting products; paragraph 1040.30" in "Regulations for the administration and enforcement of the Radiation Control for Health and Safety Act of 1968." HHS Publication FDA 88-8035, (Rockville: Center for Devices and Radiological Health).

Henderson, S. T. and Marsden, A. M. ed. 1972 "Lamps and lighting" Second Edition, Edward Arnold, London.

IES (1981) "IES lighting handbook: Reference volume", Kaufman J. E. and Haynes, H. eds. (New York: Illuminating Engineering Society of North America) ISBN 0-87995-007-2.

McKinlay, A. F. and Diffey, B. L. (1987) "A Reference Action Spectrum for Ultra Violet Induced Erythema in Human Skin" CIE Journal 6(1), 17-22.

McKinlay, A. F., Harlen, F. and Whillock, M. J. (1988) "Hazards of Optical Radiation. A Guide to Sources, Uses and Safety", Adam Hilger, Bristol and Philadelphia. ISBN 0-85274-265-7.

McKinlay, A. F., Whillock, M. J. and Meulemans, C. C. E. (1989) "Ultra-violet Radiation and Blue Light Emissions from Spotlights Incorporating Tungsten Halogen Lamps", National Radiological Protection Board (NRPB), UK, Report R228.

Mohan, K., Knight, W., Chen, K. W., Leun, I. and Heinisch, (1980) "Optical Radiation Emissions from Selected Sources. Food and Drug Administration Report, FDA 81-36, Bureau of Radiological Health, Maryland, USA.

Philips (1988) "Comprehensive Handbook (Lighting) Including Technical Section", Philips Lighting, Croydon, UK.

Piltingsrud, H. V., Fong, C. W., Odland, L. T. (1978), Am. Ind. Hyg. Assoc. J. 39, 406-13.

Whillock, M. J., Clark, I., McKinlay, A. F., Todd, C. and Mundy, S. (1988) "Ultra-violet Radiation Levels Associated with the use of Fluorescent General Lighting, UVA and UVB Lamps in the Workplace and Home", National Radiological Protection Board (NRPB), UK, Report R221

Whillock, M. J., McKinlay, A. F, Kemmlert, J. and Forsgren, R. G. (1989) "Ultra-violet Radiation Emissions from Miniature (Compact) Fluorescent Lamps", Report in preparation.

CHARACTERISTICS OF SYNCHROTRON RADIATION

E. Burattini

CNR - INFN,
Istituto Nazionale di Fisica Nucleare
Laboratori Nazionali di Frascati
Via E. Fermi 40, 00044 Frascati, Rome, Italy

1. INTRODUCTION

When relativistic electrons experience centripetal acceleration due to the presence of a magnetic field, they emit electromagnetic waves that are commonly called "synchrotron radiation" because first observed in the visible region, at the 70 MeV small electrosynchrotron, by a research group working in 1947 at the General Electric Research Laboratory (Schenectady - N.Y.).

In 1944, Ivanenko and Pomeranchuk[1,2] drew attention to the fact that the electron beam circulating in the cyclic inductive accelerator, called the "betatron", was interrupted because of the high quantity of energy irradiated by the electrons.

Originally, synchrotron radiation struck the attention of the physicists because of its very interesting and unusual properties, such as high power, high degree of collimation in the vertical plane, very broad spectral distribution and linear polarization on the orbital plane. The emission of radiation from relativistic particles traveling in a magnetic field has, more recently, attracted the attention of astrophysicists as well, since the synchrotron radiation mechanism has been found present, from the radiofrequency range to the X-ray range, also in the radio emission of Crab Nebula and several supernovae.

In 1950, synchrotron radiation became more important in connection with the development and construction of cyclic accelerators like electrosynchrotrons. This new type of accelerator was able to produce increasingly intense photon beams used not only to carry out pioneering works in several different fields,[3,4,5,6] but also as a tool to understand the dynamics of the particles in the accelerator.

In 1955 the spectral and angular distributions of the ultra violet radiation emitted by the 300-MeV Cornell synchrotron were examined by Tomboulian and Hartman for various electron energies over a wide photon energy range.

In 1960 Codling and Madden began to use the 180-MeV electrosynchrotron at NBS in Washington as a light source to perform experiments in the far ultra violet energy range. In Western Europe, first experiments with synchrotron radiation were carried out in 1965 by a French-Italian cooperation utilizing the 1.1 GeV electrosynchrotron in operation at Frascati to measure the mass absorption coefficient of gold thin films in the wavelength region 26 to 120 Å.

At the beginning of 1970, a major advance occurred due to the fact that a new kind of synchrotron radiation source, the so-called "storage ring" for electrons and positrons, became available, with a great improvement of both the electron current circulating in the accelerator and the stability of the photon beams.[7,8,9,10]

Today, several storage rings around the world have been specifically designed, built and put in operation only to serve as synchrotron radiation sources, such as - to mention only a few examples - the 2 GeV storage ring in operation at Daresbury, the Photon Factory machine (E=2.5 GeV) in operation at Tzukuba (Japan), or the two storage rings working at the NSLS - National Synchrotron Light Source at Brookhaven (N.Y.)

The unique properties of synchrotron radiation has opened entirely new horizons in the study of solid state physics, atomic and molecular physics, chemistry, material science, molecular biology,[11,12] and so on.

In 1980, work in such exciting new fields as X-ray lithography,[13] tomography, X-ray microscopy,[14] trace element analysis,[15,16] and digital subtraction angiography started in several laboratories in the world on experiments that are impossible to attempt with conventional X-ray sources.[17,18]

2. RADIATION EMITTED FROM BENDING MAGNETS

In practice, synchrotron radiation is emitted by electrons or positrons forced to move into a vacuum chamber along a closed trajectory under the action of the Lorentz force due to the presence of a magnetic field generated by dipoles or quadrupoles. As we have already mentioned, the main synchrotron radiation sources are "storage rings", whose typical assembly is reported in Fig.1 which shows the vacuum chamber, the radiofrequency apparatus to replace the energy lost from the particles, the bending magnets, and two beam lines to collect the radiation emitted tangentially to the trajectory.

2.a Nonrelativistic case: $\beta=(v/c)\ll 1$

As a starting point, to evaluate the power P radiated by a single nonrelativistic accelerating particle with charge *e*, we can use the Larmor[19,20,21,22] formula:

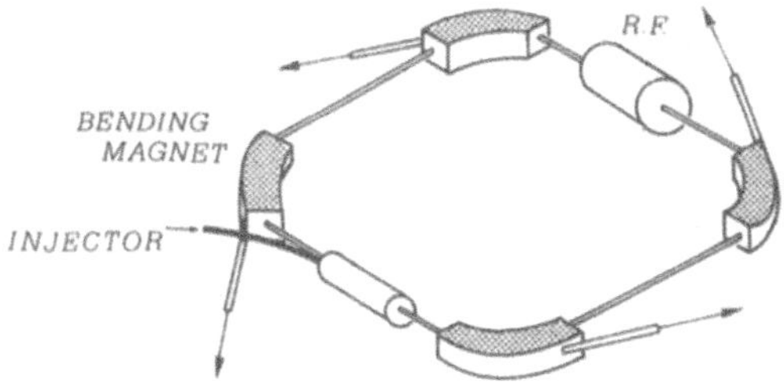

Fig.1 Sketch of an electron storage ring specifically designed as a synchrotron radiation source.

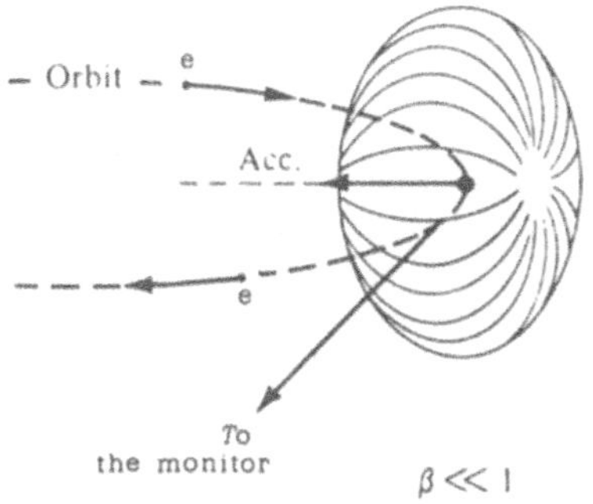

Fig.2 Angular distribution of synchrotron radiation in the nonrelativistic case.

$$P = \frac{2}{3}\frac{e^2}{c^3}\left|\frac{d\vec{v}}{dt}\right|^2. \tag{1}$$

In the special case of an electron moving along a circular orbit of radius R with constant velocity v, the power radiated is given by

$$P = \frac{2}{3}\frac{e^2}{c^3}\frac{v^4}{R^2}, \tag{2}$$

and the radiation is emitted in a dipole pattern like the one shown in Fig.2 The maximum intensity is emitted in a direction perpendicular to the orbital plane.

2.b Relativistic case $\beta=(v/c)\simeq 1$

In the relativistic case,[23,24,25,26] the radiation is emitted in a narrow cone very near to the orbital plane, as shown in Fig.3.

For a particle of mass m and kinetic energy E so that $\beta=(v/c)\simeq 1$, traveling in a circular orbit of radius R, Eq.(1) changes to

$$P=\frac{2}{3}\frac{e^2 c}{R^2}\beta^4\left(\frac{E}{m_0 c^2}\right)^4, \tag{3}$$

where E/m_0c^2 is the ratio of the particle energy and its rest energy. This quantity, indicated generally by γ, can be written for an electron as

$$\gamma=\frac{E}{m_0 c^2}=1957E\ \text{(GeV)}. \tag{4}$$

Remembering that the rest mass of a proton is roughly 2000 times greater than that of an electron, and noting that at very high energy particle momenta become practically independent of the rest mass so that electrons and protons can have comparable R value, it is clear that at the same kinetic energy an electron radiates $(2000)^4\simeq 10^{13}$ times more than a proton and, consequently, electrons, not protons, are used to obtain intense synchrotron radiation sources.

The total instantaneous power increases with the 4th power of the kinetic energy of the particles, which is completely different from the nonrelativistic case.

Finally, the instantaneous power emitted by the particles cannot be neglected in circular accelerators: consequently, suitable apparatus must be inserted into an accelerator to replace the energy lost by the particles to avoid losing the particle beam.

The energy U_0 radiated, per turn, by an electron can be evaluated multiplying Eq.(2) by the quantity $2\pi R/c\beta$ obtaining

$$U_0\ \frac{4\pi}{3}\frac{e^2}{R}\gamma^4. \tag{5}$$

In practical units (E in GeV, and R in meters) the energy U_0 can be written as

$$U_0=88.5\ \frac{E^4(\text{GeV})}{R(\text{meter})}, \tag{6}$$

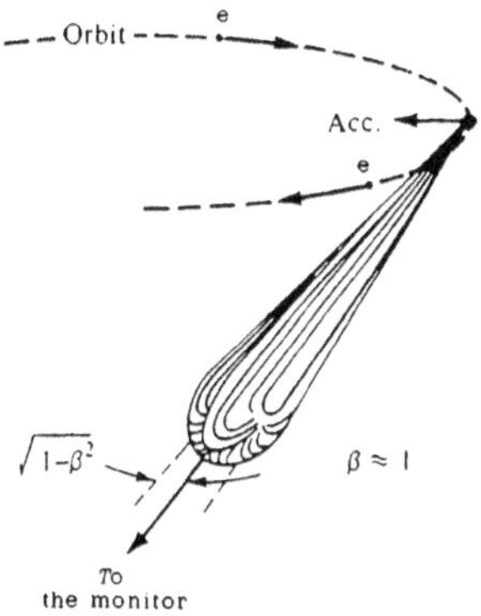

Fig.3 Angular distribution of synchrotron radiation in the relativistic case.

where U_0 is in keV per electron, per turn. For a storage ring like Adone in operation at Frascati (Italy), whose energy E is 1.5 GeV and R is 5 meters, the energy lost per electron, per turn is equal to

$$U_0 = \frac{88.5 \cdot (1.5)^4}{5} \text{ (keV)} = 89.6 \text{keV/electron turn}$$

From Eq.(6) it follows that the total power in kW radiated in a storage ring for electrons at energy E(GeV), with a circulating current I in ampere, is given by

$$W = 88.5 \ I \ \frac{E^4}{R}. \tag{7}$$

For the European Synchrotron Radiation Facility in construction at Grenoble (France), a storage ring with E=6 GeV, R=20 m and I=100 mA, Eq.(7) gives the following value for the total power emitted along all the orbit:

$$W = \frac{88.5 \times 0.1 \times (5)^4}{20} \text{ kW} = 276 \text{kW},$$

an incredible amount of energy.[27,28]

The power given by Eq.(7) is emitted over all the orbit and sometimes it could be more useful to know the power W_1 available for milliradians of horizontal angle, which is a quantity obtainable dividing Eq.(6) by $2\pi\cdot 10^3$, thus obtaining

$$W_1 = \frac{W}{2\pi\cdot 10^3} = 14.1\ \frac{IE^4}{R}, \tag{8}$$

where W_1 is given in watts for milliradians of orbit.

3. GENERAL PROPERTIES OF THE SYNCHROTRON RADIATION FROM A BENDING MAGNET

For a quantitive description of the properties of synchrotron radiation, a suitable starting point is Schwinger's formula[29,30,31] for the total power P emitted by a monoenergetic relativistic electron traveling on a circular orbit of radius R:

$$P(\lambda\cdot\psi) = \int I(\lambda\cdot\psi)d\lambda\ \ d\psi, \tag{9}$$

where $I(\lambda,\psi)$, given by

$$I(\lambda\cdot\psi) = \frac{27}{32\pi^3}\frac{e^2c}{R^3}\left(\frac{\lambda_c}{\lambda}\right)^4\gamma^8[1+(\gamma\psi)^2]^2$$
$$\left[K^2_{2/3}(\xi) + \frac{(\gamma\psi)^2}{1+(\gamma\psi)^2}K^2_{1/3}(\xi)\right] \tag{10}$$

is the instantaneous power radiated by one single electron along the whole orbit per revolution in the direction ψ, which is measured from the orbital plane, with wavelength λ. In Eq.(10), λ_c is the so-called "critical wavelength", $K_{1/2}$ and $K_{3/2}$ are modified Bessel functions of the second kind, and ξ is given by

$$\xi = \left(\frac{\lambda_c}{2\lambda}\right)[1+(\gamma\psi)^2]^{3/2}.$$

Equation (10) can be integrated over the elevation angle ψ obtaining the instantaneous power P_1 radiated by one electron per turn in the whole solid angle as a function of wavelength:

$$P_1(\lambda) = N\frac{c^2 c}{R^3}\gamma^7\left(\frac{\lambda_c}{\lambda}\right)^3\int_{\lambda/c}^{\infty} K_{5/3}(\eta)\,d\eta,$$

where N is a numerical factor. The quantity $P_1(\lambda)d\lambda$ gives the spectral distribution of the synchrotron radiation. Due to the relativistic effect, the frequency distribution of synchrotron radiation extends from the fundamental frequency $\omega_0 = v/R$ of the circulating electrons to very high harmonics. It could be demonstrated that the frequency cut-off, which corresponds to the critical wavelength λ_c, is given approximately by

$$\omega_c \cong \omega_0 \gamma^3. \tag{11}$$

For a storage ring like Adone ($\omega_0 \simeq 2\pi \times 10^7$ Hz, $\gamma = 3\times10^3$), the high frequency cut-off is roughly equal to 10^{18}, which is a value corresponding to photons whose wavelength is equal to few angstrom. Consequently, the radiation is distributed over a large energy range generally extending from the infrared to the X-ray region; the shape of the general spectrum is shown in Fig.4. The horizontal wavelength scale is defined in terms of the ratio λ/λ_c and the vertical scale in terms of the spectral flux defined as the number of the photons per second, emitted in a relative bandwidth $\Delta\lambda/\lambda = 0.1\%$, in a milliradian of horizontal angle and integrated in the vertical plane. As shown in the figure, the spectral distribution is continuous and shows its maximum for $\lambda = 0.42\ \lambda_c$. It falls off exponentially for $\lambda \ll \lambda_c$ and decreases slowly for $\lambda \gg \lambda_c$. In the low energy region, the spectral distribution is practically independent of the energy of the particles. The critical wavelength which divides the spectrum into two parts of equal radiated power is given by

$$\lambda_c = \frac{4\pi}{3}\ \frac{R}{\gamma^3}. \tag{12}$$

In practical units λ_c can be written as

$$\lambda_c = 5.59\ \frac{R}{E^3}. \tag{13}$$

Taking account that between energy E, radius R, and magnetic field B, there exists the relationship

$$R(m) = 3.33\ \frac{E(GeV)}{B(T)},$$

the critical wavelength can be written as

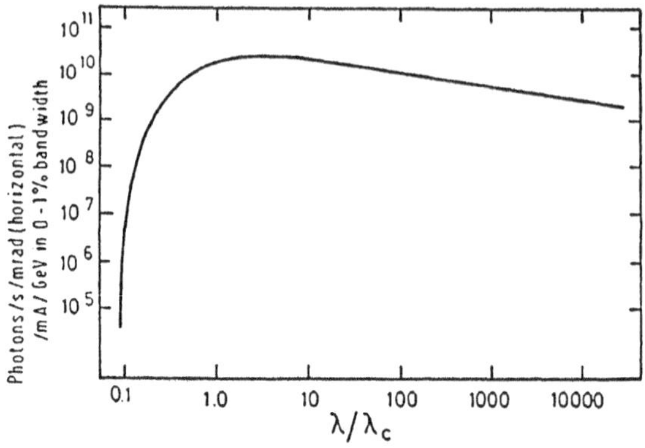

Fig.4 General shape of the spectral distribution of the synchrotron radiation per GeV.

$$\lambda_c = 18.64 \frac{1}{E^2(\mathrm{GeV})B(\mathrm{T})}, \tag{14}$$

where λ_c is in Å, R in meters, E in GeV, and B in tesla. The shape of the spectrum reported in Fig.4 indicates that in order to obtain a good spectral flux extending to the X-ray region, λ_c must be equal to 1 Å or shorter and this can be achieved with a value of E of some GeV and R no more than a few meters. The radiation emitted in a very small angular cone around the instantaneous direction of flight of the particle is shown in Fig.3. The angular distribution $\Delta\psi$ is given roughly by

$$\Delta\psi = \frac{1}{\gamma}\left(\frac{\lambda}{\lambda_c}\right)^{1/3} \qquad \text{for } \lambda \gg \lambda_c,$$

and

$$\Delta\psi = \frac{1}{\gamma}\left(\frac{\lambda}{\lambda_c}\right)^{1/2} \qquad \lambda \ll \lambda_c$$

For $\lambda = \lambda_c$, $\Delta\psi$ is equal to γ^{-1}.

For a storage ring working in the GeV region, the emission angle is of the order of 0.1 milliradians; thus, the emission in the off plane direction is typically confined in a region of only about one milliradian of angular spread. This property together with the wide spectral range makes synchrotron radiation a source with high spectral brilliance over a large energy range.

For a synchrotron radiation source, the spectral brilliance is generally defined as the number of photons per second, per milliradian square, per millimeter square in a relative bandwidth ($\Delta\lambda/\lambda$) equal to 0.1%. It follows from the definition that to achieve high brilliance, both small angular dispersion and small source size (the electron beam dimensions) are required and only synchrotron radiation shows both these properties.

The synchrotron radiation is 100% polarized in the orbital plane with the electric vector parallel to the orbital plane. Above and below this plane, the radiation is elliptically polarized due to the presence of a polarization component perpendicular to the orbital plane. The two terms present in Eq.(10) in the square brackets are associated with the two components of polarization: the first describes the component with the electrical vector E parallel to the orbital plane; the second one describes the component with E perpendicular to the orbital plane.

If we adopt as the definition for the degree of polarization the quantity given by

$$P=\frac{I_{\parallel}-I_{\perp}}{I_{\parallel}+I_{\perp}},$$

where $I_{\parallel}$ and $I_{\perp}$ are the parallel and perpendicular components respectively, for P we obtain

$$P=\frac{K_{2/3}^2(\xi)-[(\gamma\psi)^2/(1+(\gamma\psi)^2)]K_{1/3}^2(\xi)}{K_{2/3}^2(\xi)+[(\gamma\psi)^2/(1+(\gamma\psi)^2)]K_{1/3}^2(\xi)}.$$

This equation shows that for $\psi=0$, the radiation is 100% linearly polarized with the electric vector parallel to the orbital plane. Figure 5 reports the intensities of the parallel and perpendicular polarization components as a function of emission angle for different wavelengths. As can be seen from the figure, the so-called opening angle ψ is approximately γ^{-1} only for $\lambda\approx\lambda_c$. At shorter wavelengths it becomes smaller and at longer wavelengths it becomes increasingly larger.

Finally, the electrons travel in a storage ring in a compact group - so-called bunches - in synchronism with an applied radiofrequency field which replaces the energy lost by the synchrotron radiation mechanism. Thus, the radiation is emitted in short flashes of bunch length time and shows a very stable intensity from bunch to bunch. To perform particular spectroscopic experiments, the repetition frequency is an important parameter which is determined by the repetition frequency of the filled

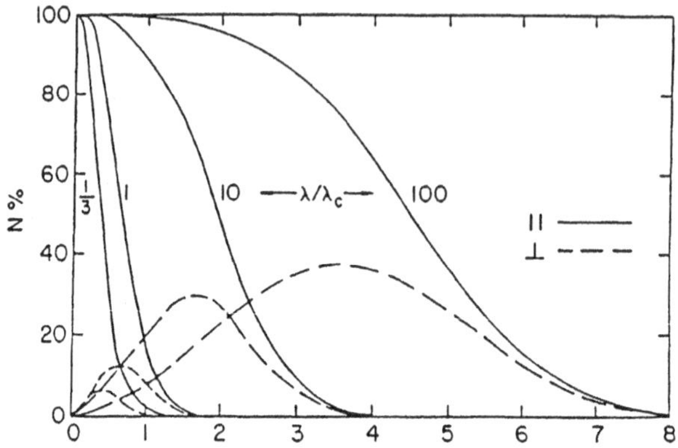

Fig.5 Intensities of parallel and perpendicular polarization components as a function of emission angle for different wavelengths.

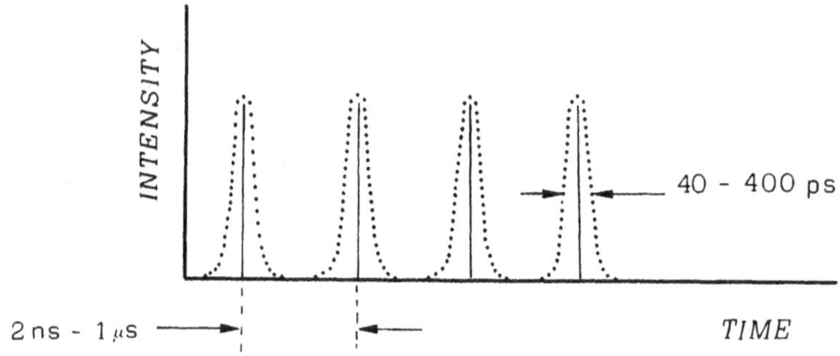

Fig.6 Intensity of the radiation reaching the detector as a function of time for a typical storage ring.

bunches; maximum separation between two contiguous pulses can be obtained working in a single bunch mode. Figure 6 reports the intensity of radiation reaching the detector as a function of time for a typical storage ring.

4. SYNCHROTRON RADIATION FROM WIGGLERS AND UNDULATORS

Up to now we have considered the radiation emitted from the circular trajectories described by electrons moving in the bending magnets of a storage ring. It is, however, possible to design different magnetic structures that are more effective sources of radiation than that coming from a bending magnet. The structure of such devices, known as wigglers[32,33] and used as a synchrotron radiation source, is reported in Fig.7. It is clear, following Eq.(14), that the critical wavelength λ_c can be decreased by increasing the magnetic field B. For a variety of technical and economic reasons, the field in bending magnets is limited to 10-14 kG, whereas wiggler fields equal to 20 kG are possible using conventional magnets and fields of 50 kG or more can be produced using superconducting magnets. Thus, standard wigglers can shift the spectrum of a relatively low-energy machine into the hard X-ray region; in addition, increasing the number N of the particle beam oscillations, the photon flux can be increased by a factor roughly equal to the number of the oscillations.

More generally speaking, similar periodic alternating magnetic field devices, used to force the electrons to make several "wiggles", as shown in Fig.8, are called "wigglers" or undulators depending on their characteristic features. In the approximation that the magnetic field is exactly sinusoidal, they are best characterized by the so-called "deflection parameter K" given by

$$K = eB_0\lambda_0/2\pi m_0 c = \alpha\gamma = 0.0934 \quad B_0\lambda_0,$$

where B_0 is the peak magnetic field in tesla, λ_0 is the magnet spatial period, and α the maximum angle deflection. Remembering that between γ and the opening angle ψ there exists the relaxation $\psi \simeq \gamma^{-1}$, it is clear that K is the ratio between the maximum deflection angle and the natural opening angle of the radiation.

The distinction between multipole wigglers and undulators is not so sharp. For $K \gg 1$, the device is called "multipole wiggler" which can, as just mentioned, decrease the critical wavelength and increase the flux by a factor approximately equal to the number of poles. The simplest standard wiggler produces just one full oscillation or "wiggle" of the beam. Generally, a device like this, shown schematically in Fig.9, is called a 1λ standard wiggler or "wavelength shifter" because it can shift the critical wavelength without flux shifting. The comparison between the flux of the radiation emitted by a bending magnet of the Adone (Frascati) storage ring with the flux emitted from the wiggler (single-pole) is shown in Fig.10.

For K less than 1 or of the order of 1, strong interference effects occur which result in a spectrum of the emitted radiation consisting of - for $K \ll 1$ - or several - for $K \simeq 1$ - quasi-monochromatic lines called harmonics. In these cases, the device is called "undulator" and the wavelength of the nth harmonic at an observing angle θ relative to the device axis is given by

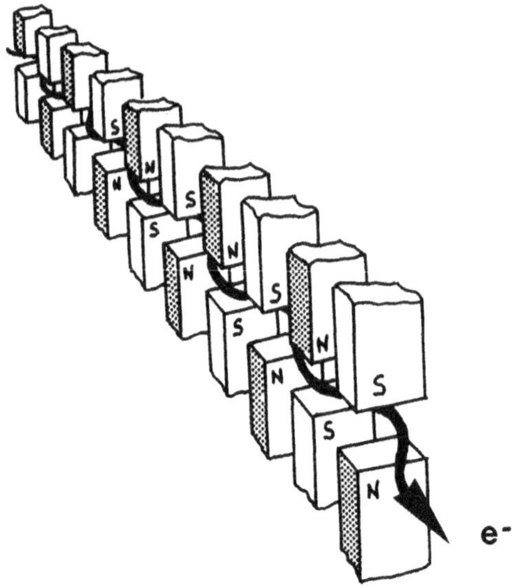

Fig.7 Typical magnetic structure (wiggler) used to obtain a more intense synchrotron radiation source.

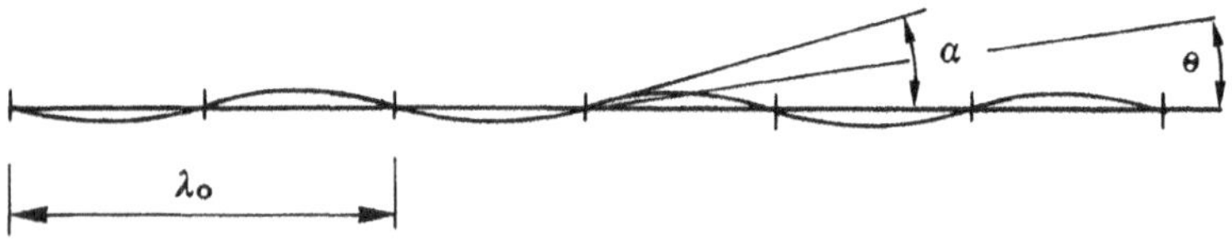

Fig.8 Trajectory of an electron beam traveling in a multipole wiggler.

$$\lambda_n = \frac{\lambda_0}{2n\gamma^2}\left(1 + \frac{K^2}{2} + \gamma^2\theta^2\right). \qquad n = 1, 2, 3\ldots$$

For $\theta = 0$, i.e., on the axis, one has

$$\lambda_n = \frac{\lambda_0}{2n\gamma^2}\left(1 + \frac{K^2}{2}\right).$$

For $n = 1$, the wavelength,

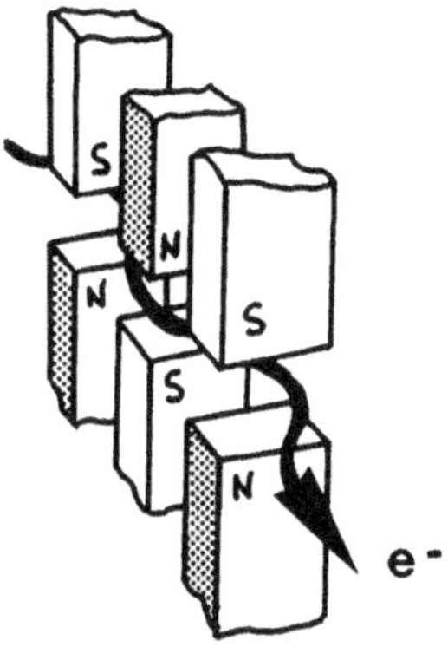

Fig.9 Schematic representation of a wiggler wavelength shifter.

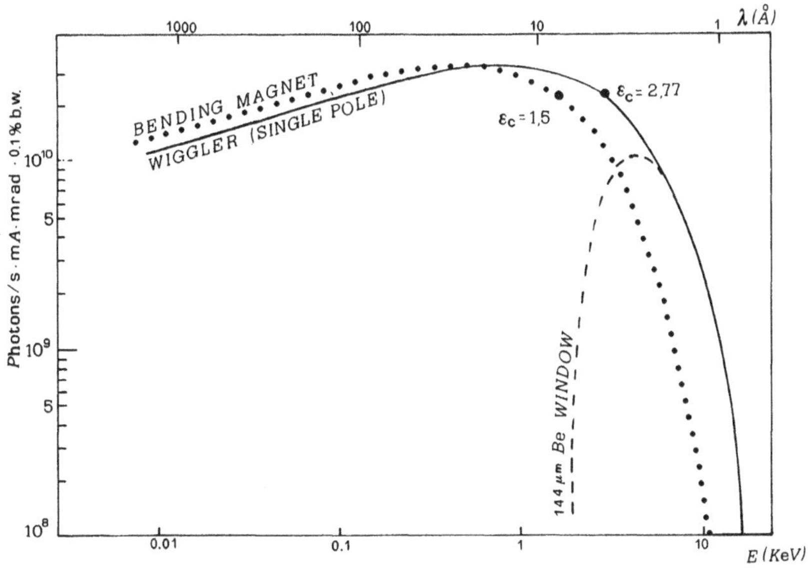

Fig.10 Comparison between the flux of the radiation emitted from a bending magnet of the Adone storage ring with the flux emitted from the wiggler (B=1.85 T - single pole).

$$\lambda = \frac{\lambda_0}{2\gamma^2}\left(1 + \frac{K^2}{2}\right),$$

is called "fundamental".

The natural relative bandwidth of the emitted radiation for an ideally parallel electron beam is given by

$$\frac{\Delta\lambda_n}{\lambda_n} = \frac{1}{nN},$$

where N is the number of poles.

To conclude, some basic synchrotron radiation relationships, which are very useful for any kind of user, are reported in Table 1.

TABLE 1

Basic synchrotron radiation relationships

Total power radiated by a relativistic electron:

$$P = \int I(\lambda, \psi) d\lambda d\psi = \frac{2e^2c}{3R^3} \frac{E^4}{(m_0c^2)^4}.$$

Energy loss per turn, per electron:

$$\delta E(\mathrm{keV}) = 88.5 \frac{E^4(\mathrm{GeV})}{R(\mathrm{m})}.$$

Total power radiated:

$$P_{tot}(\mathrm{kW}) = 26.6 E^3(\mathrm{GeV}) \cdot B(\mathrm{T}) \cdot I(\mathrm{A}).$$

Critical wavelength:

$$\lambda_c(\mathrm{\mathring{A}}) = 5.59 \frac{R(\mathrm{m})}{E^3(\mathrm{GeV})} = \frac{18.64}{B(\mathrm{T}) \cdot E^2(\mathrm{GeV})}.$$

Critical energy:

$$\epsilon_c(\mathrm{keV}) = 2.218 \frac{E^3(\mathrm{GeV})}{R(\mathrm{m})} = 2.96 \times 10^{-7} \frac{\gamma^3}{R(\mathrm{m})}.$$

Opening angle:

$$\psi \approx \frac{1}{\gamma} = \frac{mc^2}{E},$$

with

$$\gamma = \frac{E}{m_0c^2} = 1957 E \ (\mathrm{GeV}).$$

Energy of a photon:

$$\epsilon(eV) = \frac{12.40}{\lambda(\mathring{A})}.$$

For an undulator
Deflection parameter:

$$K = eB_0\lambda_0/2\pi m_0 c = \alpha\gamma.$$

N^{th} harmonic

$$\lambda_n = \frac{\lambda_0}{2n\gamma^2}\left(1 + \frac{K^2}{2} + \gamma^2\theta^2\right).$$

REFERENCES

1. D. Ivanenko and J. Pomeranchuk, Phys. Rev. 65:343 (1944).
2. D. Ivanenko and A.A. Sokolov, Doke. Akad. Neuk. 59:1551 (1948).
3. R.P. Madden and K. Codling, Phys. Rev. Lett. 10:516 (1963).
4. L.G. Parrat, Rev. Sci. Instrum. 30:297 (1959).
5. P. Jaeglé, G. Missoni, and P. Dhez, Study of the Absorption of Ultrasoft X-rays by Bismuth and Lead using the Orbit Radiation of the Frascati Synchrotron, Phys. Rev. Lett., Vol.18, 21:887 (1967).
6. P. Jaeglé, F. Combet-Farnaux, P. Dhez, M. Cremonese, G. Onori, Experimental and Theoretical Study of the Absorption of Ultra-Soft X-rays in Platinum and Tantalium, Phys. Lett. 26A, 8:364 (1968).
7. E.E. Koch, Nucl. Instrum. Methods 177:7 (1980).
8. C. Kunz, Perspective of Synchrotron Radiation in: "Vacuum Ultraviolet Radiation Physics", p.753, Viewey Pergamon Braunschweig (1979).
9. V. Rehn, Nucl. Instrum. Methods 177:193 (1980).
10. E.M. Rowe, "Phys. Today", p.28 (1981).
11. M. Belli, A. Scafati, A. Bianconi, E. Burattini, S. Mobilio, C.R. Natoli, L. Palladino, A. Reale, EXAFS and XANES Structure Determination of M_n^{+2} Binding in ATP Complexes, Il Nuovo Cimento Vol.2D, No.5:1281-1304 (1983).
12. A. Bianconi, A. Congiu-Castellano, M. Dell'Ariccia, A. Giovannelli, P.J. Durham, E. Burattini, and M. Barteri, XANES Study of Iron Displacement in the Haem of Myoglobin, FEBS Letters 1:2039 (1984).
13. A. Heuberger, H. Betz, and G. Pongratz, Present Status and Problems of X-ray Lithography, Festkörper probleme - Adv. in Solid State Phys. 19:259, Vieweg Braunschweig (1980).
14. M. Yoshimatsu and S. Kozaki, "High Brilliance X-ray Source - X-ray Optics", Springer Verlag, Berlin-Heidelberg-New York, 9, 1977.

15. J.M. Kenney, C. Jacobsen, J. Kirz, M. Rarback, F. Cinotti, W. Thomlinson, R. Rosser, G. Schidlowsky, Absorption Microanalysis with a Scanning Soft X-ray Microscope, Journal of Microscopy 138:321 (1984).

16. E. Burattini, Synchrotron Radiation Applications in Biophysics and Medicine, Physics in Environmental and Biomedical Research 23-31, World Scientific Publishing (1986).

17. A. Akisada,M. Ando, K. Hyodo, A. Maruhashi, M.A. Kisada, K. Konishi, F. Toyofuku, K. Nishimura, S. Hasegawa, A. Suwa, K. Korha, An Attempt at Coronary Angiography with a Large Size Monochromatic SR Beam, Nucl. Instrum. Methods A246:713 (1986).

18. E. Rubenstein, E.B. Hughes, L.E. Campbell, R. Hofstadter, R.L. Kirk, T.J. Krolicki, J.P. Stone, S. Wilson, H.D. Zeman, W.R. Brody, A. Macowsky, and A.C. Thompson, SPIE 314:42 (1981).

19. R. Becker, "Teoria dell'elettricitá", Vol. I,II, Sansoni Edizioni Scientifiche - Firenze (1950).

20. A. Sommerfeld, "Elettrodinamica", Sanzoni Edizioni Scientifiche, Firenze (1950).

21. R.P. Godwin, Springer Tracts, Modern Physics 51:1-73 (1968).

22. J.D. Jackson, "Classical Electrodynamics", Wiley, New York (1975).

23. A.A. Sokolov and I.M. Ternov, "Synchrotron Radiation", Pergamon Press, New York (1968).

24. A.A. Sokolov and I.M. Ternov, "Radiation from Relativistic Electrons", American Institute of Physics, New York (1968).

25. E.E. Koch, "Synchrotron Radiation", Vol.1a, North Holland Publishing Company (1983).

26. E. Burattini, "Synchrotron Radiation applied to Biophysical and Biochemical Research", Plenum Press, New York and London 1-27 (1979).

27. E. Burattini, A. Reale, E. Bernieri, N. Cavallo, A. Morone, M.R. Masello, A. Rinzivillo, G. Dalba, P. Fornasini, C. Mencuccini, Experimental Activity at the Adone Wiggler Facility, Nucl. Instrum. Methods 208:91 (1983).

28. B. Buras and S. Tazzari, European Synchrotron Radiation Facility, CERN - Geneva (1984).

29. J. Schwinger, Phys. Rev. 70:798 (1946).

30. J. Schwinger, Phys. Rev. 75:1912-25 (1949).

31. H. Winich and S. Doniach, "Synchrotron Radiation Research", Plenum Press, New York-London (1980).

32. H. Mott, J. Appl. Phys. 22:527 (1951).

33. E. Burattini, A. Balerna, E. Benieri, C. Mencuccini, R. Rinzivillo, G. Dalba, P. Fornasini, Adone Wiggler Beam Lines Progress Report, Nucl. Instrum. Methods A246;125 (1986).

CHARACTERISTICS OF COHERENT OPTICAL SOURCES

FRANZ HILLENKAMP[1] and WESLEY J. MARSHALL[2]

[1]INSTITUTE OF MEDICAL PHYSICS
UNIVERSITY OF MÜNSTER
MÜNSTER, FEDERAL REPUBLIC OF GERMANY

[2]LASER MICROWAVE DIVISION
U.S. ARMY ENVIRONMENTAL HYGIENE AGENCY
ABERDEEN PROVING GROUND, MARYLAND 21010-5422

INTRODUCTION

Coherent Optical Sources are with few exceptions (see Synchrotron radiation) *Lasers*. Therefore only lasers will be considered here as a source of coherent radiation.

RELEVANT PARAMETERS OF LASERS

Wavelength

The main range of wavelengths important for interaction with biological systems extends from 193 nanometers (nm) (border of vacuum UV) to 10 micrometers (μm) in the far infrared. To judge the influence of the emission wavelength, one has to follow the sequence of: Radiation producing absorption by a given chromophore which then produces a biological effect. The (very generalized) absorption characteristics of soft tissue as a function of the wavelength are shown in Figure 1. The solid line represents the ubiquitous absorption of the proteins and nucleic acids in the UV and the tissue water in the IR, exhibited by all tissues. In addition to these main absorbers, there are two other strong absorbers which, however, are confined to specific tissue structures: (Oxy) hemoglobin and melanin. Other absorbers, such as rhodopsin or a large variety of enzymes, have strong molar absorbency; but are not found in high enough a concentration in tissue to account for a major energy transfer.

Temporal Emission Characteristics

The temporal exposure characteristics also strongly influence the interaction between laser radiation and tissue. The following is a very condensed list of the relevant parameters and their typical ranges of variation.

Single pulse exposure:

- Relevant quantities:
  - exposure time, t: 10^{-14} s $\leq$ t $\leq$ 10^{6} s
  - energy per pulse, Q: 10^{-9} J $\leq$ Q < 10^{6} J

- Related quantities:
  - Power P: P = Q/t
  - Radiant exposure, H: H = Q/A (A=exposed area)
  - Irradiance, E: E = P/A

Exposure by repetitive pulses:

- Relevant quantities:
  - pulse repetition frequency, PRF.
  - pulse repetition period, T=1/PRF.
  - duty cycle = t/T.

- Related quantities:
  - peak power, $\hat{P}$, and average power, $\bar{P}$, as illustrated in Figure 2.
  - peak and average values of radiant exposure: $\hat{H}$, $\bar{H}$, and irradiance: $\hat{E}$, E.

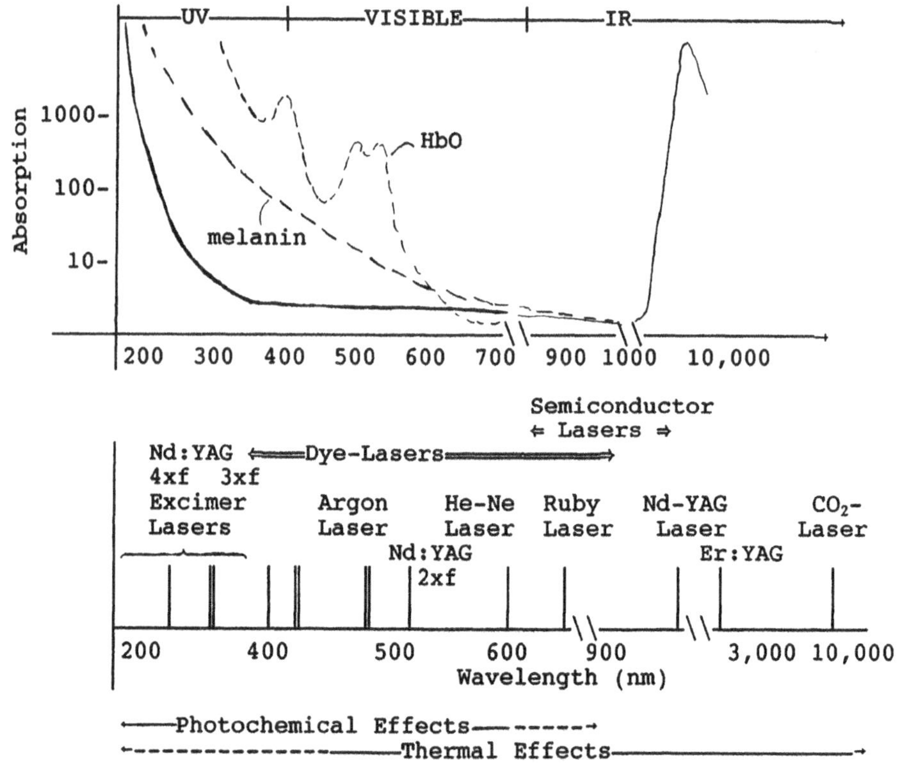

Figure 1. Comparison of the wavelength of common lasers to the spectral absorption of biological tissue. The dashed line represents a minor biological effect and the solid line represents a major effect.

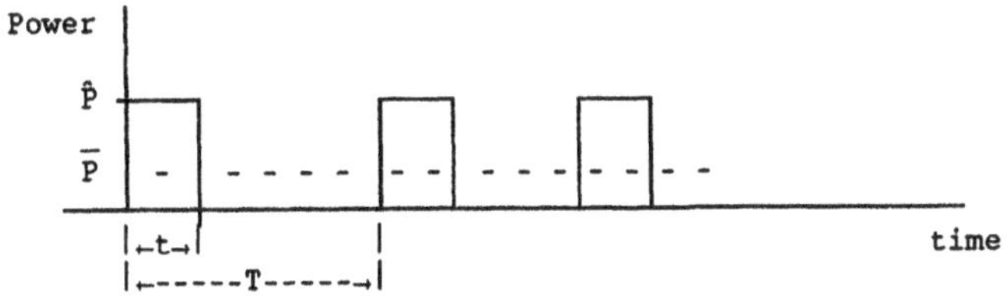

Figure 2. Graphical representation of peak power, average power, pulse width, and pulse period.

Biological effects usually depend in a complex fashion on both, wavelength, λ, and exposure time, t. No well defined parameter ranges exist for the different primary effects and, more often than not, several occur simultaneously. Plasma formation occurs for laser pulse widths less than a few microseconds (µs). Mechanical (acoustic) effects predominate for pulse widths from a picosecond (ps) to about a µs. Both acoustic effects and plasma formation are observed in the whole wavelength range. Photochemical effects predominate for lasers with a wavelength less than 400 nm and a pulse width of at least 1 nanosecond (ns). The blue light photochemical effect in the retina requires a pulse duration of a few seconds and a wavelength less than 550 nm. Thermal effects generally occur when the pulse width is at least 1 millisecond (ms) and the wavelength greater than 400 nm.

THE NATURE OF COHERENCE

Coherence is the property of radiation fields which determines their ability to form stable interference patterns. Or, in other words, it is the ability of radiation fields to get superimposed such that the *amplitudes* of the waves add (or subtract!) linearly rather than the energies, powers, or irradiances.

Example: Two waves of equal amplitude, A, and equal power, $P = \alpha\cdot A^2$, are superimposed. If they are coherent, they give (in the extreme cases) either: $A_s = A_1 + A_2 = 2A$; $P_s = \alpha\cdot A_s^2 = 4P^2$ if they are in phase, or $A_s = 0$, $P_s = 0$ if they are out of phase.

If they are incoherent, they always add up to: $P_s = 2P$ if averaged over time. It is helpful, to discuss two aspects of coherence separately: temporal coherence and spatial coherence.

Temporal Coherence

A source has temporal coherence if it emits a wave train which retains a fixed phase over a given coherence time, Δt_c. Temporal coherence is directly related to the monochromaticity, i.e., the bandwidth of the radiation $\Delta\lambda$, Δf or $\Delta\omega$:

$$\Delta t_c = 1/\Delta\omega\,, \qquad \Delta t_c = \text{coherence time}\,,$$
$$\Delta\omega = 2\pi\Delta f\,, \qquad \Delta f = \frac{c}{\lambda^2}\Delta\lambda\,. \tag{1}$$

The coherence length, Δl_c, is the distance light travels during the coherence time, Δt_c:

$$\Delta l_c = c\cdot\Delta t_c\,, \quad c = 3\cdot 10^8\ m/s\,. \tag{2}$$

Spatial Coherence

An extended source has spatial coherency if two separate points can emit waves which retain a fixed phase relationship over a given time. The smaller a source, the better its spatial coherence. Good collimation (a "parallel" beam) is equivalent to a "point" source; as will be shown later.

Relevance of Coherence for Biological Effects and Hazard Potential

Temporal coherence is of very limited relevance, if of any at all, because the absorption bands of biologically relevant molecules have bandwidths of typically tens of nanometers.

$$e.g.,\ \Delta\lambda = 20\ nm \rightarrow \Delta\omega \sim 10^{14}\ s^{-1} \rightarrow t_c = 10^{-14}\ s, \tag{3}$$

in contrast in spectroscopy:

$$\Delta\lambda \geq 10^{-7}\ nm \rightarrow \Delta\omega = 10^{6}\ s^{-1} \rightarrow t_c = 10^{-6}\ s. \tag{4}$$

Spatial coherence is of great importance because it determines the "focusability" of the radiation. i.e., the maximum energy or power per unit area (radiant exposure or irradiance, respectively) which can be attained under optimum focussing conditions, e.g., by the optics of the eye, as well as the distance, over which the energy can be transmitted without essentially losing the focusability (by real lenses), i.e., the collimation of the beam.

Temporal coherence (i.e., coherence length) and spatial coherence can be lumped together into the concept of a *coherence volume*. For a more extended discussion of coherence and its practical consequences, see Demtroeder (Demtroeder, 1988).

BEAM PROPAGATION AND BEAM TRANSFORMATION

The following discussion assumes ideal lenses without aberration.

Classical, Extended, Noncoherent Sources

These sources are mainly thermal emissions or discharge lamps. Any real lens can collect only radiation emitted from a finite area of the source, A_s, and a finite solid angle of the source, Ω_s, as shown in Figure 3. D_s represents the diameter of the object, and D_i represents the diameter of the image. α_s and α_i represent the linear angles subtended by the lens at the location of the source and the image, respectively.

$$\Omega_s = 2\pi(1-\cos\alpha_s)\ , \qquad A_s = \left(\frac{D_s}{2}\right)^2 \cdot \pi . \tag{5}$$

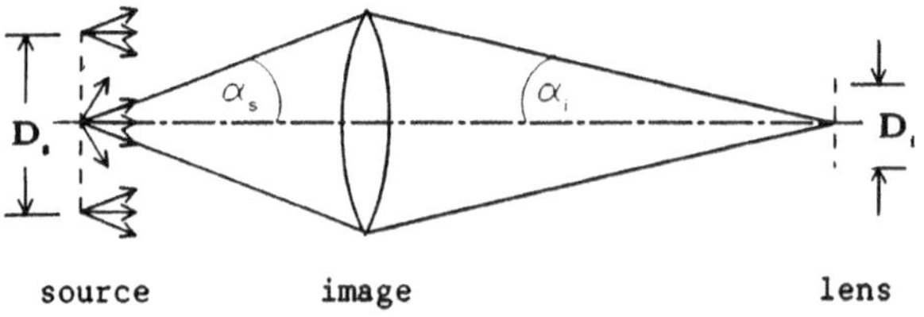

Figure 3. Illustration of an extended source transformed by a lens.

The brightness of the image (i.e., the ratio of the irradiance, E_i, and solid angle, Ω_i) is always less than the brightness of the source.

For distant nonlaser sources, it becomes increasingly difficult to collect appreciable amounts of energy in an image because of the limited size of lenses. For such sources the hazard therefore decreases typically with $1/r^2$ (r = distance from source to lens).

Sources with Limited Coherence: Lasers with Poor Mode Structure

The beams of such sources are usually described as collimated beams of diameter, D_B, typically in the millimeter range and a divergence, θ_B, typically in the 1 mrad range. Figure 4 is an illustration of a laser beam with diameter, D_L, transformed by a lens of approximately the same diameter. The smallest spot in the focal plane of the lens is:

$$d_F = \theta_B \cdot f. \tag{6}$$

The important parameter is the ratio of the irradiance in the focus, E_F, over that in the beam, E_B:

$$\frac{E_F}{E_B} = \frac{P}{A_F}\cdot\frac{A_B}{P} = \left(\frac{D_B}{d_F}\right)^2 = \left(\frac{D_B}{\theta_B f}\right)^2 \tag{7}$$

This ratio is maximized if a lens with some F# is fully filled by a beam, i.e., with $F_\# = f/D_L$ and $D_B = D_L$:

$$\frac{E_F}{E_B} = \frac{1}{(\theta_B F_\#)^2} \tag{8}$$

Galilean telescopes, as illustrated in Figure 5, can be used to adapt a given beam of D_B and θ_B to a given lens of D_L, f_L and F#:

$$\frac{\theta_2}{\theta_B} = \frac{D_B}{D_L} \quad ; \quad d_F = f_L\theta_2 = f_L\frac{D_B}{D_L}\,\theta_B \; ; \tag{9}$$

$$\frac{E_F}{E_B} = \frac{1}{(\theta_B F_\#)^2} \quad \text{instead of} \quad \frac{E_F}{E_B} = \frac{1}{(\theta_B F_\#)^2}\left(\frac{D_B}{D_L}\right) \tag{10}$$

without telescope.

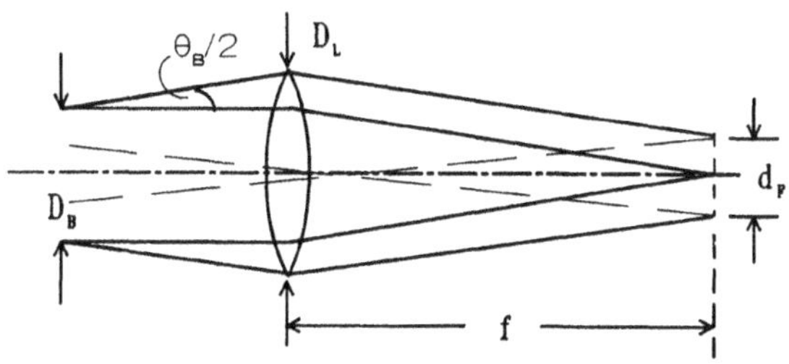

Figure 4. Collimated laser beam focussed by lens.

Example: ruby laser beam: $D_B = 7$ mm, $\theta_B = 10^{-3}$ rad, filling the 7-mm diameter iris of the eye with focal length (in air) of $f_L = 25$ mm. The ratio of the irradiance on the retina compared to that on the cornea is:

$$\frac{E_{Retina}}{E_{Cornea}} = \left(\frac{D_B}{\theta_B f_L}\right)^2 = 10^5 \tag{11}$$

Monomode Lasers Beams with Full Coherence. Generation and Transformation of Gaussian Beams

The generalized spherical resonator. The generalized spherical resonator is illustrated in Figure 6. The equation for the radius of curvature of the laser beam at all points along the Z' axis is given by:

$$R(Z') = Z'\left[1+\left(\frac{d}{2Z'}\right)^2\right]. \tag{12}$$

The local irradiance across the center of the Gaussian beam, E(r), as illustrated in Figure 7, is given by:

$$E(r) = E(o)\ e^{-\frac{r^2}{\sigma^2}} \tag{13}$$

Definition: The beam diameter 2W is defined by that radius r, at which the irradiance has fallen to $E(w) = E(o)/e^2$

$$E(r) = E(o)\ e^{-\frac{2r^2}{W^2}}\ ;\quad W = \sigma\sqrt{2}. \tag{14}$$

In order to choose a beam limiting aperture such that its effect on beam distortion is minimal [because at its edge, $r=D/2$, $E_{(D/2)}$ must be equal to $E(o)/100$], an aperture with a diameter, D, must be chosen such that $D = 3.04 \cdot W$. W is the Gaussian beam radius at the location of the aperture.

Gaussian (TEM_{00}) beam. The irradiance of a laser beam measured radially across the center is given by:

$$E(r) = E_0\, e^{-\frac{2r^2}{W^2}}, \tag{15}$$

where E(r) represents the local irradiance at any point from the center of the beam and E_0 represents the center beam irradiance.

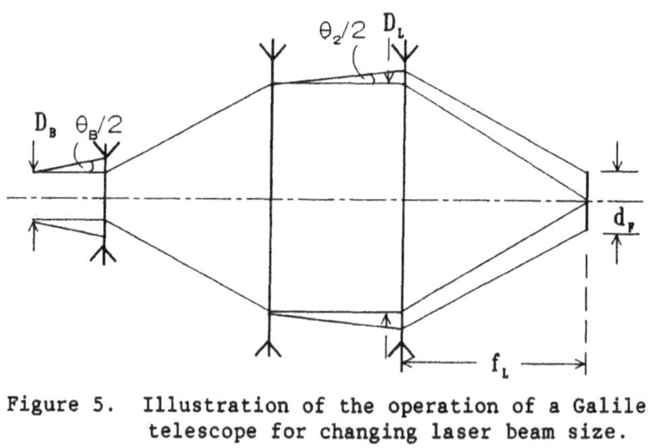

Figure 5. Illustration of the operation of a Galilean telescope for changing laser beam size.

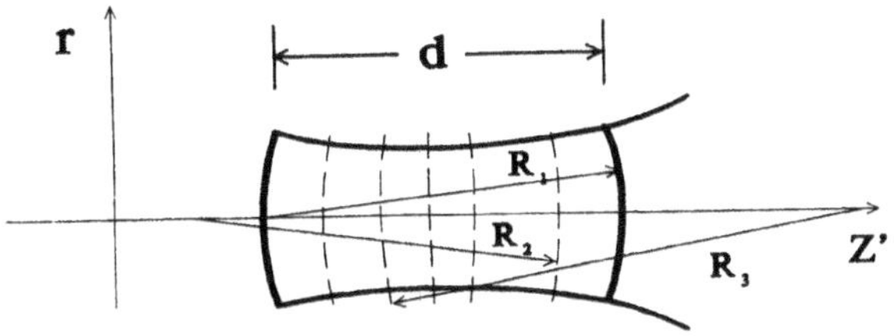

Figure 6. The generalized spherical resonator.

The irradiance at any radial point from the center of the beam with respect to a beam waist, as illustrated in Figure 8, also changes with distance and is provided by the following:

$$E(r, Z') = E_0 \, e^{-\frac{2r^2}{W^2(Z')}}, \tag{16}$$

where Z' is the distance from the beam waist, and W(Z') is the beam radius at the distance Z'. The beam waist often occurs within the resonator but this is not necessarily so. The beam radius is a function of both the waist radius, W_0, and Z', as shown below:

$$W(Z', W_0) = W_0 \sqrt{1 + \left(\frac{Z'}{Z_R}\right)^2}, \tag{17}$$

where Z_R is the Rayleigh range.

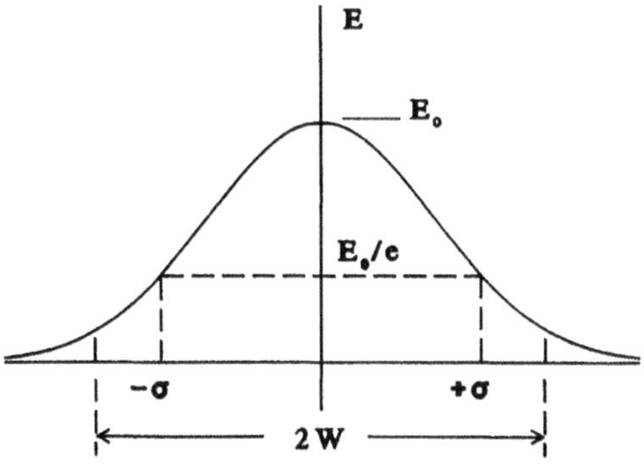

Figure 7. Gaussian (TEM_{00}) beam profile

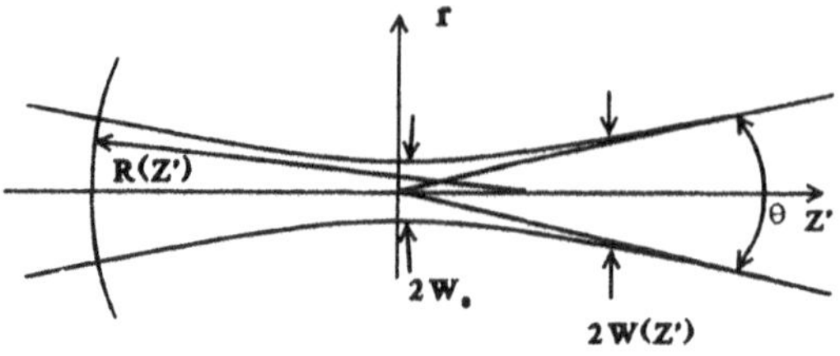

Figure 8. The formation of a Gaussian beam waist.

The radius of curvature of the beam front, R, is provided by the following:

$$R(Z', W_0) = Z'\left[1 + \left(\frac{Z_R}{Z'}\right)^2\right]. \tag{18}$$

The Rayleigh range for a TEM_{00} laser beam is given by:

$$Z_R = \frac{W_0^2 \pi}{\lambda}. \tag{19}$$

e.g.: He-Ne laser Wo = 1 mm; λ = 633 nm $\Rightarrow$ Z_R = 5 m. The beam divergence, θ, in the far field may be derived as follows:

$$\theta = 2 \lim_{Z \to \infty} \frac{d\,W(Z')}{d\,Z'} \Rightarrow \theta = \frac{2\lambda}{\pi W_0} = \frac{2W_0}{Z_R}. \tag{20}$$

Imaging of point sources by classical optics. In contrast, the formation of an image by a thin lens may be examined according to classical optics as illustrated in Figure 9. The radius of the light rays coming from the object at the surface of the lens, R_1, is the same as the distance from the object to the lens, S_1'. Likewise the radius of the rays forming the image, R_2, is the same as the distance from the lens to the image, S_2'. Only the signs of the

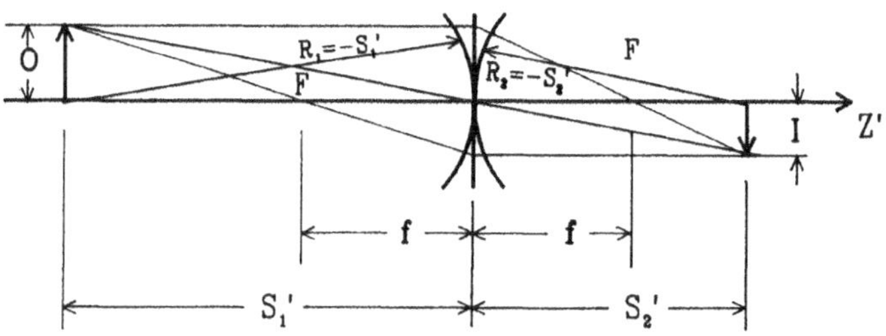

Figure 9. Imaging of point sources by classical optics.

quantities are different. These two radii are related to the focal length of the lens, f, by:

$$\frac{1}{R_1} - \frac{1}{R_2} = \frac{1}{f}, \tag{21}$$

or, in terms of object and image distances:

$$\frac{1}{S'_2} - \frac{1}{S'_1} = \frac{1}{f} \rightarrow \frac{S'_2}{f} = \frac{\frac{S'_1}{f}}{\frac{S'_1}{f} + 1}. \tag{22}$$

The image size is then related to the object size according to the following:

$$I = \frac{S'_2}{S'_1} \cdot O \quad \rightarrow I = \frac{O}{\frac{S'_1}{f} + 1}. \tag{23}$$

<u>Transformation of a Gaussian beam by a lens</u>. For a thin lens, the radius at the lens (Z=0) is the same [i.e. $W_2(0) = W_1(0)$], as illustrated in Figure 10. For Gaussian beams, lenses transform the radii of curvature of the wavefronts in exactly the same way a they do for classical beams:

$$\frac{1}{R_1(0)} - \frac{1}{R_2(0)} = \frac{1}{f} \rightarrow \frac{R_2(0)}{f} = \frac{\frac{R_1(0)}{f}}{\frac{R_1(0)}{f} + 1} \tag{24}$$

In the case of Gaussian beams, the effect of equations 17 and 18 on the transformation leads to more complicated equations relating the size and location of the beam waist as generated by the lens. It should be observed that it is not proper (and may lead to confusion) to call the waist on one

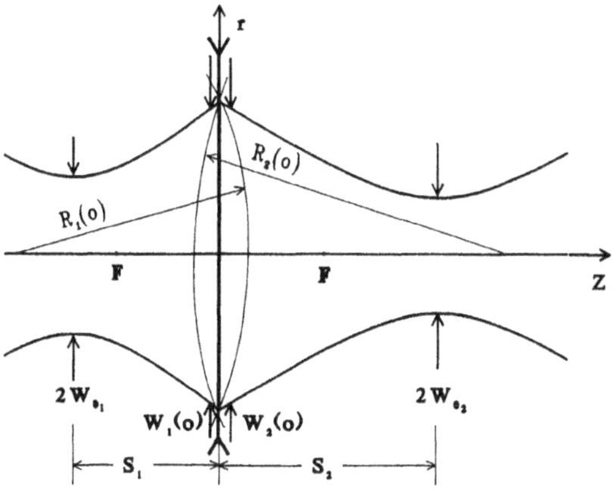

Figure 10. Transformation of a Gaussian beam by a lens.

side of the lens the "image" of the other. Imaging refers to the transformation of a set of *point* sources, constituting the "source" into another set of point sources, called the image. Since a Gaussian beam waist cannot be considered a point source, it will, therefore, be transformed, not imaged.

Equations 25 and 26 provide, respectively, the beam waist radius, W_{0_i}, and the beam waist location, S_i, on either side of the lens (i=1,2), provided the beam radius $W_i(Z)$ and the corresponding radius of curvature $R_i(Z)$ of the wave front are known at any location on that side of the lens. For i =1,2:

$$W_{o_i}{}^2(W_i(Z),R_i(Z)) = \frac{W_i^2(Z)\lambda^2 R_i^2(Z)}{\pi W_i^4(Z) + \lambda^2 R_i^2(Z)}, \tag{25}$$

$$S_i\,(W_i(Z),R_i(Z)) = \frac{R_i(Z)}{2}\left\{1+\left[1-\left(\frac{2\pi W_i^2(Z)\lambda R_i(Z)}{\pi W_i^2(Z)+\lambda R_i(Z)}\right)^2\right]^{1/2}\right\}. \tag{26}$$

With equations 17, 18, 25, and 26, it is possible, through a rather lengthy derivation, to arrive at the actual transformation equations, which relate the size and location of the beam waist on one side of the lens to those on the other. The Table lists these equations in comparison to the "corresponding" equations for classical imaging of point sources. Please observe that S_1 and S_2 are negative at the left side of the lens.

The differences between the Gaussian and the classical case are obvious if the relative location of the classical objects and images are compared to those of the two beam waists on both sides of the lens. Figure 11 is a graphical representation of these distances normalized to the focal length, f, of the lens. The two curves labeled $W_{0_1} = 0$ represent point sources (i.e. the well known classical case). The details of the Gaussian transformation depend strongly on the magnitude of the normalized Rayleigh range, Z_{R_1}/f. Two cases for $Z_{R_1}/f = 0.3$ and $Z_{R_1}/f = 1.6$ are shown. A few special cases illuminate the differences between the classical and the Gaussian case:

Special Cases

1. Whereas classically the image of objects in the focal plane, $S_1 = -f$, is located at $S_2 = +\infty$; a left side beam waist, $S_1 = -f$, always transforms into a beam waist in the right side focal plane, $S_2 = +f$. All Gaussian curves pass through the point $S_1 = -f$; $S_2 = +f$.

2. Beam waists at $S_1 = \pm\infty$ always transform into beam waists in the focal plane at $S_2 = +f$.

Cases 1 and 2 show that the beam waist position on the right side of the lens and transformed by a lens is not a single valued function of the waist position on the left side because both $S_1 = -f$ and $S_1 = -\infty$ result in $S_2 = +f$.

3. Whereas in classical imagery, an object (or its image) placed in the plane of the lens is not imaged at all (a case which is commonly made use of in illuminating optics). In Gaussian optics, a beam waist located in the plane of the lens, $S_1 = 0$, will be transformed to a location somewhere within the focal length on the right side of the lens, $0 \lesssim S_2 \lesssim f$, depending on the value of Z_{R_1}/f.

4. For $|Z_{R_1}| > |f|$, $S_2 \approx +f$ independent of S_1. Observe how close the curve $Z_R/f = 1.6$ in Figure 11 is to the $S_2 = +f$ line!

Table. Comparison of the equations for Gaussian beam transformation to those used for classical image formation.

Gaussian	Classical
$$\frac{S_2(W_{0_1}, S_1)}{f} = 1 - \frac{\frac{S_1}{f} + 1}{\left(\frac{S_1}{f} + 1\right)^2 + \left(\frac{Z_{R_1}}{f}\right)^2} \qquad (27)$$	$$\frac{S_2'}{f} = \frac{\frac{S_1'}{f}}{\frac{S_1'}{f} + 1} \qquad (28)$$
$$\left(\frac{S_1}{f}\right)_{\max} = \pm \frac{Z_R}{f} - 1 \; ; \qquad \left(\frac{S_2}{f}\right)_{\max} = 1 \pm \frac{1}{2}\frac{f}{Z_R} \qquad (29)$$	
$$W_{0_2}^2(W_{0_1}, S_1) = \frac{W_{0_1}^{\;2}}{\left(\frac{S_1}{f} + 1\right)^2 + \left(\frac{Z_{R_1}}{f}\right)^2} \qquad (30)$$	$$I^2 = \frac{O^2}{\left(\frac{S_1'}{f} + 1\right)^2} \qquad (31)$$

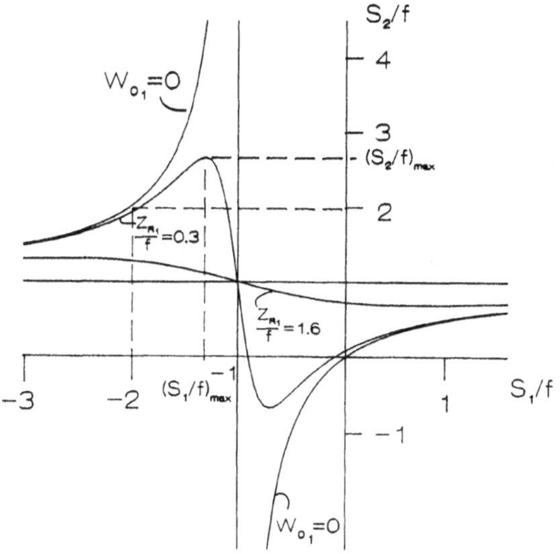

Figure 11. Graphical representation of the beam transformation equations.

5. If $Z_{R_1} \gg f$, a substitution of the identity for Gaussian beams,

$$W_{0_1} = \frac{\theta_1}{2Z_{R_1}}, \tag{32}$$

into equation 30 leads to:

$$W_{0_2} = \frac{1}{2} \frac{\theta_1 f}{\sqrt{\left(\frac{S_1}{Z_{R_1}}\right)^2 + 1}}. \tag{33}$$

If $Z_{R_1} \gg S_1$ as well, equation 33 reduces to the following:

$$W_{0_2} = \frac{\theta_1 f}{2}, \tag{34}$$

which is also valid for sources with limited coherence.

6. If f and S_1 are both larger or about equal to Z_{R_1}, as may be the case for diode lasers or monomode optical fibers,

$$W_{0_2} \lesssim \theta_1 f, \tag{35}$$

and with $S_1 \approx -f$, $S_2 \neq +f$. In these cases, the approximation for the diameter of the focus, generated by the lens, as derived for sources of limited coherence may be substantially in error.

7. For sources of limited coherence, the smallest spot size from equation 34 is always very close to the focal plane of the lens. As is obvious from Figure 11, the beam waist of Gaussian beams may be located a substantial distance from the focal plane if $f \gtrsim Z_{R_1}$. It can be shown, however, that *in the focal plane*,

$$W_2(+f) = \frac{1}{2}\theta_1 f, \tag{36}$$

independent of the values of W_{01}, S_1, and Z_{R_1}.

Tight Focussing

The above discussion also permits the determination of the smallest focus attainable for a given classical or Gaussian beam with a given lens of focal length, f_1 and clear aperture diameter, d_L, characterized by its F number, $F_\#$:

$$F_\# = \frac{f}{d_L}. \tag{37}$$

The first and main condition is, of course, that the clear aperture of the lens be fully filled by the beam. For classical beams, it may be necessary to use a telescope in front of the lens, as discussed above.

For Gaussian beams, there is no real radical cutoff of the beam profile and it must be decided at which radius, W(Z), to limit the beam profile by the aperture of the lens. To avoid excessive diffraction by the aperture of the

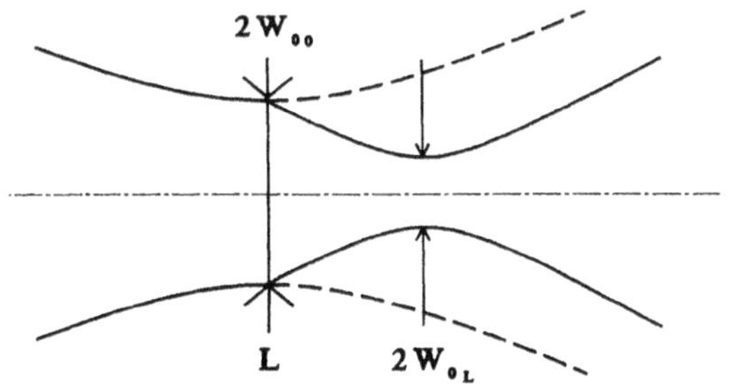

Figure 12. Focussing of Gaussian beam by a lens

lens, a typical cutoff is somewhere between E(r) = $E_0/10$ (the $1/e^2$ radius of the beam) and E(r) = $E_0/100$. It is useful to introduce a cutoff parameter, η:

$$\eta = \frac{r_L}{W(Z)} \quad where \quad r_L = \frac{d_L}{2}.$$

$$\text{For} \quad E(r) = \frac{E_0}{e^2}: \quad \eta = 1.0;$$

$$E(r) = \frac{E_0}{100}: \quad \eta = 1.5. \qquad (38)$$

Also for Gaussian beams, the smallest beam waist is generated by the lens, if the lens is placed into the beam waist of the raw beam, as shown in Figure 12. Most real Gaussian beams, however, have too small beam waists to completely fill the aperture of any real lens. Therefore a telescope must be used; or, with only very little degradation of the focus, it is possible to prefocus with a simple single lens, L_P, of reasonably short focal length as shown in Figure 13.

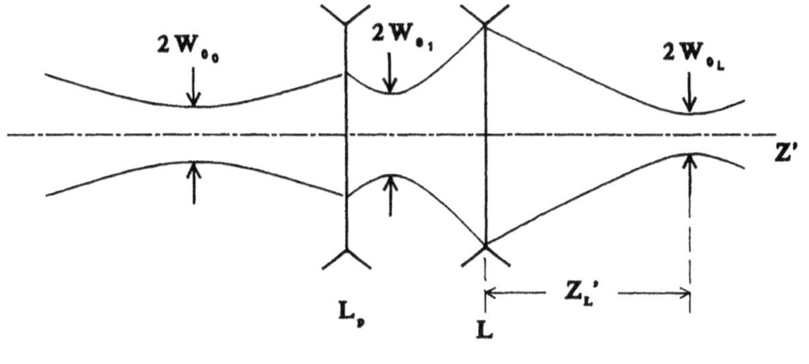

Figure 13. A lens used to prefocus a small Gaussian beam

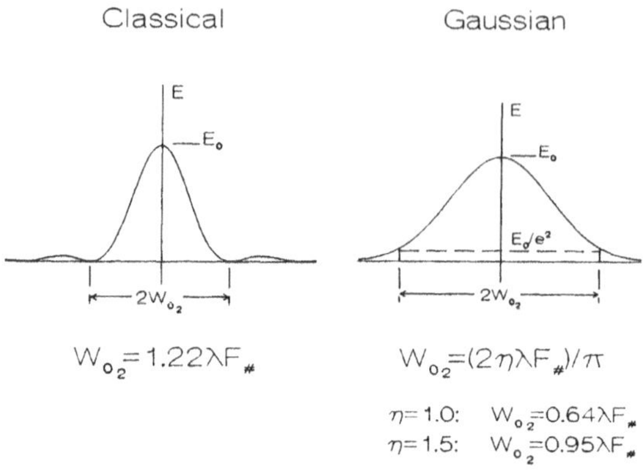

Figure 14. Comparison of the diffraction limit for a classical beam to that of a Gaussian beam.

For typical Gaussian laser beams, W_{0_0} is in the order of 0.2 - 2 mm. If a prefocussing lens is chosen, L_p with a focal length, $f_p = 20$ mm, this lens is only used with an F-number of $F_{\#} = 13$. Therefore, any simple plano-convex lens can be used for the prefocussing, instead of an expensive, convected optics. The focussing lens L must, of course, be of high quality to avoid excessive aberrations and degradation of the focus.

Figure 14 compares the minimal foci for classical beams as limited by diffraction to that for Gaussian beams. The minimal focus attainable is pretty much the same for both cases.

REFERENCES

Demtroeder, W., "Laser Spectroscopy," 1988, Basic Concepts and Instrumentation, Springer Series in Chemical Physics, Vol. 5, Springer-Verlag, New York.

Svelto, O., " Principles of Lasers," 1989, Third ed., Plenum Press, New York.

RADIATION INTERACTIONS WITH MATTER: SOME GENERAL CONSIDERATIONS ON THE IONIZING/NON-IONIZING BOUNDARY REGION

Mario Fabretto and Alessandro Rindi

Sincrotrone Trieste
Trieste, Italy

1. INTRODUCTION

The increasing interest in wide-spectrum radiation sources (like synchrotron radiation machines) brings the need to describe in an unitary way the interaction of the electromagnetic radiation with matter in the range from visible to soft X rays (i.e. the radiation energy interval between some eV up to a few keV).

This region includes effects usually described by quantum-mechanical methods as well as, at the same time, phenomena well explained in the classical electromagnetic theory.

This classical theory implies the solution of the Maxwell equations. However the classical form of those equations in the material media does not represent a closed system and to solve them we introduce a set of phenomenological parameters like conductivity, electric permittivity and magnetic permeability of the medium. These are macroscopic parameters (connected with measured optical properties) that make difficult their use when microscopic levels of interaction have to be considered.

On the other end, when one tries to extend the methods used in the high energy region to ultraviolet and visible, problems concerning the interpretation of phenomena like reflection from a surface arise.

This is principally due to an incorrect transition to the classical limit that does not retain the wave-particle duality.

We will try to show the difficulties of both interpretations at the energy boundary region, where, f.i. two effects as reflection and photoemission are simultaneously effective. In this region both interpretations shall be used to describe and interpret the interaction of radiation with matter as required for health physics purposes.

2. MAXWELL EQUATION

The starting point for the description of the interaction of electromagnetic radiation with matter are the Maxwell equations in the material media. In c.g.s. units and in differential form they are:

$$
\begin{aligned}
&\nabla \vec{D} = 4\pi \rho^{ext} \\
&\nabla \times \vec{E} = -\frac{1}{c}\frac{\partial \vec{B}}{\partial t} \\
&\nabla \vec{B} = 0 \\
&\nabla \times \vec{H} = \frac{1}{c}\frac{\partial \vec{D}}{\partial t} + \frac{4\pi}{c}\left(\vec{J}^{cond} + \vec{J}^{ext}\right)
\end{aligned}
\qquad (2.1)
$$

The (2.1) are eight equations in the fifteen variables $\vec{D}, \vec{E}, \vec{B}, \vec{H}, \vec{J}^{cond}$ so that more equations are requested for solving the system. The quantities ρ^{ext} and $\vec{J}^{ext}$ are external sources that, when considering the interaction of the radiation with a medium, can be zero. The incident radiation (in vacuum), controlled by the experimenter, is described by the fields $\vec{D}$ and $\vec{H}$, while $\vec{E}$ and $\vec{B}$ represent the fields in the medium, given by the sum of the external and the induced fields.

We can write the relation between the two pairs of fields introducing two new vectors:

$$
\begin{aligned}
&\vec{D} = \vec{E} + 4\pi \vec{P} \\
&\vec{H} = \vec{B} - 4\pi \vec{M}
\end{aligned}
\qquad (2.2)
$$

where $\vec{P}$ and $\vec{M}$ are, respectively, the polarization and magnetization density vectors in the irradiated media. Last equations are formally exact but they don't turn eq. (2.1) in a closed system. For solving the (2.1) we must, in some way, introduce a description of the medium, obtaining from this the missing equations. This is achieved by introducing phenomenological relations between the field quantities $\vec{D}$, $\vec{H}$, $\vec{E}$ and $\vec{B}$ which, usually, are linear relations defining material-dependent coefficients:

$$\vec{D} = \varepsilon \vec{E}$$
$$\vec{B} = \mu \vec{H} \tag{2.3}$$
$$\vec{J}^{cond} = \sigma \vec{E}$$

The coefficients ε (electric permittivity), μ (magnetic permeability), and σ (conductivity) are generally second-rank tensors; in the following we always will consider isotropic media so they will be scalars. The assumptions leading to the (2.3) represent an attempt to describe some properties of matter (its response to the radiation field) and are based on approximations. The introduced parameters ε, μ and σ are macroscopic quantities because they are defined in terms of macroscopic quantities (the fields in the eq.(2.1) are mean values of the microscopic fields over small but finite volumes, such to have linear dimensions smaller than radiation wavelength but still containing a great number of atoms). To understand the microscopic properties we must relate these parameters to microscopic quantities that can be expressed using the basic principles of the physics involved. The absorption and the dispersion of electromagnetic radiation by the medium are related (in non-ferromagnetic media) to the electric permittivity (or dielectric function) ε and the conductivity σ. We will consider as our basic function the electric permittivity, and the following considerations will mainly be devoted to a better knowledge of ε, even if we would get the same final results starting from the study of the conductivity.

Substituting the (2.3) in the (2.1) and after a little algebra, we obtain the equation for the electric field:

$$\Delta \vec{E} - \frac{\mu\varepsilon}{c^2} \frac{\partial^2 \vec{E}}{\partial t^2} - \frac{4\pi\sigma\mu}{c^2} \frac{\partial E}{\partial t} = 0 , \tag{2.4}$$

that has as solution plane waves:

$$\vec{E}(\vec{r},t) = \vec{E}_0 e^{i(\vec{k}\vec{r} - \omega t)} , \tag{2.5}$$

where the wave vector $\vec{k}$ is a complex quantity and, as one can immediately find by derivating the (2.5) and substituting it in the (2.4):

$$k^2 = \mu \frac{\omega^2}{c^2} \left(\varepsilon + i \frac{4\pi\sigma}{\omega} \right) . \tag{2.6}$$

The formal relation $\vec{D} = \varepsilon\vec{E}$ is preserved if we admit ε being a complex function of ω of the form:

$$\hat{\varepsilon}(\omega) = \varepsilon(\omega) + i\,\frac{4\pi\sigma}{\omega}. \tag{2.7}$$

To write the wave vector $\vec{k}$ in the complex form we must extract the square root from the (2.6). This can be done by defining $\vec{k}$ in such a way that its real part gives the usual dependence between the wave vector and the refractive index of the medium n, while the imaginary part be a quantity ξ, that we will call the "extinction coefficient", connected, as we will see later, to the energy absorption in the medium. The form of $\vec{k}$ is[1]:

$$\vec{k} = \left(\frac{\omega}{c}\right)\hat{n} = \left(\frac{\omega}{c}\right)(\vec{n} + i\,\vec{\xi}) \;. \tag{2.8}$$

The refractive index of the medium n is defined as $n = c/v$ where v is the phase velocity. Making the square of the (2.8) and comparing the result with the (2.6), considering the (2.7), we find the relation between the extinction coefficient ξ and the dielectric function $\varepsilon(\omega)$:

$$\xi = \frac{\mu}{2n_\xi}\,\mathrm{Im}\{\hat{\varepsilon}(\omega)\} = \frac{2\pi\sigma\mu}{\omega n_\xi}\;, \tag{2.9}$$

where n_ξ is the projection of the vector $\vec{n}$ in the direction of $\vec{\xi}$.

Substituting the (2.8) in the (2.5) we obtain:

$$\vec{E}(\vec{r},t) = \vec{E}_0\, e^{-\frac{\omega}{c}\vec{\xi}\,\vec{r}}\; e^{i\left(\frac{\omega}{c}\vec{n}\,\vec{r} - \omega t\right)}\;, \tag{2.10}$$

from which we see that we can consider $e^{-2\frac{\omega}{c}\vec{\xi}\,\vec{r}}$ as a term describing an exponential absorption of radiation intensity and the quantity:

$$\eta = \frac{4\pi\sigma\mu}{n_\xi c} = \frac{\omega\mu}{n_\xi c}\,\mathrm{Im}\,\{\hat{\varepsilon}(\omega)\}, \tag{2.11}$$

directly coming from writing the (2.9) in the real exponent of the (2.10), as a linear energy absorption coefficient.

We have now a relation between the imaginary part of one of the material-dependent coefficients ε, defined in the (2.3) and a quantity, the linear energy absorption coefficient, that can be studied with a quanto-mechanical approach, as we will see in section 5.

More, the real and imaginary parts of $\hat{\varepsilon}(\omega)$ are not independent from each other, being related by the well known Kramers-Kronig dispersion relations[1,2].

It is of interest to analyze the asimptotic behaviour of the dielectric function $\hat{\varepsilon}(\omega)$ for low and high values of the incident radiation frequency. The dependence of $\hat{\varepsilon}(\omega)$ from ω is called its "dispersion law" and can be expressed starting from the integral relation[3]:

$$\vec{D}(t) = \vec{E}(t) + \int_0^{\infty} f(\tau)\vec{E}(t-\tau)d\tau, \tag{2.12}$$

which indicates the most general form of a linear dependance between $\vec{D}$ and the values of $\vec{E}$ at any instant before t. Here $f(\tau)$ is a time function depending from the medium. If the time dipendence of the electric field $\vec{E}$ is of the form $e^{-i\omega t}$ the dielectric function becomes:

$$\hat{\varepsilon}(\omega) = 1 + \int_0^{\infty} f(\tau)e^{i\omega\tau}d\tau\,. \tag{2.13}$$

From this it is possible to note that the real part (ε_1) is an even function, while the imaginary part (ε_2) is an odd function of ω. For low values of the frequency we can expand both functions in power of ω. In the $\omega \rightarrow 0$ limit , for dielectrics, the function $\hat{\varepsilon}(\omega)$ obviously approaches the electrostatic value ε_0, so the expansion of the real part starts with this value. The imaginary part starts with a term proportional to ω. Also in metals it is possible to find the low frequency behaviour of $\hat{\varepsilon}(\omega)$. It is easy to verify[3] that the first term of the expansion must be purely imaginary and equal to $i\,4\pi\sigma/\omega$. This form is consistent with the zero electric field inside conductors (see 2.3).

In the $\omega \rightarrow \infty$ limit the function $\hat{\varepsilon}(\omega)$ tends to the unity (due to the rapidly oscillating exponential in the integral (2.13)). In the next section, within the framework of the Drude model for metals, we will see how to express the asimptotic form of the function in this limit.

Because in all non-magnetic media $\mu \approx 1$, from now on we will consider $\mu = 1$.

3. THE LORENTZ-DRUDE OSCILLATOR MODEL

In the preceding section we saw that the Maxwell equations in the matter are solvable if we introduce some material-dependent macroscopic parameters like ε, σ, ... Using a very simple model we will obtain an explicit, even if naïve, form for these parameters. In spite of the roughness of the assumptions, the results are useful starting points in the understanding of the optical properties.

The Lorentz model is applicable to insulators while the Drude model is applicable to free-electron metals.

We shall emphasize that many features of these classical models have quantum mechanical counterparts.

In the Lorentz model the matter is considered composed by atoms with electrons bound to the nucleus by an elastic force. Because the atomic electrons velocity v is much lower than the speed of light, the distance traveled by an electron during a wave period T can be considered small with respect to the wavelength λ. This fact allows to keep the electric field of the incident wave indipendent from the electron position vector $\vec{r}$. The motion of an electron in an external field $\vec{E}$ is then described by:

$$m\frac{d^2\vec{r}}{dt^2} + m\Gamma\frac{d\vec{r}}{dt} + m\,\omega_0^2\,\vec{r} = -\,e\,\vec{E}\ , \tag{3.1}$$

where $m\,\omega_0^2$ is the elastic constant and the term $m\,\Gamma\frac{d\vec{r}}{dt}$ represents viscous damping and provides for an energy loss mechanism. The charge e is positive, so the minus sign in the right hand side of the equation.The (3.1) neglects the small force arising from the motion of the electron in the magnetic field of the incident wave.

Assuming the time-dependence of $\vec{E}$ in the form $e^{-i\omega t}$, the solution of the (3.1) is:

$$\vec{r}\,(t) = \frac{\dfrac{-\,e\,\vec{E}\,(t)}{m}}{\left(\omega_0^2 - \omega^2\right) - i\,\Gamma\,\omega}\ . \tag{3.2}$$

The induced dipole momentum is:

$$\vec{p} = \frac{e^2\vec{E}}{m} \frac{1}{\left(\omega_0^2 - \omega^2\right) - i\Gamma\omega} \tag{3.3}$$

and the polarization density is:

$$\vec{P} = \frac{N e^2\vec{E}}{m} \frac{1}{\left(\omega_0^2 - \omega^2\right) - i\Gamma\omega}, \tag{3.4}$$

where N is the density of the oscillating electrons in the irradiated medium.

From the definition (2.2) it is possible to write the dielectric function:

$$\hat{\varepsilon}(\omega) = 1 + \frac{4\pi N e^2}{m} \frac{1}{\left(\omega_0^2 - \omega^2\right) - i\Gamma\omega}, \tag{3.5}$$

and writing in an explicit form the real and the imaginary parts we have:

$$\begin{aligned} \mathrm{Re}\left\{\hat{\varepsilon}(\omega)\right\} &= 1 + \frac{4\pi N e^2}{m} \frac{\left(\omega_0^2 - \omega^2\right)}{\left(\omega_0^2 - \omega^2\right)^2 + \Gamma^2\omega^2}; \\ \mathrm{Im}\left\{\hat{\varepsilon}(\omega)\right\} &= \frac{4\pi N e^2}{m} \frac{\Gamma\omega}{\left(\omega_0^2 - \omega^2\right)^2 + \Gamma^2\omega^2}. \end{aligned} \tag{3.6}$$

The low frequency behaviour of the dielectric function can be obtained posing $\omega << \omega_0$ in the (3.6). Under this assumption we find the following expressions:

$$\begin{aligned} \mathrm{Re}\{\hat{\varepsilon}(\omega)\} &\sim 1 + \frac{4\pi e^2}{m\omega_0^2}; \quad \omega \to 0 \\ \mathrm{Im}\{\hat{\varepsilon}(\omega)\} &\sim \frac{4\pi e^2\Gamma}{m\omega_0^4}\omega; \quad \omega \to 0 \end{aligned} \tag{3.7}$$

that have the ω dependence expected from the (2.13). The equations (3.6) are valid if the system has only one kind of electron per atom; if we have N_j electrons with resonance frequency ω_j ($j = 1,2,3,\ldots$) it will be:

$$\hat{\varepsilon}(\omega) = 1 + \frac{4\pi e^2}{m} \sum_j \frac{N_j}{\left(\omega_j^2 - \omega^2\right) - i\Gamma_j \omega}$$

$$\sum_j N_j = N \tag{3.8}$$

It is important to notice that we have expressed a complex phenomenological parameter using other unknown quantities: N_j, ω_j and Γ_j. The only relevant improvement is in the introduction of a model for the matter (nuclei surrounded by electrons bound by an elastic force).

The Drude model for metals can be derived starting from the previous analysis. It corresponds to the high frequences behaviour of the eq. (3.1) ($\omega >> \omega_0$). Due to the fact that conduction electrons in a metal can be considered as almost free, we put $\omega_0 = 0$ in the equation (3.1) and the following, obtaining:

$$\mathrm{Re}\left\{\hat{\varepsilon}(\omega)\right\} = 1 - \frac{4\pi N e^2}{m} \frac{1}{\left(\omega^2 + \Gamma^2\right)} ,$$

$$\mathrm{Im}\left\{\hat{\varepsilon}(\omega)\right\} = \frac{4\pi N e^2}{m} \frac{\Gamma}{\omega\left(\omega^2 + \Gamma^2\right)} . \tag{3.9}$$

The eq. (3.9) show how $\hat{\varepsilon}(\omega)$ tends to the unity for $\omega \to \infty$, as we predicted analyzing the (2.13).

The parameter Γ, introduced to describe a damping mechanism, can be estimated classically from the condition that the dissipated power is (dipole radiation):

$$\frac{dw}{dt} = \frac{2}{3} \frac{e^2 \dddot{r}^2}{c^3} . \tag{3.10}$$

We obtain[2]:

$$\Gamma = \frac{2}{3} \frac{e^2}{m} \frac{\omega^2}{c^3} . \tag{3.11}$$

For optical frequencies ($\omega \sim 10^{15}\ s^{-1}$) the value of Γ is of the order of $10^6\ s^{-1}$; actually its value is greatly increased by other effects than the single dipole radiation hypothesis we used to derive eq. (3.11). In a free-electron metal the origin of the viscous damping is the ordinary scattering of electrons associated with electrical resistivity. It results that $\Gamma = \tau^{-1}$, where τ is the mean free time between two collisions, which is typically of the order of 10^{-14} s^{1}.

The quantity:

$$\frac{4\ \pi\ N\ e^2}{m} = \omega_p^2 \tag{3.12}$$

appearing in the eq. (3.9) is the square value of the so called "plasma frequency". This term arises from the collective oscillatory motion of the free-electrons with respect to the positive ions of the lattice. The number of electrons involved in this motion is a function of the energy of the radiation and, as we will see in the following, it is deducible, from experimental results, using the so-called "sum-rules".

Until now we have considered both the field and the interacting atom as classical quantities. As a next step we will introduce a more realistic picture of the atom, based on quantum mechanic.

Frequencies ω_j must be considered as the transition frequencies of an electron between two states separated in energy by $\hbar\omega_j$. The quantity N_j, representing the electron density corresponding to the resonant frequency ω_j, must be replaced by a term Nf_j, where N is the total electron density and f_j, called the oscillator strength, is a measure of the relative probability of a quantum mechanical transition. It is possible to demonstrate that[1]:

$$\Sigma_j f_j = 1 \ , \tag{3.13}$$

which is the quantum mechanical analogy of the second term of eq. (3.8). In the section 5 we will go further with the quantum description introducing the radiation field quantization.

Clearly Γ, N, f_j and all the related phenomenological quantities introduced until now have to be deduced from microscopic theories.

4. OPTICAL PROPERTIES OF MATTER

Starting from the Lorentz-Drude model we want to rapidly analyze the connection between the frequency dependence of $\hat{\varepsilon}$ and the optical properties of solid. The relevant equations are the (3.6) and the (3.9) and starting from them it is possible to describe the frequency-dependent behaviour of a solid in terms of whether it is primarily reflecting, absorbing, or transparent.

From the (2.7) and (2.8) we have, for the refractive index n and the extinction coefficient ξ:

$$n = \{\frac{1}{2}[(Re^2\{\hat{\varepsilon}\} + Im^2\{\hat{\varepsilon}\})^{\frac{1}{2}} + Re\{\hat{\varepsilon}\}]\}^{\frac{1}{2}},$$

$$\xi = \{\frac{1}{2}[(Re^2\{\hat{\varepsilon}\} + Im^2\{\hat{\varepsilon}\})^{\frac{1}{2}} - Re\{\hat{\varepsilon}\}]\}^{\frac{1}{2}}. \tag{4.1}$$

In figure 4.1 we show the spectral dependence of $\varepsilon_1 = Re\{\hat{\varepsilon}\}$ and $\varepsilon_2 = Im\{\hat{\varepsilon}\}$ as from (3.6), for an insulator, in the case $\hbar\omega_0 = 4$ eV, $\hbar\Gamma = 1$ eV and $\hbar^2 4\pi N e^2/m = 60$ (eV)^2.

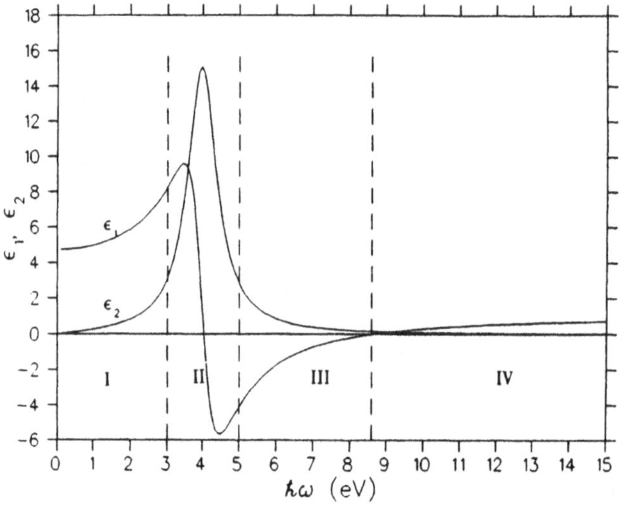

Fig. 4.1 Spectral dependence of Re $\{\hat{\varepsilon}\}$ (ε_1) and Im $\{\hat{\varepsilon}\}$ (ε_2). The curves are calculated with $\hbar\omega_0 = 4$ eV, $\hbar\Gamma = 1$ eV, and $\hbar^2 4\pi N e^2/m = 60$ (eV)^2.

Fig. 4.2 Spectral dependence of n and ξ from the values of ε_1 and ε_2 in fig. 4.1.

From these curves one can plot the spectral dependence of n and ξ as in figure 4.2 and, defining r as the reflected fraction of the electric field (reflectance)[3], deduce the quantity R (reflectivity):

$$R = \frac{(n-1)^2 + \xi^2}{(n+1)^2 + \xi^2} = |r|^2 \tag{4.2}$$

In the figure 4.3 we can see the normal reflectivity versus energy curve obtained from the previous data.

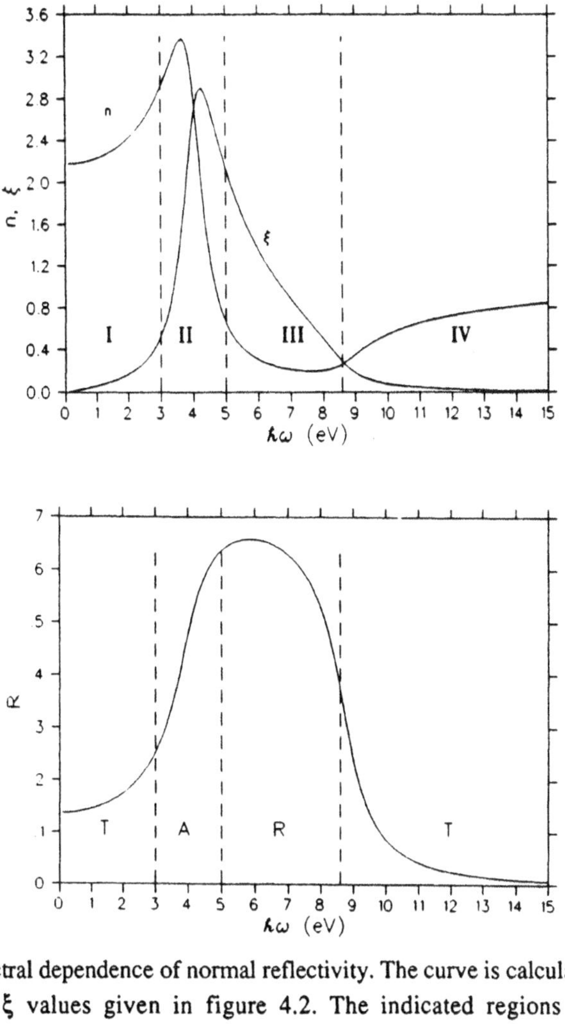

Fig. 4.3 Spectral dependence of normal reflectivity. The curve is calculated from the n and ξ values given in figure 4.2. The indicated regions are primarily transmitting (T), absorbing (A) and reflecting (R).

The region I of fig. 2, where $\omega<\omega_0$, has a very small value of ξ, so the substance is mainly transparent, with no absorption and small reflectivity. This corresponds to an incident radiation with frequency less than the transition frequency ω_0. The region II is characterized by strong absorption, evidenced by the high peak of ξ, but also reflection may be appreciable. In region III $\omega>\omega_0$ and the electrons of the insulator act as free electrons, giving to it a metallic reflectivity; the refraction index n is less than unit. At higher frequencies (region IV), ξ is again very small, and we have transparence. Regions III and IV for an insulator are usually in the ultraviolet and cannot be observed visually. For semiconductors like Ge and Si the forbidden band gap lies in the infrared and the region of metallic reflectance falls in the visible.

Developing the same analysis for a free electron metal, starting from the (3.9), we obtain the spectral behaviours illustrated in figs. 4.4-6, all calculated for $\hbar^2\omega_p^2 = 30\ (\text{eV})^2$ and $\hbar\Gamma = 0.02$ eV.

Comparing the figures 4.1-3 with the 4.4-6 it appears that an insulator responds like a metal to photons of energy $\hbar\omega > \hbar\omega_0$.

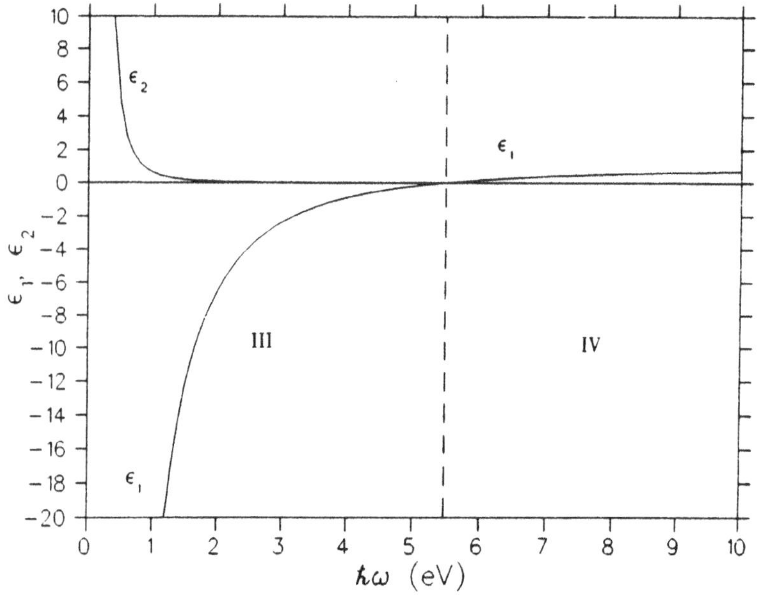

Fig. 4.4 Spectral dependence of Re $\{\hat{\varepsilon}\}$ (ε_1) and Im $\{\hat{\varepsilon}\}$ (ε_2) for a free-electron metal. The curves are calculated with $\hbar^2\omega_p^2 = 30\ (\text{eV})^2$ and $\hbar\Gamma = 0.02$ eV.

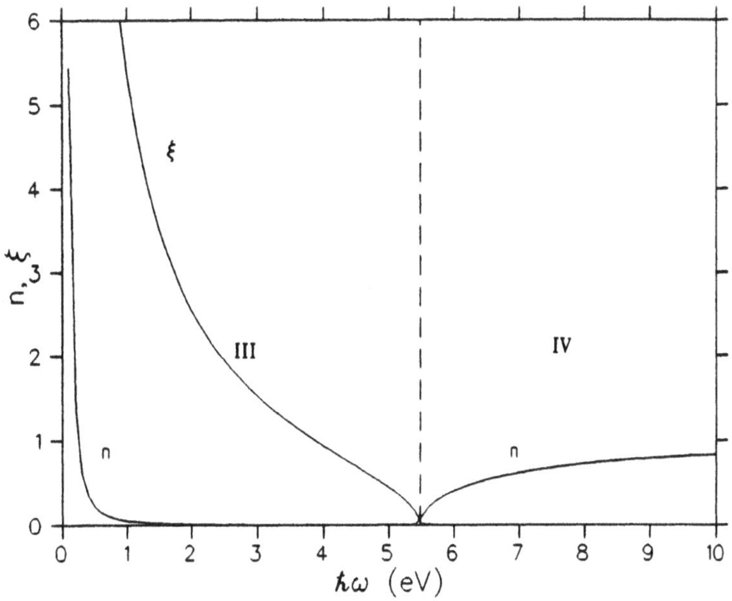

Fig. 4.5 Spectral dependence of n and ξ for a free-electron metal. The curves are calculated from the values of ε_1 and ε_2 given in figure 4.4.

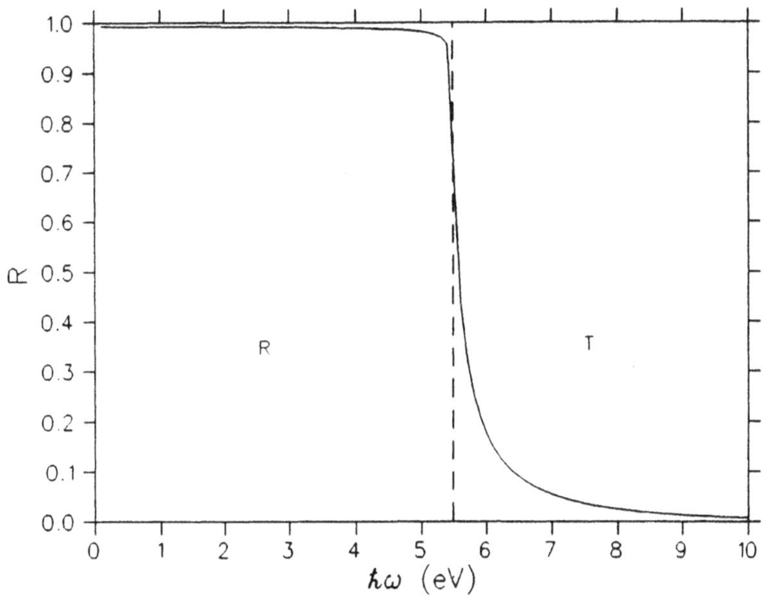

Fig. 4.6 Spectral dependence of normal reflectivity for a free-electron metal. The curve is calculated from the n and ξ values given in figure 4.5. The sharp decreasing of the curve corresponds to the plasma frequency (≈ 5.48 eV).

Some words have to be spent about the plasma frequency we introduced in the previous section. From the (3.9) we see that ω_p defines (in the $\Gamma \rightarrow 0$ limit) the value for which the real part of the dielectric function is zero. The plasma frequency typically lies in the visible or ultraviolet spectral region, corresponding to $\omega > 10^{15}\ s^{-1}$ (E > 1 eV). The mean free collision time for free electrons in metal is $\tau \approx 10^{-14}\ s$ so the product $\omega\tau \gg 1$ and the eq. (3.9) can be written:

$$\mathrm{Re}\left\{\hat{\varepsilon}(\omega)\right\} = 1 - \frac{\omega_p^2}{\omega^2}$$

$$\mathrm{Im}\left\{\hat{\varepsilon}(\omega)\right\} = \frac{\omega_p^2 \tau}{\omega\left(1 + \omega^2 \tau^2\right)} \tag{4.3}$$

and, because $\omega > \omega_0$:

$$n \approx 1 - \frac{1}{2}\frac{\omega_p^2}{\omega^2} \quad . \tag{4.4}$$

For X-rays it is usual to write[4]:

$$\hat{n} = 1 - \delta - i\beta$$

$$\hat{\varepsilon} = 1 - 2\delta - i2\beta \tag{4.5}$$

where:

$$\delta = \frac{\omega_p^2}{2\omega^2} = \frac{N\ e^2\ \lambda^2}{2\ \pi\ m\ c^2}$$

$$\beta = \frac{\eta\lambda}{4\ \pi} \tag{4.6}$$

(remember 2.11) are both small quantities ($\sim 10^{-5}$).

From these results it follows that, for X-rays, n<1 and, from Snell law we can find a critical grazing angle:

$$\theta_c = \sqrt{2\delta} \tag{4.7}$$

below which total reflection for non-absorbing media appears. For absorbing media we can obtain the spectral behaviour of reflectance starting from (4.5).

Defining the two quantities:

$$\begin{aligned} 2a^2 &= \left[\left(\sin^2\theta - 2\delta\right)^2 + 4\beta^2\right]^{1/2} + \left(\sin^2\theta - 2\delta\right) \\ 2b^2 &= \left[\left(\sin^2\theta - 2\delta\right)^2 + 4\beta^2\right]^{1/2} - \left(\sin^2\theta - 2\delta\right) \end{aligned} \tag{4.8}$$

we have, for an incident wave normally polarized to the incidence plane:

$$R(\theta) = \frac{(\sin\theta - a)^2 + b^2}{(\sin\theta + a)^2 + b^2} \tag{4.9}$$

where θ is the grazing angle. For an ideal free-electron media the angular behaviour of R is showed in figure 4.7, while the energy dependence is that illustrated in figure 4.6. Actually this is rarely the case. Figure 4.8 shows the experimental reflectance of Ag at normal incidence[5].

In the free-electrons model, for silver, the plasma energy has a value near 9 eV (corresponding approximately to one free-electron per atom). The sharp decrease at 3.9 eV visible in the fig. 4.8, has to be understood as an effect due to "*d*" states which lie 4 eV below the Fermi energy. The effect of the *d* states is to make a positive contribution to Re $\hat{\varepsilon}(\omega)$ and so to shift to lower energies the point at which Re $\hat{\varepsilon}(\omega)$ passes through zero[1] (see figs. 4.4-4.6).

Some final words have to be spent about the density of the electrons involved in the optical phenomena. As we already stated, this density depends on the energy of the incident radiation. In metals, with the valence band uncompletely filled, it would be immediate to assume that all the valence electrons are effective, at least until the incident radiation energy is not such to make possible excitations of the inner levels. This assumption is rather close to the real behaviour, but it happens that the straightforward application of this rule does not lead to a correct description of the optical effects. An example of this can be seen in the fig.4.9, where the gold reflectivity versus energy curve, obtained from this model (solid curve), is compared with the experimental one (dashed curve)[6].

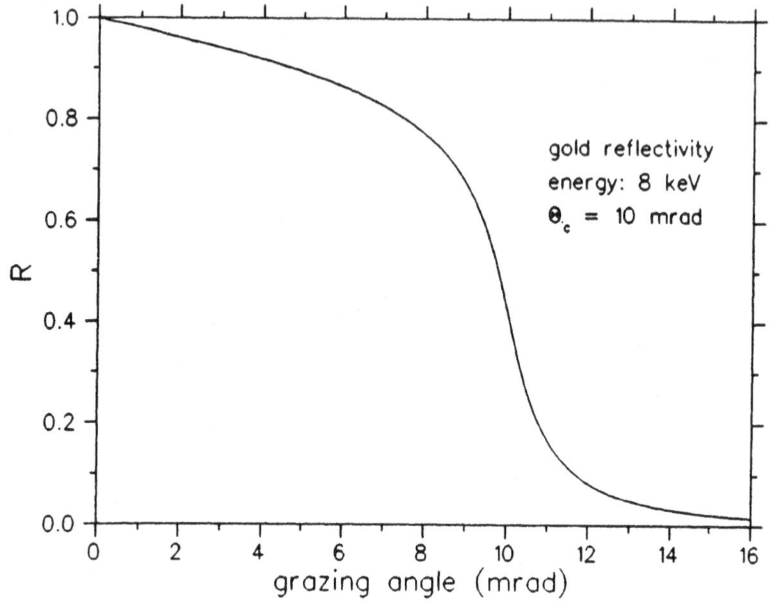

Fig. 4.7 The gold reflectivity versus the grazing angle calculated from the eq. (4.9) at 8 keV. The grazing angle is the complementary of the incidence angle.

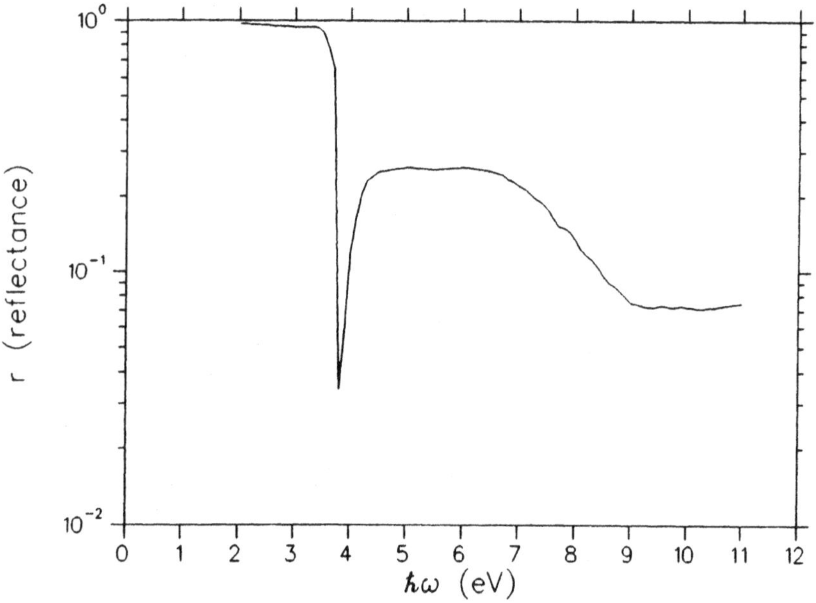

Fig. 4.8 Reflectance of silver.

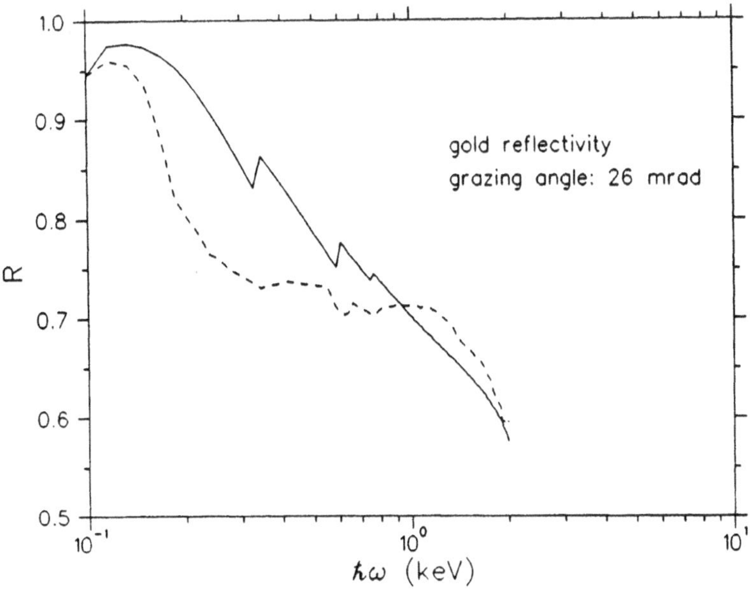

Fig. 4.9 The gold reflectivity versus energy curve for a grazing angle of 26 mrad (1.5°). The dashed curve fits experimental data while the solid one represents the result of the simple model discussed in the text.

A way to handle the problem is to find some functions of measurable quantities where the only unknow parameter is the electron density and to fit the experimental points using the effective number of optical electrons per atom n_{eff} as a fitting parameter. An alternative choice is to keep this quantity fixed but to consider an effective electron mass different from its vacuum value. Among these techniques, of great importance are the so called "sum rules"[1]. As an example let us consider the rate of energy absorption from an electric field per unit volume:

$$\frac{dw}{dt} = \frac{1}{4\pi} \mathrm{Re}\left(\hat{\varepsilon}\vec{E}\,\frac{\partial\vec{E}}{\partial t}\right) = \frac{1}{4\pi}\omega\,\mathrm{Im}\left\{\hat{\varepsilon}(\omega)\right\}|\vec{E}|^2 \tag{4.10}$$

The integral:

$$\int_0^\infty \omega\,\mathrm{Im}\left\{\hat{\varepsilon}(\omega)\right\}\,d\omega \tag{4.11}$$

is a measure of the energy absorption for all frequencies.

Using the expression (3.6) for Im $\{\hat{\varepsilon}(\omega)\}$ we can write:

$$\int_0^\infty \omega \operatorname{Im}\left\{\hat{\varepsilon}(\omega)\right\} d\omega = \frac{4\pi N e^2 \Gamma}{m} \int_0^\infty \frac{\omega^2 \, d\omega}{\left(\omega_0^2 - \omega^2\right)^2 + \Gamma^2 \omega^2} \tag{4.12}$$

For small Γ significant contributions come only for $\omega \approx \omega_0$; thus:

$$\int_0^\infty \omega \operatorname{Im}\left\{\hat{\varepsilon}(\omega)\right\} d\omega = \frac{1}{2}\pi \left(\frac{4\pi N e^2}{m}\right) \tag{4.13}$$

Writing $N = N_a\, n_{eff}$ where N_a is the density of atoms and n_{eff} a fitting parameter with the meaning of an effective number of electrons per atom contributing to the optical properties over a finite frequency range, eq. (4.13) becomes:

$$\int_0^{\omega_c} \omega \operatorname{Im}\left\{\hat{\varepsilon}(\omega)\right\} d\omega = \frac{1}{2}\pi \left(\frac{4\pi N_a e^2}{m}\right) n_{eff}(\omega_c) \; . \tag{4.14}$$

From the experimental knowledge of the left hand side of eq. (4.14) it is possible to deduce $n_{eff}(\omega_c)$ as showed in figure (4.10) for metals like Ag and Cu[7].

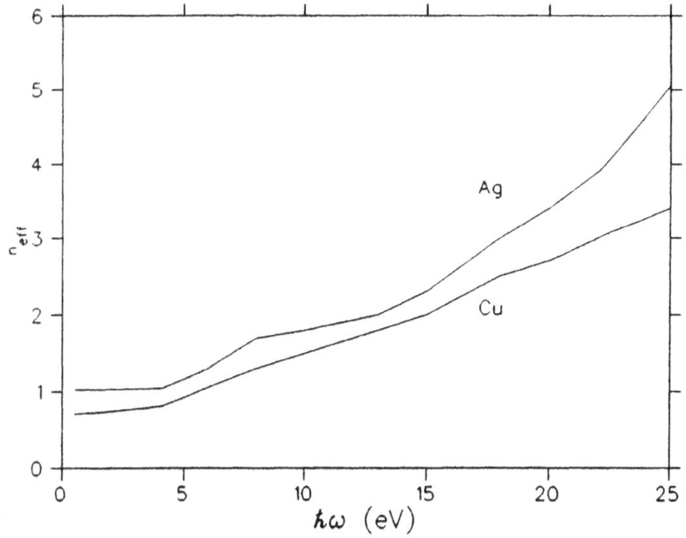

Fig. 4.10 Effective numbers of electrons per atom contributing to the optical properties over a given energy range versus energy.

5. QUANTUM DESCRIPTION OF THE OPTICAL CONSTANTS

In the preceeding sections we described the incident radiation as traveling electric and magnetic fields in the framework of the classical electrodynamics point of view. The irradiated medium had been supposed to be constituted by negative electrons, bounded to positive ions or free, as in the case of metals. We arrived at a semi-classical model, where the radiation power is mainly absorbed only for some particular frequences, corresponding to transitions between quantized energy levels in the matter. Some experimental facts, as the photoemission of electrons from a surface or the Compton scattering, occuring at energies up from some eV, cannot be explained using this model. It is necessary to assume that also the energy of the incident radiation be quantized and it could be excanged only by discrete quantities called "photons"[8]. The theory allowing us to handle in a full satisfactory way the problem in both, low and high energy regions, is the so called quantum electrodynamics (QED). This theory, on the other side, makes it very difficult to treat in a simple way coherence phenomena as reflection. A complete QED description of the optical properties of matter is beyond the scope of this paper, so we will limit ourselves to show the way to relate the macroscopic optical constants of a medium (ε,σ,μ,..) to its microscopic structure within the framework of quantum theory. Further developments and more detailed treatments can be found in literature[1,2,8].

The interaction of the radiation with matter can be described with the hamiltonian formalism. The hamiltonian function must contain two terms describing the "free" radiation and matter fields and an interaction term "weighted" with the electromagnetic coupling constant. Because of the low value of the electromagnetic coupling ($\frac{e^2}{4\pi\hbar c} \approx \frac{1}{137}$), it is possible to treat this interaction with the time dependent perturbation theory. It is convenient to consider the electromagnetic field in a box of finite volume V and to proceed with the second quantization formalism (see f.i.[9]). In this way the free electromagnetic field is completely described by specifying the number of photons with given wave-vector $\vec{k}$ and polarization α. This means that any state can be written as a linear combination of states with definite number of photons; these are eigenstates of the energy and we indicate them by $| n_{\vec{k}\alpha} \rangle$. We will see in the following that a more convenient set of base functions has to be used to handle coherence phenomena.

The material system is composed by electrons and nuclei. To understand how to write the wave functions relative to it let us proceed in the following way. Let H be the complete hamiltonian of the system:

$$H = H_R + H_M + H_I = H_0 + H_I \tag{5.1}$$

with H_R the radiation hamiltonian, H_M the matter hamiltonian and H_I the interaction term; $| n_{\vec{k}\alpha} >$ and $| \psi_n >$ are eigenvectors of, respectively, H_R and H_M, so:

$$| \phi_n > = | n_{\vec{k}\alpha} > | \psi_n >$$

is an eigenvector of H_0. Assuming the system to be in the state $| \phi_n >$ at time t=0, its wavefunction at later times can be written:

$$| \phi (t) > = \sum_{n'} c_{n'} (t) | \phi_{n'} (t) > \tag{5.2}$$

with $|c_{n'} (t)|^2$ the probability to find the system in the state $|\phi_{n'} >$ at the time t. The $c_n (t)$ coefficients obbey to the equation:

$$i\hbar \frac{dc_m (t)}{dt} = \sum_{n'} < \phi_m | H_I | \phi_{n'} > c_{n'} (t) \tag{5.3}$$

with initial condition $c_{n'}(0) = \delta_{nn}$. The matrix element $< \phi_m | H_I | \phi_{n'} >$ can be evaluated using the following non-relativistic form of the interaction hamiltonian:

$$H_I = - \sum_i \left[\frac{e_i}{m_i c} \vec{A} \left(\vec{r}_i\right) \vec{p}_i + \frac{e_i^2}{2 m_i c^2} \vec{A}^2 \left(\vec{r}_i\right) \right] \tag{5.4}$$

where the index i labels the material components, $\vec{A}\left(\vec{r}_i\right)$ is the vector potential of the radiation field at the point $\vec{r}_i$ and $\vec{p}_i$ is the impulse of the i-th particle. Without going deeper in the formalism, eq. (5.3) leads to evaluate integrals of the form:

$$\int_V \psi_f^* (\vec{p} \, \vec{\varepsilon}) \, e^{\pm i (\vec{k} \vec{r})} \psi_i \, d\vec{r} \tag{5.5}$$

where $| \psi_i >$ and $| \psi_f >$ are the initial and final states, eigenfunctions of the imperturbated material hamiltonian H_M, $\vec{\varepsilon}$ is the polarization vector of the radiation and the two signs at the exponent keep into account, respectively, emission and absorption of one photon with wavevector $\vec{k}$. Integrals like (5.5) come from the action of the linear term, with respect to $\vec{A}$, of the interaction hamiltonian (5.4).

At first order in perturbation theory, this term describes transitions where the final state differs from the initial by only one photon (emission or absorption).

To describe scattering phenomena, where two photons are involved, it is necessary to consider higher orders in the linear term and the contribution coming at first order from the quadratic one, as we will early see.

States $|\psi>$, as we said, describe the material system and must contain all the microscopic information to allow a quantum description of the optical constants.

It can be demonstrated that[2]:

$$\mathrm{Im}\,\hat{\varepsilon}(\omega) = \frac{1}{\omega t} \sum_{n,n'} |c_{n'}(t)|^2 \, [f(n) - f(n')] \tag{5.6}$$

where f(n) is the probability to find the system in the state n and the involved process is the absorption of a photon. The determination of Im $\hat{\varepsilon}$ (ω) is sufficient to determine all the optical constants of the system as shown in the previous chapters.

The problem is now to give explicit form to wavefunctions $|\psi>$.

As a first step let us return to eq. (5.5). If we denote with $\boldsymbol{a}$ a length comparable with the linear dimensions of the region where $|<\psi|\psi>|^2$ is significantly greater than zero, then, if

$$k\boldsymbol{a} \sim \frac{\boldsymbol{a}}{\lambda} \approx 0, \tag{5.7}$$

the exponential in the (5.5) is almost a constant and can be carried out from the integral:

$$e^{\pm i(\vec{k}\vec{x})} \int_V \psi_f^* \, (\vec{p}\,\vec{\varepsilon}) \, \psi_i \, d\vec{r} \; . \tag{5.8}$$

The condition (5.7) is equivalent to say that the radiation wavelength must be greater than the dimension of the interacting particle (or system), and is classically known as the "dipole approximation". Considering the atomic dimension to be of the order of 10^{-10} m, it follows that the energy must be less than ~ 24 keV, including in this way the whole border region from visible up to soft X-ray. For well localized electrons (as core electrons are) the integral (5.5) reduces to a sum over atomic indexes:

$$\sum_j e^{\pm i(\vec{k}\vec{x}_j)} \int_{V_j} \psi_f^* \, (\vec{p}\,\vec{\varepsilon}) \, \psi_i \, d\vec{r}_j \tag{5.9}$$

where V_j is a finite volume around the j-th atom, $|\psi_i>$ and $|\psi_f>$ are atomic states.

This result still remains valid also if one of the two states is relative to delocalized electrons. This is the case describing the non-relativistic photoelectric effect. In this case we can calculate the integral using for ψ_i the k-shell electron wave function, while for ψ_f the free electron wave-function.

If electrons are delocalized (valence band electrons) but the irradiated medium has well defined symmetry properties, we can make use of Bloch's theorem and consider $|\psi_n>$ as a many-particle state satisfying:

$$T_{\vec{R}}\,|\psi_n> = e^{i\vec{k}_n\vec{R}}\,|\psi_n> \tag{5.10}$$

where $T_{\vec{R}}$ is a translation by a lattice vector $\vec{R}$. In this case, if the wavelength is large compared with the unit cell (i.e. $\vec{k}$ is small with respect to the dimension of the Brillouin zone), the dipole approximation is still valid[2].

At the first order in perturbation theory the previous assumptions allow to describe transition between atomic levels, direct interband transitions, non-relativistic photoelectric effect. To describe scattering phenomena we must consider second order processes in which two photons are involved. These processes need an intermediate state. We can represent this processes by the two Feynman diagrams:

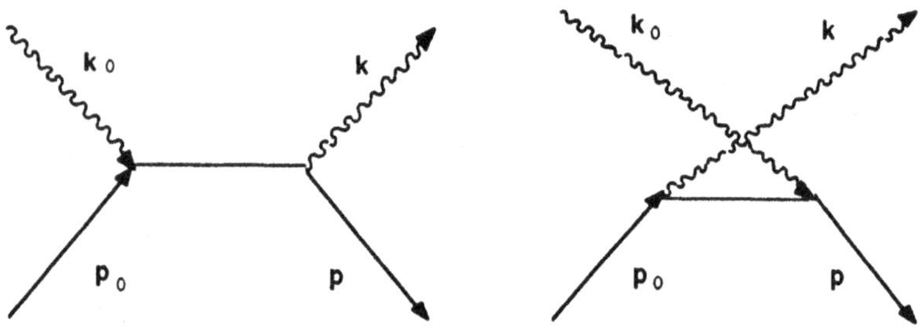

The corresponding matrix elements are:

$$\sum_i <\psi(\vec{p})\ \vec{k}\ |H_I^L|\ \psi(\vec{p}_{i1})><\psi(\vec{p}_{i1})\ |H_I^L|\ \psi(\vec{p}_0)\ \vec{k}_0>$$
$$\sum_i <\psi(\vec{p})\ \vec{k}_0\ |H_I^L|\ \psi(\vec{p}_{i2})><\psi(\vec{p}_{i2})\ |H_I^L|\ \psi(\vec{p}_0)\ \vec{k}> \tag{5.11}$$

with H_I^L the linear part of the interaction hamiltonian.

The final state can be the same as the initial one (Rayleigh scattering) or different (Raman scattering).

Also the quadratic term in the (5.4) gives a small contribution to two photons phenomena. Under the same condition leading to dipole approximation (λ and $\lambda_0 \gg a$), it contributes to Rayleigh scattering only[8].

We have, in fact, to evaluate the integral:

$$\int_V \psi_f^* \, e^{i(\vec{k}-\vec{k}_0,\,\vec{r})} (\vec{\varepsilon}\,\vec{\varepsilon}_0)\, \psi_i \, d\vec{r} \tag{5.12}$$

If $\lambda = \frac{2\pi}{|\vec{k}|}$ and $\lambda_0 = \frac{2\pi}{|\vec{k}_0|}$ are both larger than the atomic dimension, then it is possible, as before, to write:

$$e^{i(\vec{k}-\vec{k}_0,\,\vec{x})} (\vec{\varepsilon}\,\vec{\varepsilon}_0) \int_V \psi_f^* \, \psi_i \, d\vec{r} = e^{i(\vec{k}-\vec{k}_0,\,\vec{x})} (\vec{\varepsilon}\,\vec{\varepsilon}_0)\, \delta_{fi} \tag{5.13}$$

In the opposite limit, for λ and λ_0 small, the exponential function cannot be considered as a constant in the integrals.

The numerical value of the matrix elements decreases when the exponential function vary appreciably inside the atom, and so does the scattered intensity. At very high energies, corresponding to $k_0 \gg \frac{\hbar c}{a} \approx 2 \times 113 \, \frac{I}{Z}$ (I is the ionization energy), the intensity of the scattered wave vanish[8].

If the final state is a free electron with momentum $\vec{p}$ and energy E, it introduces a new exponential term in the integral (5.12), just compensating the factor $e^{i(\vec{k}-\vec{k}_0,\,\vec{r})}$ and making the result of integration large, however small the primary wave-length may be. In this case we have:

$$\begin{aligned} \vec{p} &\approx \vec{k}_0 - \vec{k} \\ E &= E_0 + \hbar\omega_0 - \hbar\omega \end{aligned} \tag{5.14}$$

corresponding to momentum and energy conservation. For a given scattering angle we obtain, at the values fixed by (5.14), a line whose sharpness increases as radiation wavelengths decreases. For very short wavelengths the process considered here becomes identical with the scattering of a photon by a free electron (Compton scattering).

The picture arising from the previous discussion shows that radiation tends to interact with objects having dimensions comparable to its wavelength. Starting from very short wavelength, where all electrons are seen as "free" electrons (remember the effective number of electrons per atom in the sum rule of section 4), Compton scattering dominates the absorption mechanism in eq. (5.6). Increasing the wavelength, Raman scattering (anelastic) decreases its importance while Rayleigh diffusion (elastic) become more and more intense. At the same time the photoelectric effect becomes the most important cause of absorption. At lower energies, below the ionization energy, transitions between atomic levels appear, producing an intense scattered radiation called "resonance fluorescence"; at still lower energies interband and intraband (in metals) transitions are the main responsibles of energy absorption[2].

This description comes from a model in which only electrons are coupled to the radiation field.

A more realistic model for matter must include, in the function ψ, both electrons and nuclei coordinates. Since nuclear masses are much greater than the electronic mass, the Born-Oppenheimer approximation can be used, writing:

$$\psi_{nv}(\vec{r}, \vec{R}) = \psi_n(\vec{r}, \vec{R})\, \chi_{nv}(\vec{R}) \tag{5.15}$$

where $\vec{r}$ are $\vec{R}$ are referred to all electrons and nuclei coordinates, respectively. The ψ_n in eq. (5.15) are eigenstates of the electronic hamiltonian in which nuclear coordinates are parameters and the corresponding eigenvalue $\varepsilon_n(\vec{R})$ represents the effective potential in the nuclear hamiltonian. Writing the nuclear kinetic energy:

$$-\frac{\hbar^2}{2\, m_N} \Delta_R\, \psi_{nv} = -\frac{\hbar^2}{2\, m_N}\left(\psi_n\, \Delta_R\, \chi_{nv} + 2\nabla_R\, \psi_R \cdot \nabla_R\, \chi_{nv} + \chi_{nv}\, \nabla_R\, \psi_n\right) \tag{5.16}$$

the form (5.15) for ψ_{nv} follows from approximating eq. (5.16) with just the first term on the right hand side. The second term is not totally negligible and can be treated as a correction to the Born-Oppenheimer approximation, resulting in electron-phonon interaction. These corrections become very important and produce the dominant features of optical processes in the spectral regions where the stronger direct transition are absent, i.e., at frequencies below the direct interband gap.

The electron-phonon interaction also accounts for indirect interband absorption which is observed in semiconductors. Another important mechanism of radiation absorption in semiconductors is connected to exciton states. We can describe an exciton as the interaction between an excited electron and the hole it left in the valence band. An electron-hole pair acts as a kind of positronium atom with hydrogen - like energy levels where the lowest state corresponds to the recombination of the pair, while ionization corresponds to a free conduction electron and a free hole in the valence band. The excition levels lie in the forbidden gap of dielectrics.

Although short and incomplete, this qualitative description of the relevant mechanisms in the interaction of the radiation with matter will allow, in the next section, to describe the "border zone" between ionizing and non-ionizing effects.

6. FROM IONIZING TO NON-IONIZING

The aim of this section is to follow, step by step, what happens when irradiating a material medium with photons of decreasing energy.

Let us start from an energy of 1.021 MeV, just below the pair production threshold (almost until the end we don't need to specify whether the medium is a metal or a dielectric).

The world seen by a photon of this energy is essentially a kind of gas of electrons, each one indipendent of the others, and the result of every interaction is an electron-photon pair with energies and momenta as expected after an elastic electron-photon collision. An interaction of this kind is known, as we have seen, as Compton scattering. The Compton scattering is considered as an inelastic interaction in spite of the conservation of kinetic energy because the incoming and scattered photons have different frequencies and are considered as "different " quanta: we saw that the interaction takes place through the absorption of a photon and the re-emission of a new one.

Electrons, however, can be seen as completely free only when their binding energies are negligible compared to the energy of the incoming photon: decreasing the photon energy electrons appear to be bound to (invisible) nuclei. The conservation laws are so only approximately satisfied (because also nuclei enter in the balance) and elastic interactions are not completely negligible. We call the inelastic interaction Raman scattering and it is clear that Compton scattering is its limit for photon energies much higher than K-shell ionization energy. If the energy of the diffused photon is the same of the incoming one (elastic scattering) we call it Rayleigh diffusion. Elastic scattering becomes more important than inelastic scattering at energies below some keV in water and below some hundred of keV in lead. Before this happens, however, a more important effect comes to dominate the scene.

Just below 1 MeV in lead and at ≈ 50 keV in water, the photon has a high chance to disappear being absorbed by a K-shell electron that is extracted from the atom.

This is the well-known photoelectric effect and it is the main responsible for photon absorption form hard X-rays to ultraviolet. In this energy region, radiation "sees" the atom like a single particle and the momentum conservation law is completely violated if we consider only one electron involved in the interaction. The right way to see the process is assuming that the photon interacts with the atom as a whole. At an energy of 0.5 MeV the photon wavelength is 0.025 Å, while the radius of the 1s electron orbit in lead is ≈ 0.006 Å; this justifies the use of the dipole approximation, that we discussed in the previous sections, for the theoretical calculations. As the energy still decreases, shells with higher n are excited by the radiation and the ionization thresholds give the photoelectric (and total) cross-section the characteristic saw-toothed form. This until the lowest ionization energy is reached which usually happens at energies of some eV, corresponding to ultraviolet frequencies.

From this point further we are completely in the non-ionizing domain. Superimposed on elastic scattering (always present if intermediate states are available), transitions between atomic, molecular and cristalline energy levels are now the ways of interaction and absorption of the radiation. It is at this point that we must use different models for dielectrics and for metals. So exciton states have to be considered in insulators and semiconductors while absorption of light with successive electron-phonon interactions, leading to the heating of the cristal by lattice mode excitation, are fundamental in metals. It is beyond the scope of this paper to further investigate these processes that we can generally classify as absorption with heat formation. More important is the nature of reflection. The wave picture, using the classic macroscopic Maxwell equations, explains very well the existence of the reflected wave but, how to explain this phenomenon using the quantum description adopted in section 5? The frequency of the reflected light is the same of the incoming one, so the scattering must be elastic. But the Rayleigh cross-section per atom for non-polarized light is (at low frequences)[10]:

$$d\sigma = \frac{1}{2}\left(\frac{e^2}{mc^2}\right)^2 Z^2 (1 + \cos^2\theta)\, d\Omega \tag{6.1}$$

where Z is the number of electrons per atom and θ is the angle between the scattered and the incident wavevectors. The expression (6.1) refers to the scattering from a single atom and does not allow us to deduce any reflection from an ordered layer of atoms. Reflection is generated by coherent emission of light from an aggregate of atoms. In the classical theory it has origin by the fact that the phase of the scattered wave is the same of the phase of the incoming beam. The same must be true also in the quantum theory but the application of the idea of coherence requires some care. Let us consider N photons of frequency ω interacting with a medium. From the uncertainty relation:

$$\Delta E \, \Delta t \geq \hbar \qquad (6.2)$$

because $E = N\omega\hbar$ we can write:

$$\omega \, \Delta N \, \Delta t \geq 1 \qquad (6.2)$$

or

$$2\pi \frac{\Delta t}{T} \Delta N \geq 1 \qquad (6.3)$$

We can interpret (as it can be demonstrated in a more rigorous way[8]) $2\pi \, \Delta t/T$ as an uncertainty in the determination of the phase:

$$\Delta \phi \, \Delta N \geq 1 \ . \qquad (6.4)$$

This means that treating the electromagnetic field in the second quantization picture, using the "number of occupation" base (the base in which the number of photons with a given energy is fixed; so $\Delta N \rightarrow 0$), makes us to lose all the knowledge about the phase and, as a consequence, about coherence. In the case of the scattering of a single photon, as treated for Compton, Raman and Rayleigh scattering, we cannot know anything about the phase of the quantum scattered. Inversely, if we use a base in which the wave functions are eigenfunctions of the phase we don't know how many quanta are involved in the process, so it is hard to keep on speaking about scattering.

To find the phase difference between scattered photons when two atoms are involved is possible even if the number of quanta is definite; the price to pay is that we will not know from which atom the light quantum is scattered. Recalling the (5.8) we find, in the matrix element in the (5.6), a term:

$$e^{i(\vec{k}\vec{X}_A)} + e^{i(\vec{k}\vec{X}_B)} = e^{i(\vec{k}\vec{X}_A)}\left(1 + e^{i\left(\vec{k}(\vec{X}_B - \vec{X}_A)\right)}\right) \qquad (6.5)$$

and for the transition probability:

$$\left|1 + e^{\left(i\vec{k}(\vec{X}_B - \vec{X}_A)\right)}\right|^2 = 2\left\{1 + \cos\left[\vec{k}(\vec{X}_B - \vec{X}_A)\right]\right\} \ . \qquad (6.6)$$

The scalar product $\vec{k}\left(\vec{X}_B - \vec{X}_A\right)$ represents the difference of light path for the two scattered wave, i.e. the classical difference of the phases. This result is analogous to the classical one leading to the reflected beam.

Instead of using one of the alternate rappresentations for the electromagnetic field (number of occupation $| n >$ or phase $| \phi >$ eigenstates), to handle coherence phenomena it is better to use the so-called "coherence states", which are eigenstates of the positive frequency part of the electric field operator. The coherent state $|\alpha>$ is intermediate between $|n>$ and $|\phi>$ in the sense that neither the amplitude, nor the phase of the state is precisely defined.

Both quantities have an uncertainty spread with a root-mean-square deviation. They are defined as[11]:

$$| \alpha > = e^{-\frac{1}{2}|\alpha|^2} \sum_n \frac{\alpha^n}{(n!)^{\frac{1}{2}}} | n > \tag{6.7}$$

It is easy to verify that states (6.7) are normalized, not orthogonal and overcomplete (there are more $|\alpha >$ states than $|n>$ states, this is a consequence of the lack of orthogonality). Coherent states are the best way to perform the limit transition from quantum to classical mechanics in the case of the electromagnetic field. We must remember that, in particle mechanics, the quantum state most similar to a classical particle is a wave packet with uncertainty spread in its momentum and position.

From these considerations it results that a "mixed" approach could be the best way to analyze the interaction of a wide energy spectrum radiation field with the matter. The classical results from Fresnel equations (section 4) can be used to separate the reflected beam spectrum from the transmitted one. The interaction of the penetrating component with matter can then be handled with the usual absorption cross-sections. More, from the dependence of reflectivity from the linear absorption coefficient, quantum results enter in the classical Fresnel formulas too. This can be obtained in two ways: by the direct use of the experimental reflectivity curve (phenomenological approach) or by calculating this curve from the absorption coefficients using Kramers-Kronig dispersion relation[6]. In both cases an useful concept is that of the "effective number of electrons per atom", described in the previous sections.

From the radioprotection point of view a rough approximation made with a step function defined considering all the electrons with binding energy lower than the radiation energy gives a meaningful starting point and is rather good for this purpose, as outlined in the section 4.

REFERENCE

1. F. Wooten, "Optical Properties of Solids", Academic Press, New York, London (1972).
2. F. Bassani, M. Altarelli, "Interaction of Radiation with Condensed Matter", on Handbook on Synchrotron Radiation, Vol.1, ed. by E.E. Koch, North Holland P.C. (1983).
3. L.D. Landau, E.M. Lifšits, "Elettrodinamica dei mezzi continui", Editori Riuniti, Roma (1986).
4. B.L. Henke, Phys. Rev. A 6, 94 (1972).
5. G.B. Irani, T. Muen, F. Wooten, Phys. Rev. 3B, 2385 (1971).
6. B.L. Henke, P. Lee, T.J. Tanaka, R.L. Shimabukuro, B.K. Fujikawa, At. Data Nucl. Data Tables 27,1 (1982).
7. H. Ehrenreich, H.R. Philipp, Phys. Rev. 128, 1622 (1962).
8. W. Heitler, "The Quantum Theory of Radiation", Oxford University Press (1944).
9. L.D. Landau, E.M. Lifšits, "Teoria quantistica relativistica", Editori Riuniti, Roma (1978).
10. L.D. Landau, E.M. Lifšits, "Teoria dei campi", Editori Riuniti, Roma (1976).
11. R.J. Glauber, Phys. Rev. 131, 2766 (1963).

RADIOMETRIC MEASUREMENTS FOR BROADBAND OPTICAL SOURCES

Jürgen H. Bernhardt

Institute for Radiation Hygiene
of the Federal Health Office
D 8042 Neuherberg

PRINCIPLES AND REQUIRED RADIOMETRIC DATA

To evaluate a broadband optical source, it is normally necessary to determine the spectral distribution of optical radiation emitted from the source at the point of nearest human access. This accessible emission and its spectral distribution for a lighting system may differ from that actually being emitted by the lamp alone due, e.g., to the filtration in the light path.

It is desirable to use the correct terminology and definitions referring either to the source or to the surface on which the radiation is incident. The definitions of quantities and units are treated in another chapter in this book. Radiant emittance (watts per square metre), radiant intensity (watts per steradian), radiance (watts per steradian per square metre) and integrated radiance (joules per steradian per square metre) all refer to the source. Irradiance (watts per square metre) and radiant exposure (joules per square metre) both apply to the surface on which the radiation is incident. Thus radiant emittance and irradiance have the same dimensions but the former is power per unit area of source and the latter per unit area of surface. The term spectral placed before any of the quantities implies restriction to a unit wavelength band, e.g. spectral irradiance (watts per square metre per nanometre).

There are three important factors to be considered when interpreting radiometric measurements: characteristics of the source, response of the detector and biological effect (Moseley, 1988).

(a) The source - dimensions, spectral power distribution and radiance (power per unit area per unit solid angle).
(b) Detector - spectral response, power range, accuracy, precision and angular response.
(c) Biological effect -action spectrum, irradiance and dose requirements.

There are some general desirable features in a metering system. One is stability with time, so that sensitivity does not change during a measurement. Another is a constant, low (preferably zero) reading when no source is present. It is unlikely that the sensor will have a uniform

Light, Lasers, and Synchrotron Radiation
Edited by M. Grandolfo *et al.*, Plenum Press, New York, 1990

response over its surface and so if calibration is performed by filling the receiver the instrument should be used in the same way. It is also useful to know where the limiting aperture is in the measurement system since the inverse square law relates to irradiance at this position. Its location should be verified by investigation of the irradiance-distance relation. It may be that the sensing element is the limiting aperture. Determination of the wavelength response of a detector is often inaccurate. It is usually accomplished by passing light from a high-power xenon lamp through a monochromator. The difficulty with this method is that a wavelengths where the detector has a low sensitivity special precautions must be taken to ensure that the reading obtained is more sensitive. Finally, the response time of the detector must be suitable for the source and type of measurement intended. For example, if the sensor is to respond to pulses of short duration, a short time constant is required; if it is the average output from an AC source which is important, then a long time constant is more suitable.

The spectral irradiance E_λ should be measured as completely as possible from 200 nm to at least 1400 nm. Fluorescent lamps and many commonly used arc lamps emit little infrared radiation beyond 1200 nm and the spectral region between 1200 nm and 1400 nm often need not be measured if a suitable instrument is not available. For incandescent lamps, however, it is important that the near-infrared (IR-A) radiation be measured (Sliney and Wolbarsht, 1980). The spectral irradiance E_λ at the nearest point of access is of interest in assessing the potential hazards to both the skin and eye from the ultraviolet component in the spectrum, and the potential hazards to the skin form the entire spectrum. The spectral irradiance in the infrared is also needed to assess the potential hazards to the anterior part of the eye. For wavelengths beyond 1400 nm the spectral irradiance is of marginal interest; although total irradiance in the IR-B and IR-C should be known.

BIOLOGICAL WEIGHTING OF SPECTRORADIOMETRIC DATA

For the assessment of detrimental health effects from radiant exposure of a broadband optical source it is necessary to know the photobiologically effective irradiance E_{biol} of a radiation source:

$$E_{biol} = \sum_\lambda E_\lambda \cdot s\,(\lambda)_{biol,\ rel} \cdot \Delta\lambda \qquad \text{in W/m}^2 \qquad (1)$$

E_λ: spectral irradiance in $W/m^2 \cdot nm$

$s\,(\lambda)_{biol,\ rel}$: relative spectral effectiveness

For the wavelength dependent evaluation of the spectral irradiance two methods are generally applied: either a direct one by appropriate evaluating measurement procedures, or the indirect way by calculation from spectral measurement data.

Wide-range radiometers adapted to the relevant biological effect function are often easy to operate and rapidly provide measurement results (Goldberg and Klein, 1974; Mountford et.a., 1984, Nagpal, 1978; Pepper and Mountford, 1983; Schreiber and Ott, 1984). The photobiologically effective irradiance can be approximated using suitable filters which adapt their transmittance to the effect functions. One disadvantage of this procedure is that these approximations only refer to one single effect function and that filters have to be changed when another photobiologically effective evaluation is planned.

This disadvantage may be avoided by means of spectral measuring procedures which enable evaluations using different effect functions independent of the measurement (McKenzie, 1982; Schreiber and Ott, 1984). With this technique the spectral irradiance E_λ as a function of wavelength is measured. For this purpose a monochromator with rotating grating and minimum band width of 2.5 nm with an appropriate detector may be used. Supported by a small computer it is possible to store measuring data, to calibrate the instruments and to calculate various biological effects using specific effect functions $(s(\lambda)_{biol, rel}$. However, since the intensity of both the emitter and the effect function often varies by several orders of magnitude in different wavelength regions, this requires much in terms of equipment (e.g. double monochromator) in order to achieve the required accuracy of measurement.

The spectrum of an arc lamp, a gas discharge lamp, or a fluorescent lamp consists of both line structure and a continuum. Significant errors can be introduced in the representation of the spectrum and in the weighting of the spectrum against biological or safety action spectrum if the fraction of energy in each line is not properly added to the continuum (see, e.g., Sliney and Wolbarsht, 1980).

The recommended units for presenting the spectrum in tabulated form should have the measured spectral irradiance in microwatt-per-square-centimeter-per-nanometer [$\mu W/(cm^2 \cdot nm)$] of the continuum at regular intervals (e.g., every 5 nm) with a separate list of the irradiance in microwatts-per-square-centimeter ($\mu W/cm^2$) for each line. This procedure is particularly important in critical spectral regions where biological weighting functions vary rapidly (e.g., 300 - 320 nm; 395 - 500 nm). Narrow bandwidth recording (e.g., 1 nm) is also of obvious importance in such transition regions. The procedure of radiometric measurements are described in more detail by Sliney and Wolbarsht (1980).

The spectral distribution of a lamp or arc can be graphically illustrated when the continuum and line structures are recombined in a histogram. The spectral division of the histogram should accurately reflect the spectral resolution of the data. To prepare a histogram from tabulated data the separate line contributions must be added back to the continuum.

From the time integral of the photobiologically effective irradiance, the photobiologically effective radiant exposure (dose) H_{biol} can be determined:

$$H_{biol} = \int_0^{t_1} E_{biol} \, dt, \text{ in } J/m^2 \qquad (2)$$

For assessing a radiation dose, those values for the photobiologically effective radiant exposure may be used which produce a just noticeable threshold reaction (e.g. erythema). Numerical values for threshold radiant exposures $H_{s,biol}$ are for instance:

erythema:	100 - 500 J/m^2	(max 297 nm), value dependent on skintype
conjunctivitis:	50 J/m^2	(max 260 nm)
keratitis:	40 J/m^2	(max 270 nm)
direct pigmentation	10^5 J/m^2	(max 340 nm)

With $H_{s,biol}$ and E_{biol} the exposure time may be determined which leads to a just noticeable threshold reaction:

$$t = H_{s,biol}/E_{biol} \qquad (3)$$

SPECTRORADIOMETER

A spectroradiometer is an instrument for measuring irradiance within a narrow bandwidth centred at a wavelength which is continuously variable and selectable by the user. It comprises input optics, a monochromator and a detector plus readout. A microcomputer is often incorporated. Radiation is dispersed (separated according to wavelength) in the monochromator by a quartz or water-filled prism or a grating.
Radiation enters the monochromator through an entrance slit which is at the focal point of a collimating mirror. Radiation is reflected from the mirror as a parallel beam incident on a grating which separates the radiation into parallel beams with wavelength dependent on angle. A telescope mirror collects radiation at one particular angle (and therefore wavelength) and focuses it into the exit slit. Radiation emerging from monochromator is detected by a sensing element. The wavelength presented to the detector is altered by angular rotation of the grating.

The flux which passes through the exit slit depends among others on the angular dispersion and the square of the bandwith. The latter is thus an important parameter in determining the throughput of radiation flux.

The shape of the bandpass function is dependent on the widths of the entrance and exit slits (for more details see, e.g. Johns and Rauth, 1965, and Handbook of Optics).

Monochromators exhibit a strong angular dependency, in many cases with little response to radiation beyond 10° off-axis. Also, they possess polarisation sensitivity of up to 40 %. It is, therefore, necessary to employ a depolariser/diffuser (see section input optics). In addition temperature control or compensation is required for precision measurements, since even the best spectroradiometers have a wavelength temperature coefficient of 0.1 nm per degree Celsius.

Accurate spectroradiometry is very difficult when spectral irradiance varies rapidly with wavelength. An example of this is the measurement of shortwave radiation at the surface of the earth, where irradiance changes by about 3 % per 0.1 nm at 305 nm and about 10 % per 0.1 nm at 295 nm. Errors arise from the fact that the triangular-shaped slit function has tails which extend to wavelengths far removed from the selected pass band. These errors produce an apparent shift of several nm towards larger wavelengths.

Another the error in a spectral irradiance measurement is due to stray light. The rapidly increasing spectral irradiance of sunlight near 300 nm is poorly characterized by a system with single grating which has significant stray light. A plateau at shorter wavelengths is present. A double-monochromator significantly improves the situation. However there exist a limited sensitivity due to the limited throughput characteristic of all double monochromators.

The presence of stray light can be checked during a measurement by placing a short-wave cut-off filter in front of the monochromator. The filter's cut-off should be at a wavelength just above wavelength being measured. Hence, a series of filters are really needed. If the instrument reading is reduced to zero then the reading should be reasonably correct, provided, of course, that there is not significantly more radiation at still shorter wavelengths, If the stray light contribution cannot be reduced to zero, the subtractive technique can be applied to obtain an estimate of the spectral irradiance at that wavelength. Obviously all of the facts of monochromator performance must be known and understood before accurate evaluations of broad-band optical sources can be performed. Problems associated with measurements are described more detailed in the book of Sliney and Wolbarsht (1980).

DETECTORS

The detector attached to a spectroradiometer may be a thermal detector or a quantum detector like a photodiode or a photomultiplier. Mainly these two types and their properties are described in this chapter. Further details of detectors are described, e.g., in "Handbook of Optics" (W. Driscoll, ed., 1983).

Thermal detectors

Thermal detectors sense the change in temperature produced by absorption of incident radiation. Their spectral response can therefore be as flat as the absorption spectrum of their blackened coating and window will allow. This makes them useful for spectroscopy and radiometry.

Pneumatic

The receiver is a thin blackened membrane placed in a small airtight, gas-filled chamber. Heat absorbed by the foil is transmitted to the gas, causing a pressure rise, which is measurable in various ways (by mirror motion; electrically, by change in capacitance). A commercially available form is the Golay cell, which has found wide use in far-infrared spectroscopy.

Thermocouple

The receiver is a thin blacked flake connected thermally to the junction of two dissimilar metals or semiconductors. Heat absorbed by the flake causes a temperature rise of the junction, and hence a thermoelectric emf is developed which can be measured, for example, with a potentiometer or by a high-impedance digital voltmeter. A thermopile comprise several such thermocouples, connected in series for improved sensitivity. Since the functioning principle is a rise in temperature based on absorption of energy, thermopiles have a flat spectral response within their region of operation. For example, a thermopile with a fused silica window has a near-uniform response over the range 180 to 3400 nm. This makes it especially useful when a wide range of wavelengths is to be covered, although the sensitivity may be less than other types of detectors at specific wavelengths. The receiver may be small, less than 1 mm in diameter. Fine unsupported wires cause thermopiles to be fragile but recent developments in thin-film technology have led to the production of devices which are much more robust.

Bolometer

The receiver is a thin blackened slab whose impedance is temperature-dependent. The dc impedance change is generally sensed using a bridge circuit. The ac operation requires only a single bolometer element in series with a load and voltage source. The signal is the change in voltage across the load.

Most bolometers in use today are of the thermistor type. Their sensitivity closely approaches that of the thermocouple. Thermistor elements are made of polycrystalline Mn, Ni, and Co oxides. In their final form they are semiconductor flakes 10 μm thick, which undergo a temperature resistance change of ~4 percent/K. Since thermistor resistance changes with ambient temperature enough to alter the biasing significantly, it is usually operated in a bridge circuit, with a nearly identical thermistor shielded from signal radiation and used for a balance resistor. Some extremely sensitive low temperature bolometers are available commercially.

Pyroelectric

Ferroelectric material (e.g., a lithium tantalate crystal or Triglycine sulfate) exhibits a residual polarization in the absence of any electric field. Dipole moments, initially aligned by applying an external field, result in a surface charge which is normally slowly neutralized by leakage. This polarization is temperature-dependent (pyroelectric effect), and when a pulse of incident radiation heats a portion of an electroded sample, there is a change in surface charge (open-circuit voltage) which is proportional to the incident radiation power. Speed of response is inherently not limited by thermal inertia but by charge-reorientation time. Electrically, the device behaves like a capacitor, requiring no bias and therefore exhibiting no current noise. The receiver may again be small, for example 1 mm in diameter, but with a faster response than a thermopile.

QUANTUM DETECTORS

Photoemissive Detectors

Photoemissive detectors are generally the detector of choice in all spectral regions where high quantum efficiency is available. In the spectral region $\lambda < 600$ nm, the photomultiplier, or multiplier phototube, has close to ideal sensitivity; i.e., selected photomultiplier tubes (PMT) are capable of detecting single photon arrivals and amplifying the photocurrent (pulse) enormously without seriously degrading the signal-to-noise ratio. Time resolution can be as short as 0.1 ns.

The radiation is absorbed by a photosensitive surface which usually contains alkali metals (cesium, sodium, or potassium). Incident quanta release photoelectrons, by the photoelectric effect, which are collected by a positively biased anode. This is called a diode phototube. A photomultiplier tube consists of a photocathode in which incident light is converted into a flow of electrons. A series of electrodes, called dynodes, at progressively higher potential accelerate and amplify the electron stream to produce a measurable current. The dynode potentials are established using a resistor chain which acts as a voltage divider. In spectral regions where quantum efficiency is high ($\lambda < 550$ nm), the photoemissive detector is very ideal. Since the sensitivity in red-sensitive tubes is limited by thermally generated electrons, sensitivity can be improved by cooling.

Wavelength response is determined by the composition of the cathode and the window. The extension of sensitivity into the UV is dependent on the composition of the window. For UV operation down to about 220 nm, a quartz window is required but, for even shorter wavelengths (down to 110 nm), windows made of magnesium fluoride must be used. Photocathodes of caesium telluride or caesium iodide will not respond to wavelengths above 320 and 200 nm, respectively, whereas bialkali (such as antimony-rubidium-caesium) will extend the range to 620 nm.

A measure of cathode sensitivity is the quantum efficiency, defined as the ratio of the number of photoelectrons emitted from the photocathode to the number of incident photons. A figure of 20 % is fairly typical for a bialkali cathode. Quantum efficiency and sensitivity are related via the photon energy at each wavelength.

Photomultiplier tubes are susceptible to magnetic fields, for example from lamp running gear or electric motors. To overcome this effect Mumetal shields are available.

Photomultipliers are of optimal value where low light levels are encountered. This may arise as a result of the incident light being passed through a low-efficiency, narrow-bandpass monochromator. Photomultiplier suffer from their fragility and the requirement for a high kV stabilised power supply.

Photoconductive

The radiation is absorbed by a photoconductive material either in thinfilm or bulk form. Each incident quantum may release an electron-hole pair, thereby increasing the electrical conductivity. The usual readout circuit is the same as that for the ac bolometer.
Photoconductors can have a high quantum efficiency from the visible region out to the far infrared but lack the nearly ideal high amplification of photomultipliers. They are therefore chiefly useful in the spectral region beyond 1 µm, where efficient photoemitters are unavailable. In general, lower-temperature operation is associated with longer-wave-length sensitivity in order to suppress noise due to thermally induced transitions between close-lying energy levels.

Photovoltaic

The most widely used photovoltaic detector is the p-n junction type, where a strong internal electric field exists across the junction even in the absence of radiation. Radiation quanta incident on the junction of this film or bulk material produce free hole-electron pairs which are separated by the internal electric field, causing a change in voltage across the open-circuit cell or a current to flow in the short-circuit case.
As with the photoconductor, quantum efficiency can be high from the visible to the far infrared. Response times can be as short as several nanoseconds, which is much faster than photoconductors. The advantages of nearly ideal internal amplification have become available in avalanche photodiodes sensitive out to 1.55 µm.

The term photodiode has been applied both to vacuum- or gasfilled photoemissive detectors (diode phototubes, or photo cells) and to photovoltaic detectors (semiconductor p-n junction devices).

Photoelectromagnetic

A thin slab of photoconductive material is oriented with radiation incident on a large face and a magnetic field perpendicular to it. Hole-electron pairs generated by the incident photons diffuse through the material and are separated by the magnetic field, causing a potential difference at opposite ends of the detector.

These detectors require no cooling or biasing electric field but do require a permanent magnetic field. Photoelectromagnetic InSb has response out to 7.5 µm, where it is as sensitive as a thermocouple of equal size, and has a response time less than 1 µs. Another competing uncooled detector is InAs, which is far more sensitive out to 3.5 μm. Cooled infrared detectors are one to two orders of magnitude more sensitive.

Photographic

The receiver is an emulsion containing silver halide crystals. Incident photons are absorbed by the halide ion, which subsequently loses its electron. This electron eventually recombines with a silver ion and reduces it to a neutral silver atom. As more photons are absorbed, this process is repeated until a small but stable cluster of reduced silver atoms is formed within the crystal (latent image). Internal amplification is provided by introduction of an electron donor (photographic developer), which, using the latent image as a catalytic center, reduces to silver all the remaining silver ions within the exposed crystal. The density of reduced crystal is a measure of the total radiation exposure.
The spectral region of sensitivity for photographic detection coincides rather closely with that of the photoemissive detector. For $\lambda > 1.2$ μm there is too little energy in each photon to form a stable latent image. The photographic process is an integrating one in that the output (emulsion density) measures the cumulative effect of all the radiation incident during the exposure time. The efficiency of the photographic process can be very high, but it depends upon photon energy; e.g., in the visible region it takes only 10 to 100 photons to form a stable latent image (developable grain). The photographic process enjoys a large and efficient internal emplification ability (development) wherein the very small energy of the photons' interaction is converted into readily observed macroscopic changes.

FILM BADGE DOSIMETERS

Many substances exhibit a measurable change in one or more of their properties on exposure to Ultraviolet Radiation (UVR) and this enables them to be used as chromophores in film badge dosimeters.

One approach has been to observe thermoluminescence of MgO single crystals (Takeuchi et al. 1976). Thermoluminescence dosimetry (TLD) is an established technique for measuring the dose from ionising radiation. The crystal exhibit luminescence when their temperature is raised to a few hundred degrees Celsius and the intensity of the emitted light is dependent on the radiation dose previously received by the crystal. When MgO single crystals were exposed to UVR from a deuterium lamp, thermoluminescent intensity was shown to be proportional to exposure time. Another substance, silver-activated metaphosphate glass, has been shown to exhibit luminescence after exposure to UV at 253.7 nm, the emission intensity being proportional to UV dose (Nagpal et.al. 1976). The induced photoluminescence may be increased by heating to a temperature of 350°C. Lithium fluoride TLDS are widely employed for the detection of ionising radiation. It has been shown that they may also be used as ultraviolet radiation dosimeters. None of the TLD substances described above, nor others published in the literature, have found widespread application in UV dosimetry.

One of the first materials to be used successfully was polysulphone (Davis et.al. 1976). It was observed that polysulphone darkened on exposure to UVR. To exploit this effect a polymer film between 36 and 44 μm thick is used. Absorbance of the film changes with UV dose for wavelengths less than 250 nm up to 335 nm and exhibits a smooth response to varying dose. The attractive feature of polysulphone is that its spectral response bears a resemblance to the erythema action spectrum although it does extend too far into the UVA. Other substances have also been used in the production of UV-sensitive film (for references see Moseley, 1988).

A promising material recently investigated is diazo film (Jackson 1980, Moseley et al. 1984), which is widely used in photographic departments for producing coloured backgrounds in transparencies of diagrams. It consists of a sensitive layer containing a diazonium salt and coupling agent deposited on a 0.135 mm thick sheet of clear acetate. In the presence of ammonia, brightly coloured (usually blue) azo compounds are produced but prior exposure to UVR destroys the diazonium salt and reduces the intensity of the colour when the film is developed. Thus, the film's optical density decreases as UV dose increases. Standardisation of development procedure is essential for accuracy. Provided this is done the film is a suitable chromophore with measurable sensitivity down to 254 nm. Relative density, in which optical density is determined relative to that of fully exposed and unexposed film, changes with dose. Unfiltered, a dose range between 0 and 200 mJ cm^{-2} can be measured at 365 nm. With filtering, incident radiation could be attenuated to extend the useful range. The sensitivity of the film is the reciprocal of the dose required to produce a given relative density and varies with wavelength. Diazo film is widely available and easily processed. Suitably filtered it could provide the basis for a UV film badge with many applications.

INPUT OPTICS

Irradiance (W m^{-2}) on a surface depends on the angle between the beam and the surface. This is because rotation of the surface alters the area presented to the beam and hence the total power per unit area of the surface. Stated mathematically, the irradiance, I_s, on a surface from a uniform beam is given by the equation

$$I_s = I_d \cos\theta \tag{4}$$

where I_d is the power per unit cross sectional area in the beam at a distance d from source and θ is the angle between the normal to the surface and the axis of the beam from surface to source. This relation, often called the cosine law, also applies to an extended source where each element of the source may be considered to be a point source. A detector correctly measuring irradiance will obey the cosine law. For a good angular response, the detection area should be as close to the surface as possible. The use of filters or quartz windows will degrade the response by refraction. In the case of monochromators, appropriate input optics must be used.

Monochromators are particularly sensitive to the angle of incident radiation and the position of the beam on the entrance slit. Also, the surface of the detector, e.g. the photocathode of a photomultiplier tube, may be non-uniform in its response. A suitable diffuser must, therefore, be used. Radiation may be scattered either on transmission through a material or by surface reflections. The latter is preferable and usually takes the form of an integrating sphere. This is designed so that there is no direct path between input and output apertures. Instead, radiation undergoes multiple scattering before reaching the output so that the range of angles presented to the aperture is constant irrespective of the direction of radiation incident on the input.

DIRECT READING INSTRUMENTS

Spectroradiometric measurements are laborious and time consuming, and the interpretation of the data has the same difficulties. Although rapid-scan spectrometers have shortened the process and permit measurements of some pulsed and time-varying sources, they are, nevertheless, quite complex and expensive and the data reduction remains difficult. It is easy to understand, therefore, that there have been many efforts to develop instruments with spectral responses incorporating the several biologic weighting functions especially in the UV-range. There are several filters on the market which follow the UV action curve reasonably well. These filters, in combination with a broad-band UV detector will function reasonably well as a UV hazard monitor.

The principal difficulty in developing an ultraviolet instrument has been to develop a means for measuring only the UV-B and UV-C radiation with sufficient sensitivity while still rejecting all of the UV-A and visible light. Few instrument designers can believe at first that this design task is as difficult as it really is. They fail to appreciate the critical importance of proper spectral response in the 300 to 320 nm band and the need for near total rejection of longer wavelengths. Most ultraviolet sources for which this type of hazard monitor instrument would be used (such as plasma torches, arc lamps and welding arcs) emit far more radiation in the visible and near-ultraviolet (UV-A) regions than in the actinic (UV-B and UV-C) regions. Conventional instruments which were designed to measure actinic ultraviolet radiation would also respond to some degree to the longer wavelength, and therefore could provide an incorrect (larger) spectral irradiance value in the ultraviolet. Efforts to develop ultraviolet photochemical dosimeters for safety purposes have also met the same problem.

A relatively inexpensive instrument for measuring germicidal lamps (254 nm) is the Robinson-Berger meter with a calcium tungstate ($CaWO_4$) phosphor which has a fluorescence spectrum similar to the erythema action spectrum from 290-315 nm.

The measurement of the actinic ultraviolet radiation from the sun at sea level nicely illustrates the UV measurements problems (Sliney and Wolbarsht, 1980). At noontime in June, the irradiance for wavelengths below 315 nm would be of the order of 0.1 mW/cm^2 vs. 100 mW/cm^2 for the entire spectrum. At noontime in December, the irradiance for wavelengths below 315 nm is of the order of 0.01 mW/cm^2 vs. 80 mW/cm^2 for the entire spectrum. Therefore, to measure the actinic radiation component of the terrestrial solar spectrum, an instrument must be many thousands of times more sensitive to wavelengths below 315 nm than to wavelengths in the visible. The measurement of a welding arc would obviously require less rejection of longer wavelength radiation.

CALIBRATION

Regular, careful calibration of UV detectors is imperative. Unfortunately, it is not sufficient to accept data supplied by the manufacturer unless they can be supported by a calibrated standard.

Calibration is usually accomplished by the method of substitution by which one source of known spectral distribution is used to calibrate the other, unknown source. For this to be accomplished with complete accuracy, the two sources would need to be identical. The differences between the sources, and the extend to which the measuring system responds to their different characteristics, will determine the magnitude

of the error due to the methodology. Ultimately, a calibration is traceable to a primary standard, usually a black-body cavity radiator to which Planck's equations apply, a hydrogen arc or synchrotron radiation. The primary standards are used to calibrate lamps which become secondary standards. These are either quartz halogen tungsten filament lamps or deuterium lamps. New errors are introduced in the transference from primary to secondary standard.
The relation between the secondary and primary standards and the way in which the secondary standard is subsequently used restricts the accuracy of any laboratory calibration system. It is essential that there is a proper understanding and correct appraisal of the limitations of a detecting instrument. Among the many possible sources of error, one which might be overlooked is due to the fact that a tungsten filament is usually wound in an approximately cylindrical fashion. In some orientations, the surface of the filament at the front and back of the lamp will be visible while in other positions the filament at the back of the lamp will be obscured leading to a reduced irradiance. This can introduce an error of the order of 5 % (Moseley, 1988).

Stray light can also cause errors in the instrument calibration process in addition to those in the measuring process. A 1000-W tungsten lamp is typically used to calibrate a spectroradiometer. The spectral irradiance values from 250 nm to 400 nm differ by a factor exceeding 100. Hence the spectral calibration function of the spectroradiometer can be in error in the UV-B and UV-C regions if there is inadequate stray light rejection. The measurement of a relatively uniformly varying spectrum at a high irradiance, such as from a welding arc, will have minimal stray light errors (probably undetectable) to place the field measurements in question. But short wavelength measurements could be seriously in error as a result of stray light error introduced during calibration (Sliney and Wolbarsht, 1980).

Because of this problem a deuterium lamp calibration standard has been introduced in recent years. Stray light errors are not limited to the UV. Similar errors can arise in the infrared where a detector's sensitivity rolls off rapidly.

At least two, preferably three, calibrated sources, should be available for comparison, and a regular calibration programme should be part of any laboratory or department in which optical sources and detectors are used. Reliance should not be placed on a single, one-time-calibrated light source. Generally it is recommended, to use a calibration source similar to the source to be measured.

REFERENCES

Davis A., Deane, G.H., and Diffey, B.L., 1976, Possible Dosimeter for Ultraviolet Radiation, Nature, 261:169.

Driscoll, W. (ed), 1983, Handbooks of Optics, McGraw Hill.

Goldberg, B., Klein, W.H., 1974, Radiometer to Monitor Low Levels of UV Irradiance, Appl Optics, 13:493-496.

Jackson, S.A., 1980, A film Badge Dosimeter for UVA Radiation, J. Biomed. Eng. 2:63.

Johns, H.E., and Rauth, A.M., 1965, Theory and Design of High Intensity UV Monochromators for Photobiology and Photochemistry Photochem. Photobiol., 4:673.

Kiefer, J., (ed.), 1977, Ultraviolette Strahlen, Verlag de Gruyter, Berlin.

Koller, L.R., 1965, Ultraviolet Radiation, 2nd ed., Wiley, New York.

Mackenzie, L.A., 1982, Monitoring of Ultraviolet Radiation. In: Practical Aspects of Nonionising Radiation Protection, pp. 73-79. Conference Report Series 36, (London: Hospital Physicists Association).

Meyer, A.E.H. & Seitz, E.O., 1949, Ultraviolet Radiation, Berlin, Walter de Gruyter & Co. (in German).

Moseley, H., Robertson, J., and O'Donoghue, J., 1984, The Suitability of Diazochrome KBL Film for UV Dosimetry, Phys. Med. Biol. 29:679.

Moseley, H., 1988, Non-Ionizing Radiation: Microwaves, Ultraviolet and Laser Radiation, (Medical Physics Handbook, 18) A. Hilger, Bristol and Philadelphia. In Collaboration with the Hospital Physicists Association.

Mountford P.J., Pepper, M.G., Goldin, D., 1984, An On-line-UV-radiation Monitoring System for Control of Photosensitivity Test Dose, Phys. Med. Biol. Vol. 29, 4:407-417.

Nagpal, J.S., Ramanathan, G., and Gangadharan, P., 1976, Thermal and Optical Bleaching of Radiation Effects in Silver Activated Metaphosphate Glass - its Use in UV Dosimetry, Health Phys. 31:157.

Nagpal, J.S., 1978, Ultraviolet Radiation Dosimetry. A review, Med. Phys. Bull., 3:126-130.

Naumann, H., Schröder, G., 1987, Bauelemente der Optik, Taschenbuch für Konstrukteure, Carl Hanser-Verlag, München, Wien.

Pepper, M.G., Mountford, P.J., 1983, A Meter for UV-Dose and Irradiance. Phys. Med. Biol. 28/3:257-267.

Schreiber P., and Ott, G., 1984, Schutz vor ultravioletter Strahlung. Schriftenreihe der Bundesanstalt für Arbeitsschutz, Dortmund, Sonderschrift S 14.

Sliney, D., Wolbarsht, M., 1980, Safety with Lasers and Other Optical Sources, Plenum Press, New York and London.

Takeuchi, N., Inabe, K., Yamashita, J., and Nakamura, S., 1976, Thermoluminescence of MgO single crystals for UV dosimetry, Health Phys. 31:519.

LASER BEAM DIAGNOSTICS*

Wesley J. Marshall

Laser Microwave Division
U.S. Army Environmental Hygiene Agency
Aberdeen Proving Ground, MD 21010

INTRODUCTION

The evaluation of the direct beam laser hazards primarily consists of the determination of the NOHD with and without the use of optical devices, and the necessary optical density (OD) required for laser eye protection. The required OD would depend on whether the viewer was located at the laser exit or at a location downrange from the laser. Intrabeam viewing of a laser with optical devices generally requires a higher OD than viewing with the naked eye since more laser energy is collected by the optics. The NOHD of a specularly reflected laser beam must be determined if specular reflectors may be located where the laser beam is pointed.

LASER BEAM MODELS

Constant Irradiance Model of the Laser Beam

Although it is common knowledge that the center of a laser beam is usually more intense than the edges, scientists still tend to accept a model of the beam for calculations which assumes a constant irradiance within a diameter known as the beam diameter and zero outside this circle. The beam diameter is chosen as the radial distance where the local irradiance in the beam drops to 1/e times the peak irradiance. The reason that this diameter is chosen is that the peak irradiance is obtained from dividing the total beam power by the beam area. This relationship is only strictly applicable when the laser beam operates in the lowest transverse mode, TEM_{00}, and has a gaussian shape.

Expensive design flaws have occurred in safety equipment by using too simple a model. For instance, consider a safety filter

*The opinions or assertions contained herein are the private views of the author and are not to be construed as reflecting the views of the Department of the Army or the Department of Defense.

Light, Lasers, and Synchrotron Radiation
Edited by M. Grandolfo *et al.*, Plenum Press, New York, 1990

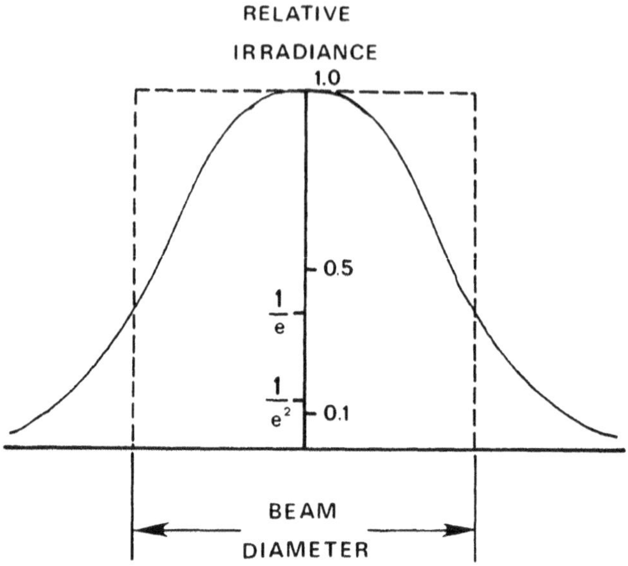

Figure 1. Comparison of the Beam Profiles of the Constant Irradiance Model of a Laser Beam (dashed line) to the Profile of the Gaussian Beam Model.

designed to provide a reduction factor of 10,000 and made only slightly larger than the beam diameter. The edges of the laser beam which had a local irradiance of only about 1-percent of the center beam irradiance are now much more intense than the center of the laser beam which is covered by the filter. Figure 1 illustrates a gaussian shaped beam with the constant irradiance equivalent model superimposed.

The main problem with the constant irradiance model occurs when the laser beam is small since the peak irradiance is calculated rather than the amount of energy that passes through the 7-mm, pupillary diameter of the eye; hence, the hazard is overestimated. The safety exposure limits are based on the total power or energy which a person could receive into the eye for visible and near-infrared ocular exposures. When the laser beam is smaller than 7 mm, the true beam irradiance at the laser exit port really overstates the hazard. The irradiance averaged over 7-mm is the desired quantity.

<u>Gaussian beam model</u>

In order to easily perform safety calculations, a model of the laser beam is necessary which will provide valid safety information but will not unnecessarily hamper the safety officer with the complexities of laser beam propagation. The exposure limit (EL) is generally given in terms of irradiance or radiant exposure on the cornea (ANSI 1986). Although the exposure limits appear to be based on corneal irradiance and corneal radiant exposure, the true hazard of a laser beam is really based on the accessible emission limit (AEL). The AEL is the total amount of power or energy which will pass through the limiting aperture, D_f (0.7 cm for visible and near infrared, and 0.1 cm for most other wavelengths) (ANSI 1986).

The gaussian beam model for hazard evaluation is based on a gaussian beam shape and measurements through apertures in order to provide accurate safety information regardless of beam size or viewing aids.

LASER BEAM CALCULATIONS

Laser Beam Diameter

Laser safety officers and laser manufacturers have historically performed laser calculations differently. Manufacturers and developers of laser systems often refer to laser beam diameters in terms of $1/e^2$ peak irradiance points (e.g. the radial distance between two points across the center of a laser beam where the local irradiance is $1/e^2$ times the central beam irradiance). If the total energy in the laser beam is divided by the area defined by the $1/e^2$ beam diameter, the average irradiance (one-half the maximum irradiance) is obtained.

Safety standards commonly refer to 1/e peak irradiance points instead, since for beams larger than 7-mm, the maximum irradiance is necessary for hazard evaluation (ANSI 1986). The 1/e definition for the beam diameter of a gaussian beam is simply a factor of $\sqrt{2}$ smaller than the $1/e^2$ definition. Hereafter, the term "beam diameter" refers to the beam diameter at 1/e points.

Having measured the power of the laser and the power through an aperture, the effective beam diameter, D_L, may be calculated by (Sliney and Marshall 1979):

$$D_L = \sqrt{\frac{-d^2}{\ell n\left(1 - \frac{\Phi_d}{\Phi_0}\right)}}, \tag{1}$$

where Φ_d represents the measured power through an aperture of size, d; and Φ_0 represents the total power of the laser at the measurement location. The aperture diameter should not pass more than 90-percent of the beam in order to provide accurate results. For very small visible and near-IR beams, an accurate determination of exit beam diameter is unnecessary for safety evaluation, since the entire beam could enter a person's eye.

For instance, consider a He-Ne laser having an output power of 5 mW and an observed beam diameter of about 2 cm at the laser exit port. The power measured through a 7-mm (0.7 cm) aperture is 1 mW. The exit beam diameter at 1/e points from equation 1 is 1.5 cm assuming a gaussian shaped beam. If a detector with a 7-mm aperture located downrange measured 0.1 mW, the calculated beam diameter is 4.9 cm. A larger aperture may be necessary to determine the diameter precisely for a nongaussian shaped beam.

Laser Beam Divergence

The beam divergence, ϕ, is the rate of beam spreading in radians and is approximated by the formula (Sliney 1974a):

$$\phi = \frac{D_L - a}{r}, \tag{2}$$

where "a" represents the exit beam diameter and "r" represents the distance between the laser and a distant measuring point.

A typical He-Ne laser beam with an exit beam diameter of 0.8 mm which has a measured diameter of 10 cm at 100 m would have a beam divergence of about 1 mrad. Note that the 100 m must be changed to 10,000 cm.

Beam waist and Rayleigh range

The minimum beam diameter is called the beam waist and may be at the laser exit, inside the laser cavity, or located some distance in front of the laser. For lasers which have a beam waist in front of the laser, a more accurate determination of beam expansion in the near field is provided by (Kogelnik and Li 1966 and Self 1983):

$$D_L = D_w \sqrt{1 + \frac{(r - r_0)^2}{R_R^2}} \tag{3}$$

where D_w is the diameter of the beam at the waist, r_0 is the distance from the laser to the waist, and R_R is the Rayleigh range. The Rayleigh range is the distance past the beam waist where the beam has expanded to $\sqrt{2}$ times D_w. In the far field, it can be seen from equations 2 and 3 that the beam divergence is equal to D_w/R_R.

For example, a laser has a beam waist of 7 mm located 10 m in front of the laser, and a Rayleigh range of 10 m. The beam diameter at the laser exit port and at 20 m is 9.9 mm, and the beam divergence is 0.7 mrad.

Calculations of power or energy through an aperture

Once the beam diameter has been determined, the power or energy which passes through an aperture of size d, may be calculated by the following equation (ACGIH 1976):

$$\Phi_d = \Phi_0 \left[1 - \exp\left(\frac{-d^2}{D_L^2}\right)\right], \tag{4}$$

For a 50 mW, far-infrared laser with a beam diameter of 5 cm, 20 μW of the laser beam would pass through a 1-mm aperture. For a pulsed laser, Φ may be replaced by the energy per pulse, Q.

Effect of Optical Viewing Devices

When optical viewing devices are used at a particular distance, the beam diameter of the laser which passes through the optical system and is incident on the cornea of the eye is reduced by the magnifying power of the optics, M. Also, the power is reduced by the transmitted fraction, t, of the optics at the laser wavelength. Therefore the power through an aperture, d (which is located in the beam path after passing through an optical device), is given by: (Marshall and Conner 1987)

$$\Phi_d = \Phi_o t \left[1-\exp\left(\frac{-d^2M^2}{D_L^2}\right)\right]. \tag{5}$$

The size of the laser beam falling on an individual's eye is physically limited in diameter by the exit aperture of the magnifying optics, D_e. However, if the exit aperture is larger than the pupil, then the laser energy is limited by the dark-adapted pupillary diameter of 7 mm. When the limiting aperture, D_f, is different from D_e, the smaller of the two apertures determines the value of d in equation 5.

For a 50 mJ, Q-switched, erbium YAG laser with a 5-cm beam diameter which passes through a pair of 7 X 50 binoculars, the amount of energy which could pass through a 1-mm aperture is 290 μJ assuming 30-percent transmission of the optics.

With the constant irradiance model, incorrect results may be obtained where the beam diameter is smaller than the entrance aperture of the optics if the beam irradiance is simply multiplied by the square of the magnifying power.

OPTICAL DENSITY

The necessary optical density (OD) for eye protection is the log of the ratio of the power or energy transmitted through the limiting aperture and the AEL. Therefore, the power or energy passing through the limiting aperture may be compared directly to the AEL to determine the hazard of the laser. The OD is the number of log units that the laser exceeds the AEL.

Two types of optical density are normally desired: one for unaided viewing and one for any possible contingency including the use of optical aids. For unaided viewing, the amount of power or energy passing through a 7-mm aperture is compared to the AEL. When optical viewing aids are used the total output power or energy is usually compared to the AEL. For a more sophisticated evaluation, the power and transmission of the optics and size of the entrance and exit aperture may be taken into account.

Nonlinear effects of goggle materials, breakage, or the accuracy of the specified OD are not considered here but must be considered when actually selecting eye protectors. The formula for OD is (Marshall and Conner 1987):

$$OD = \mathrm{Log}_{10}\left(\frac{\Phi_d}{AEL}\right). \tag{6}$$

The AEL for the erbium YAG laser mentioned above is 79 μJ. A reduction factor of 3.67 is necessary to provide protection, or 0.57 OD.

An alternative approach for calculating the OD is to compare the irradiance averaged over the limiting aperture to the EL. The average irradiance over the limiting aperture is given by (ACGIH 1976):

$$E_{avg} = \frac{4\Phi_d}{\pi D_f^2}. \tag{7}$$

The equation for calculating OD by this method is (Sliney 1974b):

$$OD = Log_{10}\frac{E_{avg}}{EL}. \tag{8}$$

NOMINAL OCULAR HAZARD DISTANCE (NOHD)

Calculation of NOHD in a vacuum

When Φ_d is set equal to the AEL and the atmospheric attenuation of the beam is neglected, the distance, r, is then the NOHD in a vacuum, which is designated as R. Equations 2 and 5 may be rearranged to provide this quantity (Marshall and Conner 1987):

$$R = \frac{1}{\Phi}\left[\sqrt{\frac{-M^2d^2}{\ell n\left(1-\frac{AEL}{\Phi_0 t}\right)}} - a\right], \tag{9}$$

where d is the minimum of D_f and D_e. When the beam waist is located in front of the laser exit port, the distance to the beam waist must be added to the value of R in order to determine the NOHD of the laser. For most laser systems, this correction is unnecessary. Except for very low power lasers, the above equation may be simplified to the following:

$$R = \frac{1}{\Phi}\left(\sqrt{\frac{\Phi_0 t M^2 d^2}{AEL}} - a\right). \tag{10}$$

For a 50 mJ, Q-switched, neodymium:YAG laser with an initial beam diameter of 5 cm and a divergence of 0.1 mrad, the value of R when viewed with 7 X 50 binoculars is 66 km, assuming 70 percent transmission of the optics.

Atmospheric Attenuation

Only a portion of the laser beam is transmitted through the atmosphere and reaches the laser target (or detector). For path lengths exceeding 1 km, the effect of the atmosphere should be considered. For most safety calculations with either a ruby or neodymium laser, an attenuation factor, μ, of 5×10^{-7} cm^{-1} may be used corresponding to an extremely clear atmosphere at sea level (RCA 1974). Much more involved calculations would be necessary to yield precise information on the effects of the atmosphere, but only a worst case analysis is necessary for safety calculations. The reduction of the beam power at a distant location is given by the following (ANSI 1986):

$$\Phi = \Phi_0 e^{-\mu r}. \tag{11}$$

Approximations of NOHD

When the atmospheric attenuation is taken into account, the NOHD cannot be solved in closed form. Therefore, the following formula may be used as a very close approximation (Marshall 1980):

$$\text{NOHD} = r \cdot e^{-\frac{\mu \cdot \text{NOHD}}{2}}. \tag{12}$$

An iterative solution of equation 12 is possible with an electronic calculator. As an alternate solution, the following equation provides a closed form approximation for NOHD values up to 90 km for ruby and neodymium lasers (Marshall and Conner 1987):

$$\text{NOHD} = \frac{\sqrt{(4+\mu r)^2+16\mu r+0.16(\mu r)^2} - (4+\mu r)}{2\mu}. \tag{13}$$

For the 66 km value of R mentioned above, the attenuated NOHD is 30.8 km according to equation 13 with an atmospheric attenuation of 5 X 10^{-7} cm^{-1}. An iterative technique with a programmable calculator yields 30.7 km.

Determining the NOHD from Reflections

When a laser beam is reflected from a surface, the reflected beam may be treated as a new laser source. Only a percentage of the laser beam will be reflected depending on the material and angle of incidence of the laser beam. The divergence of the

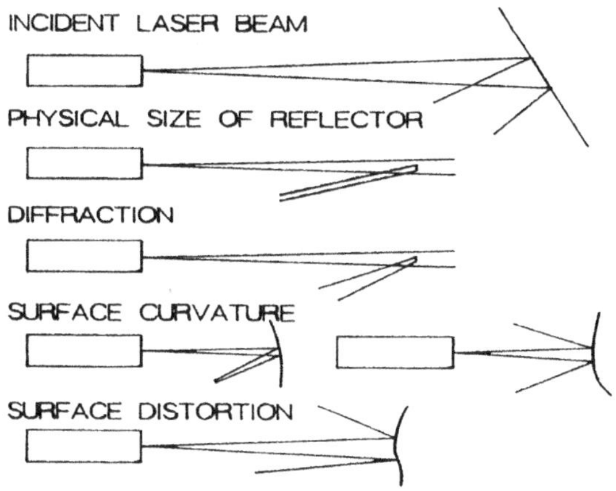

Figure 2. Graphical Illustration of the Effects on the Beam Divergence from a Laser Beam Reflected from a Relatively Flat Specular Reflector.

reflected beam will be affected by the divergence of the incident beam, size of the reflector, diffraction effects, and surface curvature (Van De Merwe and Marshall 1986). The divergence of the reflected beam is usually smaller than the incident laser beam since only the central portion of the incident beam is intercepted. The apparent size of the reflector from the incident laser position, which changes with angle, determines the diffraction effect. Surface distortion may add additional effects to the reflected laser beam. These effects are graphically illustrated in Figure 2.

CONCLUSIONS

The preceding aperture method provides a means for accurately comparing laser output parameters to exposure limits as well as determining the laser parameters through measurements. Although some of the equations may be quite lengthy, programmable calculators or computers can obtain safety information in a straight-forward manner. Several sample calculations should be performed by the user in order to fully understand the preceding equations before relying on the safety information for personal protection.

REFERENCES

American Conference of Governmental Industrial Hygienists, 1976, "Guide for Control of Laser Hazards," ACGIH, 6500 Glenway Ave., Bldg. D-5, Cincinatti, OH 45211.

American National Standards Institute, "American National Standard for the Safe Use of Lasers, ANSI Z136.1-1986," ANSI, 1430 Broadway, New York, NY 10018.

Kogelnik, H., and Li, T., 1966, Laser Beams and Resonators, Proc. IEEE, 54:1312-1329, (jointly published in Applied Optics, 5:1550-1567).

Marshall, W.J., 1980, Hazards analysis on gaussian shaped laser beams, AIHA Journal, 41:547-551.

Marshall, W.J., 1989, Laser reflections from relatively flat specular surfaces, Health Physics, 56(5):753-757.

Marshall, W.J. and Conner, P.W., 1987, Field Laser Hazard Calculations, Health Physics, 52(1):27-37.

RCA Corporation, 1974, "RCA Electro-optics Handbook," (2nd edn), Technical Series EOH-11, RCA Commercial Engineering, Harrison, NJ.

Self, S. A., 1983, Focusing of Spherical Gaussian beams, Applied Optics, 22:658-661.

Sliney, D.H., 1974, The Development of laser safety criteria -- biological considerations, in: "Laser Applications in Medicine and Biology 1," M.L. Wolbarsht, ed., Plenum Press, New York, NY.

Sliney, D.H., 1974, Laser protection eyewear, in: "Laser Applications in Medicine and Biology 2, M.L. Wolbarsht, ed., Plenum Press, New York, NY.

Sliney, D.H. and Marshall, W.J., 1979, Laser Safety. Part IV: Measuring the beam diameter, Electro-Optics System Design, 11(11):31-37.

Van De Merwe, W.P. and Marshall, W.J., 1986, Hazardous ranges of laser beams and their reflections from targets, Applied Optics, 25(5):605-611.

LASER RADIATION DETECTORS*

James Franks

US Army Environmental Hygiene Agency
Aberdeen Proving Ground, MD

INTRODUCTION

Laser safety standards in the United States and other countries use a hazard classification system to avoid the necessity for laser users to make radiometric measurements. Manufacturerers in most Western nations are required to affix a warning label to the laser housing that gives the hazard classification. This classification gives users and laser safety professionals the basis for deciding appropriate control procedures to be used in the safe use of this equipment. However if you are a manufacturer or a laser safety professional and need to verify manufacturer's specifications you may need to use a laser radiometer in your hazard classification or hazard evaluation. During a laser accident investigation measurements may be required to determine if an individual received an exposure above the protection standard (PS). Radiometers used in hazard classification and evaluation must respond to levels as small as the Accessible Emission Limit (AEL). AELs range in value from tenths of microwatts or microjoules in the visible and near infrared (400-1400 nm) to millijoules in other parts of the spectrum. Hazard classification at the Class 4 level requires that cw radiant powers of at least 500 mw be measured, however it may be necessary on rare occassions to measure cw output powers as great as 100W to 1kW.

RADIOMETERS

The radiant power or energy output from a laser is measured in absolute units of watts or joules. The devices used to measure these quantities are called, in general, radiometers and the quantities themselves are radiometric quantities.

----

*The opinions or assertions contained herein are the private views of the author and are not to be construed as reflecting the official views of the U.S. Department of the Army or the U.S. Department of Defense

Radiometers usually consist of a detector that responds to the input laser radiation, a method of reading and/or recording the detector response and a method of modifying the laser input to suit a particular measurement requirement. These modifications may use known aperture sizes to pass the laser beam through, attenuators to assure detector operation in a linear region, filters to improve signal to noise ratio, or diffusers to average sensitivity changes with location on the sensing element over the whole detector area. The heart of every radiometer is the detector.

DETECTORS

Detectors can be divided into two categories; thermal and quantum detectors. Thermal detectors absorb optical radiation and the temperature of the absorbing element changes. They are not as sensitive as quantum detectors but they exhibit a broad, relatively flat spectral response. Quantum detectors are also refered to as photon detectors and they respond to individual photons by generating a current or voltage in an external circuit. Quantum detectors are characterized by high sensitivity and fast response, but respond only to a rather narrow spectral range while exhibiting a marked wavelength dependency. Photographic emulsions could also be considered as quantum detectors since individual quanta are absorbed by silver halides and a latent image is formed by a photochemical process. A photographic detector would measure total exposure or energy. This technique is not widely used for evaluation of laser hazards because it has a limited spectral response, is linear only over a small dynamic range and is not real time. Beam detectors, such as photoluminescent materials, have been used to "see" laser beams operating outside the visible portion of the electromagnetic spectrum. These "beam visualizers" have not been used for radiometric measurements, only as an aid in locating a normally invisible beam.

Quantum Detectors

Photoemissive detectors. Photoemissive detectors utilize the photoelectric effect as the operating principal. A photocathode surface is struck by incoming photons and emits electrons. These electrons are accelerated across a potential difference and collected at the anode while generating a current in an external circuit. These detectors must therefore have an external power supply. Photoemissive detectors exhibit little sensitivity beyond 1200 nm.

Vacuum photodiodes are photoemissive detectors that function similarly to the old vacuum tube diode except that secondary are generated by the photoelectric effect rather than thermionic emission. A vacuum photodiode has only one electron generating element i.e. the photocathode and does not internally amplify current. Vacuum photodiodes exhibit fast response (1 ns risetime), high responsivity (50 mA/W at 500 nm for an S-11 photocathode) but to cover the ultraviolet and visible region with equal responsivity would require more than one detector. A photocathode that exhibits the broadest spectral response (300 nm to 1100 nm) is the S-1 but peak responsivity is down to 4 mA/W at 350 nm. These detectors can be the basis of a rugged, portable, and versatile radiometer system. They are capable of measuring radiant energy as well as radiant power with a suitable readout unit and can easily measure radiant exposure or irradiance levels at or below the PS throughout the range of 300 nm - 1100 nm. These detectors are saturated easily with short pulse lasers (<10ns), however, and tend to be rather expensive.

Photomultipliers are photoemissive detectors that operate on the same principal as the vacuum photodiode except that secondary surfaces called dynodes amplify the photoelectrons generated by approximately 6 orders of magnitude. These detectors are fast with risetimes of about 3 ns but require very stable high voltage power supplies for reliable measuremnts. Microchannel-plate photomultipliers have exhibited risetimes as short as 0.27 ns but do not have the gain of a conventional, electrostatic photomultiplier.(Leskovar, 1984) An RCA Type 7102 photomultiplier with an S-1 cathode has a peak responsivity of 660 A/W at 350 nm. This is an increase of 5 orders of magnitude over the S-1 photodiode. Laser measurements for hazard evaluation do not require this high sensitivity and for this reason photomultiplier tubes are most often used with spectroradiometers in the evaluation of potential hazards from broadband, non-laser sources.

Semiconductor Detectors. Photoconductors are semiconductor detectors whose electrical resistance changes by the creation of additional current carriers when exposed to optical radiation. An external power suppply is necessary but the detector can be biased in either direction. A voltage is normally the measured electrical quantity for photoconductors and these detectors are most often used in the mid to far infrared where photoemissive or other solid state detectors such as silicon have no response. These detectors are often cryogenically cooled to improve detectivity. In general, these detectors are slow (1ms response time) when compared to other quantum detectors but exhibit high detectivities in the mid to far IR.

Photodiodes are semiconductor optical radiation detectors that are normally operated in a reverse bias manner but may be operated with no external bias. The reversed bias mode of operation is called the photoconductive mode, while the no bias mode is called the photovoltaic mode.

Modern semiconductor photodiodes are normally PIN devices with an intrinsic (no doping) material sandwiched between P and N junctions. However, a photodiode that has been designed to operate in the photovoltaic mode only, is usually only a PN junction device with no intrinsic layer. The most common type of photovoltaic detector is the silicon detector or solar cell. These devices have their greatest sensitivity in the visible and near infrared, exhibit about 1 microsecond response times and little dark current.

The photoconductive mode is the most common mode of operation for a modern photodiode. This method of operation of a photodiode should not be confused with the bulk photoconcuctors described above. Photodiodes operated in this mode exhibit 1 ns risetimes, and about a 50 per cent increase in sensitivity with respect to the photovoltaic mode. These devices, with cryogenic cooling, can operate out to wavelengths of 12 micrometers. The most common type of photodiode operated in the photoconductive mode is the silicon photodiode and these typically have a peak responsivity of 0.5 A/W.

Avalanche photodiodes can cause an internal gain in current and sometimes are referred to as solid state photomultipliers. Photodiodes operated in the avalanche region typically have a factor of 100 increase in sensitivity and a decrease in rise time from approximately 1ns to 0.5ns.

Photoelectromagnetic Detectors. Detectors of this type are also identified as PEM detectors. Laser radiation strikes the detector and frees positive and negative charges within the semiconductor material. These charges are separated by a magnetic field, and an output voltage, proportional to the input radiation is generated. These detectors require no cooling or bias voltage but do require an external magnetic field. These detectors are most useful in the midinfrared (3000nm-6000nm), exhibit 1 microsecond response times and have detectivities equal to thermopiles.

Thermal Detectors

Thermal detectors absorb incident laser radiation, the temperature increases, and an electrical signal is generated proportional to the laser input. The newest thermal detector is the pyroelectric.

Pyroelectric detectors. Pyroelectric detectors are made from a material which responds only when it's temperature is changing. It also responds or has an output when it is mechanically shocked, so that the detector should not be subject to sudden movement or vibration during measurement. Laser radiation incident on the detector surface is absorbed by the material and converted into heat; this results in a temperature change in the crystal. A change in temperature alters the lattice spacing of the pyroelectric crystal producing a change in the spontaneous electric polarization in the material. A current is generated to balance the polarization change which is proportional to the rate of change of temperature of the crystal. The current generated can be converted to a voltage which is proportional to total radiant power. It is important to realize that pyroelectrics only respond to pulsed, modulated or chopped radiation sources. Continuous souces can be measured but are normally chopped mechanically. These detectors exhibit a broad spectral response that, like other thermal detectors, is limited by the absorbance characteristics of the absorbing material and the transmission of any window material present. These detectors also exhibit fast response (resolution < 1 ns) and are the most sensitive of the thermal detectors.

Thermoelectric Detectors. The principle of operation os these detectors is the Seebeck effect; when a junction of dissimilar metals is heated, a voltage is produced. A single junction detector is known as a "thermocouple," and one containing more than one junction is a "thermopile."

Thermopiles exhibit a broad spectral response that is usually limited by the transmission characteristics of any window material present. For example, a thermopile with a sapphire window has a flat response from 300nm to about 6500nm. Thermopiles typically exhibit a responsivity two orders of magnitude less than a pyroelectric. In modern detectors that use thermopiles, it is the flow of heat that is measured rather than the quantity of heat as in traditional calorimetry. If the absorbing element is exposed to a single laser pulse the temperature of the absorber will rise causing a temperature difference between the absorber and surroundings. The maximum temperature difference will cause a maximum votage output which wll be proportional to the energy incident. Upon cw or repetatively pulsed laser exposure the temperature difference will be measured after a steady state has been achieved. This temperature difference is proportional to the input power. The response time of these detectors can be seconds and their useful energy range is > 10 mJ while power levels must be greater than 1mW. For power levels of from 100W to 300W air cooling is sufficient while at levels greater than this water cooling is required. These devices may be calibrated electrically rather than optically and form the basis of secondary standards of calibration for laser power and energy in many laboratories.

Bolometers are detectors that employ detecting elements that sense a temperature change by changing it's electrical resistance. These temperature sensitive resistors may be metallic or semiconductors (thermistors). The bolometer is usually operated in a bridge circuit with an active element which is exposed to laser radiation and a compensating element which is exposed to ambient conditions.

BEAM VISUALIZATION TECHNIQUES

"Image Converters"

Image converters employ a photoemmissive cathode surface that emits electrons which are focussed onto a phospor screen. These devices typically have an S-1 photocathode so that they are sensitive to visible and near infrared lasers. The phospor may typically be a P-20 phospor which emits green light. These devices are useful for viewing a laser beam profile on a diffuse surface. Laser intrabeam exposure may damage the photocathode and is not recommended.

Luminescent Detectors

Luminescence is a process where light is emitted from a compound as a result of exposure to an external energy source. When the light is emitted only in the presence of the exiting source it is called fluorescence. Phosphorescence continues after the exciting source is removed. Stokes law states that the wavelength of emitted radiation is longer than the exciting wavelengths. Fluorescence in violation of Stoke's law (anti-Stokes radiation) is produced when a material is excited by visible or near ultraviolet radiation and remains in a metastable state until excited by longer infrared wavelengths. Then the material will emit light of shorter wavelength than the exitation wavelengths. Most laser beam finding materials work on this principle. Materials that are in a metastable state as a result of exposure to shorter wavelengths of optical radiation and then emit visible light as a result of being heated are thermoluminescent materials. Useful beam finders for far infrared lasers like carbon dioxide lasers that operate at 10.6 micrometers have been developed using thermoluminescent materials.

Liquid Crystal Sheets

Encapsulated sheets of liquid crystal material are available that are useful for locating normally invisible continuous wave laser beams. These materials change their reflectance with temperature so that under white light illumination they change color as their temperature changes.

"Burn Papers"

Several manufacturers in the US provide papers which are coated with a black absorbing ink that vaporizes when exposed to Q-switched laser pulses. These materials have thresholds of about 10-200 mJ/cm2 in a single q-switched pulse(Sliney, 1989). The approximate beam size and location can be determined using this paper. Other types of "thermal" papers such as those used in thermal printers or some strip chart recorders will also work as a "beam finder".

References

B. Leskovar, " Microchannel plate photomultipliers", Laser Focus/Electro-Optics, pg 347, Feb 1984.

D. H. Sliney, "Radiometry and laser safety standards", Health Physics, 5(56), pg 723, May 1989.

QUANTITATIVE DETERMINATION OF ULTRAVIOLET RADIATION*

Wesley J. Marshall

Laser Microwave Divsion
U.S. Army Environmental Hygiene Agency
Aberdeen Proving Ground, Maryland 21020-5422

INTRODUCTION

Evaluating the safety of ultraviolet radiation (UVR) directly from ultraviolet lamps, industrial sources or the sun with a simple, easy-to-use instrument has not been feasible until the last few years when quality instruments have become accessible. The problem has historically been the separation of the UVR from the visible and infrared portions of the optical spectrum and then accurately weighting the UVR to the documented human biological response curve. Many instruments designed to measure UVR from a particular source will not accurately measure the UVR from another broadband source due to spectral differences. The most accurate method for measuring the actinic UVR (UV-B and UV-C) is still to tediously measure the UVR in each wavelength region with a double monochrometer and then manually weight these values to the human response spectrum. For safety evaluation, several direct reading instruments may be used when high accuracy is really not necessary. Polysulphone film detectors may be used in those situations where direct reading is not feasible, several simultaneous measurements are desired, or in a harsh environment where electromagnetic interference is present. The exposed detectors can be evaluated in a less harsh environment later.

MEASUREMENT PITFALLS OF MONOCHROMETERS

Many pitfalls exist when measuring UVR such as geometry problems, detector field of view, nonlinearity of the detector, detector aging, to mention only a few (Marshall 1979 and Henderson 1987). Stray light is probably the most significant problem in measuring UVR.

---

*The opinions or assertions herein are those of the author and should not be construed as a reflection of the official position of the U.S. Department of the Army or U.S. Department of Defense.

Stray light

For many ultraviolet sources, the amount of ultraviolet is almost negligible compared to the visible and infrared energy emitted, even though the biological effect is substantial. When the visible and near infrared energy is a million times greater than the ultraviolet, even a small fraction of this unwanted energy striking the detector will invalidate the ultraviolet measurement.

Sources of Error. Stray light may be transmitted through a monochrometer system and detected by the detector. Even though the instrument is designed to reject all other wavelengths except the one for which the monochrometer is tuned, energy at other wavelengths may add to the total measurement. Sources of this contribution are reflections inside the monochrometer housing, particles of dirt on the grating, imperfections in the grating, quality control in the manufacture of the grating, and second order transmissions of the grating.

Solutions. Methods to reduce these sources of error include using a double monochrometer system, keeping the grating clean, and using a solar blind detector. Much of the stray light which is transmitted through the first monochrometer system can be eliminated by passing the UVR through a second monochrometer grating. The use of high quality holographic gratings which have been kept free of dust will also reduce stray light problems. Clean flowing nitrogen over the gratings may be advisable in a dirty environment. Storing the gratings in a dust free environment or keeping the gratings wrapped in plastic will help prevent dust, which will act as scattering centers, from forming on the gratings. A solar-blind detector system which is relatively insensitive to visible and near infrared radiation will eliminate much of the stray light from the UV part of the spectrum.

Filters. A filter which transmits UVR and blocks visible and near-infrared wavelengths may be used when a solar blind detector is not available. A sharp cutoff UV blocking filter may be used as a check for the presence of stray light. For instance a filter which has 50-percent transmission at 305 nm can be used to check for stray light at 250 -280 nm (Schott 1982). When the filter is inserted in the optical train, only radiation at about 300 nm and above will contribute to the reading. A small indication may be subtracted from the measurement. If the reading is large, it must be remembered that about 8-percent of the stray light is reflected off the inserted filter. A significant reading with this filter in place indicates that major problems are present in the measurement system. An order sorting filter is usually used in spectral measurements to block second order transmissions of the monochrometer grating from reaching the detector. Such a filter is usually not required in the ultraviolet because second order transmissions would have to come from the vacuum ultraviolet portion of the spectrum. If an order-sorting filter has been used for some portion of the spectrum during the calibration of the system, then the same filter must be used during measurement.

Calibration

Poor intensity or wavelength calibration of a monochrometer system may introduce errors. Stray light problems during the calibration process can carry errors into the measurement of an unknown source.

Sources of Error. An attempt to calibrate an ultraviolet monochrometer system with a source rich in visible and near-infrared energy will result in tremendous errors at the shorter wavelengths (Sliney and Wolbarsht 1980). The amount of ultraviolet energy produced by a tungsten lamp at 200 nm is so little that any measurement will usually be completely stray light. The wavelength indicator on a monochrometer system may be in error by a few nanometers even though the indication may be in tenths of nanometers.

Solutions. A much better intensity calibration may be obtained by using a calibration source such as a deuterium lamp which has a low emission in the visible and near-infrared wavelengths and a substantial UVR emission. By carefully eliminating stray light from the calibration process, better measurements can be obtained on an unknown source. A low pressure mercury arc source or a UV laser can be used to check the wavelength calibration of the monochrometer system at various wavelengths throughout the UV portion of the spectrum.

Detector Saturation

When measurements are to be attempted on a strong UVR source, care must be exercised not to damage the equipment before measurements have begun. An expensive instrument may be totally destroyed in a matter of seconds with a common UVR source such as a 1000 W tungsten lamp, if measurements at a very close distance are attempted.

Sources of Error. A detector usually produces either a current or voltage directly proportional to the incident radiation over a certain range. When this range is exceeded, an increase in radiation no longer produces the same increase in current or voltage. When the source of UVR is pulsed, the detector may actually saturate on each pulse even though the meter indication still shows operation well below the maximum limit.

Solutions. Since UVR irradiance is reduced with distance, begin measurements at a far distance and turn slowly through the wavelength range carefully watching that the limits of the instrument are not exceeded. A variety of detectors may be necessary to handle the full range of intensities encountered. A less sensitive detector may be used for high intensity lamps. If the measurement of a bank of similar lights is desired, the measurement of one lamp may be adjusted to account for the contribution from all lamps. A filter for UVR may be constructed from screen wire. Since only a fraction of the total irradiance would then reach the detector, such a filter could be used to check for saturation. If the full reduction factor of the filter is not observed, then the higher measurement is most likely saturated.

Geometry Problems

Measurements on a UVR source may be completely invalidated simply due to the geometry of the measurement process. Either an unusually high or low reading may be obtained simply from the relative location of the measuring equipment to the source of radiation.

Sources of Error. When a UVR source is also very bright, the actual source of the emission may be difficult to determine. Physical structures next to the source may only allow emission in a narrow cone. If the measurement aperture is located outside this cone, an unusually low reading will result. Conversely, if the measurement aperture is located in the most intense region, an abnormally high exposure dose will be assessed for workers in the area. Simply having a body present may reduce multiple reflections and therefore reduce actual exposure (Levin 1986). If no injuries have resulted due to overexposure but the measurement indicates a severe hazard, geometry may be the explanation. For large sources of UVR, the field of view of the detector may not be large enough to encompass the entire source.

Solutions. One method for checking for the existence of geometry problems is to check all possible exposure angles with a direct reading instrument. Even if the instrument measures visible light instead of UVR, large fluctuations in the relative irradiance will indicate a geometry problem. Another way to curb measurement error is to visually examine the source with neutral density filters to see if the brightest area is visible from the measuring position. When assessing the hazard from the standpoint of a person's true exposure, exposures from the most intense regions must be added to those from the less intense regions.

Cosine Correction

The biological response to UVR received from a variety of directions is generally considered to be proportional to the cosine of the angle that the radiation strikes the human body. Measurement systems must be able to simulate this biological response.

Sources of Error. A good diffuser on the front of a monochrometer system or direct reading instrument can produce a cosine correction during measurement. However, for some arrangements of a monochrometer system, the detector may only receive energy from a limited field of view. For instance, a hood may be employed to reduce the effects of room light.

Solution. A direct reading photometric instrument may be used to obtain a cosine corrected measurement. A comparison between the direct reading instrument placed in the limited field of view of the UV monochrometer system and the unrestricted reading will allow an adjustment to be made for cosine correction. This type of correction may also be used when the hazard from a bank of similar lamps is desired, and detailed measurements were made on only one lamp.

Mechanical Problems

Instruments that produce valid measurements in the

laboratory may sometimes give false indications when taken into the field. Rough handling of airline baggage may cause a loss of credibility for the scientist.

Sources of Error. Loose connections inside the indicator, or of the probe to the indicator can cause a loss of signal. Background noise may provide a nonzero indication even though the system is not working. If mechanical gears are used to provide data or scale factors, a slip in the knob can cause order of magnitude errors.

Solutions. Before collecting data on an unknown source, a check reading against a source of known hazard such as a fluorescent fixture will verify the system's performance. Covering the detector with a hand or opaque material will also verify that the meter is responding to the surrounding illumination.

Large spectral peaks

The presence of large spectral peaks in the source may make data collection much more difficult when using a monochrometer system and may invalidate data taken with a direct reading instrument.

Sources of Error. Large spectral peaks in the optical source can cause errors from periodic monochrometer readings. When a spectral peak is much more intense than the radiation in neighboring wavelengths, abnormally high values will appear in the neighboring periodic readings. In addition, the contribution from a spectral peak may be entirely overlooked from data taken only at specific wavelengths.

Solution. The spectral peaks must be treated separately from the slowly varying contribution. The monochrometer system bandpass can be checked by measuring a known peak accurately. The bandpass of the monochrometer is obtained by measuring the wavelength interval between readings of one-half the peak reading. The handle which drives the monochrometer grating must be turned only in one direction for this measurement. When measurements are taken at wavelength intervals, the measurements between can be checked visually for any gross deviations from normal. When a sudden increase is noted, the reading is maximized, and a peak reading is obtained. This value is recorded separately. Any regular interval measurement close to the spectral peak must be adjusted so that a portion of the peak is not included in the measurement. The hazard from both portions must then be recombined. A decrease in the bandpass of the monochrometer system by using the narrowest slits available for both entrance and exit apertures will simplify and greatly improve data gathering.

Direct Reading Instruments

Direct reading instruments are by far the easiest to use. An entire evaluation may be performed in seconds that would take hours when performed by a normal monochrometer system. However, sources of error will be present in the readings which can not be eliminated even if more time is spent in taking the measurements.

Sources of Error. The direct reading instrument must solve all the previously mentioned sources of error simultaneously and quickly. Stray light entering the detector can not usually be corrected by a simple subtractive technique. Even readings taken with a variety of instruments from the same source may yield readings that vary by factors of 2 or 3 (Roelandts 1987). Since a filter is used to match the detector response to the accepted human response curve, mismatch errors are difficult or impossible to correct. When using special purpose instruments designed to measure one particular type of source such as the sun or germicidal lamps, large errors will result when trying to use these instruments on a source which they are not designed for. Large errors would occur frequently when the source is rich in blue light and near ultraviolet, but is shielded from emitting actinic UVR through the use of a sharp cutoff UV filter. Even though the instrument generally provides good data, a reading from a source which emits primarily at one particular wavelength will generally be in error. The peak sensitivity of the instrument may not always be in the direction that one would expect from casual viewing of the instrument, or the instrument may not be cosine corrected.

Solutions. The only true solution to the myriad of problems that could occur with direct reading instruments is to choose a very good instrument and then treat it gently. When measurement at only one spectral line is desired, the detector may be calibrated specifically for that line, or type of source. With some detectors, a series of calibration factors may be necessary for each type of UVR source to provide accurate results. At the U.S. Army Environmental Hygiene Agency, we have had good success in making actinic UVR measurements with the International Light, Model IL 730. Other quality instruments may be available for measuring the actinic UVR from a wide variety of UV sources, but most instruments are designed for one specific type of source. The IL 730 detector has good rejection of UV-A, visible, and near-infrared. The detector is cosine corrected. A sharp cutoff UV filter (supplied with the instrument) is used in a subtractive technique for low-level actinic UVR sources. When very accurate data is desired, a good monochrometer system is best.

MONOCHROMETER SYSTEMS

Several monochrometer systems are available to the Laser Branch, USAEHA. The Oriel Model 77250, a small 1/8 meter monochrometer, is a compact and precision instrument. Two monochrometers may be hooked in series for a double monochrometer system. Accessaries are available for filter holders, and a cosine correcting diffuser. The model EG&G Model 585 monochrometer system is a precision instrument, but larger. A double, in-line, Model 585, monochrometer system is available at USAEHA but is rather long and unwieldy for sources located in the field. The Li-cor Model LI 1800 portable spectrometer is very portable but does not measure UVR less than 300 nm.
A Tracor Northern rapid scan spectroradiometer is also available. This unit does not have a good and easy way of determining an intensity calibration, and is rather bulky for taking field measurements, but will measure pulsed sources which may be difficult with the others. Other monochrometer systems not

available at USAEHA may work as well or better (Kaufmann and Hartman, 1989 and Killick, et al., 1988)

PRACTICAL CONSIDERATIONS

Many times, a survey officer does not have all the equipment necessary to evaluate a particular source, and therefore accurate measurements simply are not available. In these cases, a knowledge of the potential of various types of UVR sources can assist in the proper control of personnel hazards. If people have been working around a source extensively for years, and no one has received any noticeable effect, then this information should be taken into account. On the other hand if the people working around the source have a deep tan and possibly peeling skin on their face and arms, then that is also a good indication of a problem situation.

Sometimes geometry so complicates a measurement that a true hazard evaluation is almost impossible. When evaluating welding sources for UVR hazards, the personnel exposure changes tremendously from one piece of work to another. In a situation of a deep weld, the weld is down in a narrow crevice for the first few passes. During this time, the UVR is radiated only straight overhead, but on the last pass, the UVR could radiate everyone in the area. In another situation, certain reflecting paints or aluminum metal sheets may reflect UVR to a person's backside where there is no protection (Philips 1983).

NEW PROPOSED STANDARD FOR UVR

A change in the UV action spectra has been published as a proposed standard in this year's ACGIH threshold limit values (ACGIH 1989). This changed standard will probably also be published by the International Radiation Protection Association. The proposed change is to combine the effects of actinic UVR and near ultraviolet in one standard. This change will not significantly change the hazard evaluation process for most common actinic UVR sources since the spectral effectiveness values for the near UV are so low. For near UVR sources such as UV-A tanning beds and Dental curing lights, actinic radiometers can not now be used to evaluate these hazards. An improvement in accuracy of current direct reading actinic UVR instruments may actually occur since all leak a small percentage of UV-A.

MEASUREMENTS WITH POLYSULPHONE FILM

Polysulphone film detectors can be used in a harsh environment, be located in several locations at once, and be collected and evaluated at some time after the exposure. The peak response of the film is at 300 nm rather than 270 nm (Diffey, 1984 and 1987 and Ferenczi et al., 1987). The formula used for evaluating UVR exposure is (dIFFEY 1984):

$$D = 4.3\,\Delta OD^{1.38}\ \mathrm{J/cm^2}. \qquad (1)$$

However, reasonably accurate actinic UVR measurements were desired from a variety of sources.

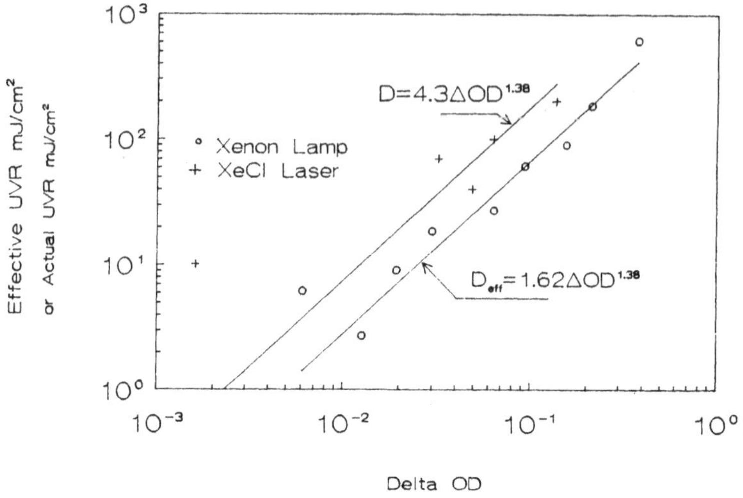

Figure1. Effective UVR Exposure from a Xenon Lamp and Actual UVR Exposure from a XeCl, Eximer Laser, Plotted as a Function of the Change in Optical Density of the Polysulphone Detectors.

For making effective actinic UVR measurements, the film was calibrated against a 1000 W, xenon arc source. The film was then checked against sources which had very different spectral distributions from the calibration source. It was discovered that the film read 30% high from a 1000 W, tungsten lamp, and 30% low against a deuterium standard lamp (Marshall 1988). The formula used to evaluate these measurements for an effective actinic UVR dose is:

$$D_{eff} = 1.62\,\Delta OD^{1.38}\ \ J/cm^2. \qquad (2)$$

This formula is different than published earlier due to the shift in spectral response between the polysulphone film and the biological effect. A more elaborate formulation for high doses is available but was not necessary for the doses examined at USAEHA (Kolari et al., 1986).

John Reid of Lumonics in Kanata, Ontario, Canada kindly supplied some polysulphone film to the USAEHA that had been exposed to a 308 nm, XeCl, eximer laser. When the same calibration was used for these exposures that had worked well with the broadband sources, the calculated effective exposure dose was off by a factor of 14. This difference is understandable since the 308 nm wavelength is near the peak response of the film but far from the peak of the biological curve. However, when the actual UVR exposure was calculated from the formula of Diffey, accurate results were obtained. Therefore the effective formula should only be used for broadband sources, and a correction factor may be applied based on the spectral characteristics of the source to provide reasonably accurate results. In the Figure, the effective UVR exposures from the Xenon lamp and the actual UVR exposures from the XeCl, eximer laser are plotted against the resulting optical density changes.

The film is evaluated by accurately measuring the optical density at 330 nm before and after exposure. At USAEHA, exposed and unexposed adjacent samples on the same piece of film were evaluated. Due to thickness variations in the film, effective actinic UVR could not be accurately measured for exposures less than 9 mJ/cm^2 . However, good results were possible at higher doses. It was found that exposed film could be cleaned with alcohol, if necessary, before evaluation without loss of accuracy.

CONCLUSIONS

Monochrometer systems still provide the most accurate actinic UVR hazard evaluation, although several measurement pitfalls must be avoided to achieve acceptable results. Direct reading instruments may also provide good results as long as the instrument is used for those sources which it has been designed. Polysulphone film provides reasonably accurate measurements on many sources, especially in a harsh environment, in spite of its differing spectral response curve.

REFERENCES

ACGIH, 1989, "Threshold Limit Values and Biological Exposure Indices for 1989-1990," available from ACGIH, 6500 Glenway Ave., Bldg. D-7, Cincinnati, OH 45211.

Diffey, B.L., 1984, Personal ultraviolet radiation dosimetry with polysulphone film badges, Photodermatology 1:151-157.

Diffey, B.L., 1987, A comparison of Dosimeters used for solar ultraviolet radiometry, Photochemistry and Photobiology, 46(1):55-60.

Ferenczi, L.Z., Hill, G. B., and Scrimger, J.W., 1987, Quantification of Human Exposure to UVR in: "Human Exposure to Ultraviolet Radiation: Risks and Regulations," W.F. Passchier and B.F.M. Bosnajakovic, ed., Elsevier Science, Amsterdam: 407-413.

Henderson, A.R., 1987, UV Radiation Hazards; Principles and Problems of Measurement, in: "Human Exposure to Ultraviolet Radiation: Risks and Regulations," W.F. Passchier and B.F.M. Bosnajakovic, ed., Elsevier Science, Amsterdam: 333-337

Kaufmann, W.F., and Hartmann, K.M., Low Cost Digital Spectroradiometer, Photochemistry and Photobiology 49(6):769-774.

Killick, H.J., Primarolo, A.A., Mackenzie, R.M., and Forkner, A.C., An Inexpensive Automatic Recorder for Biologically Important Solar UV, Photochemistry and Photobiology 48(4):555-557.

Kolari, P.J., Hoikkala, M., and Lauharanta, J. 1986, Assessment of the dose response of polysulphone film badges for the measurement of UV-radiation, Photodermatology 3:228-232.

Levin, R.E., 1986, Radiometry in Photobiology in: "The Biological Effects of UVA Radiation," F. Urbach and R.W. Gange, ed., Praeger, New York: 30-38.

Marshall, W.J., 1979, Optical Measurement Techniques in: "Non-ionizing Radiation, Proceedings of an ACGIH Topical Symposium," 175-179.
Marshall, W.J., 1988, Actinic Ultraviolet Safety Measurements with Polysulphone Film, Proceedings of Health Physics 22nd Midyear Topical Meeting on Instrumentation.
Philips, R., 1983, "Sources and Applications of Ultraviolet Radiation," Academic Press, New York: 371.
Roelandts, R., and Diffy, B.L., 1987, The Status of UVA Dosimetry in PUVA Therapy, in: "Human Exposure to Ultraviolet Radiation: Risks and Regulations," W.F. Passchier and Bosnajakovic, B.F.M., ed., Elsevier Science, Amsterdam: 391-393.
Schott Color Filter Glass Technical Information, 1982, available from Schott Opttical Glass Inc., Duryea, Pennsylvania.
Sliney, D.H., and Wolbarsht, M.L., "Safety with Lasers and Other Optical Sources," Plenum Press, New York: 456-467.

MEASUREMENTS OF WELDING ARCS AND PLASMA ARCS

Paul Eriksen

National Institute of Occupational Health
Lersø Parkallé 105
DK-2100 Copenhagen
Denmark

INTRODUCTION

Welding and plasma arcs are among the most powerful emitters of optical radiation. The temperature of welding and plasma arcs is extremely high and the ultraviolet output is therefore potentially hazardous. As is well known the visible radiation emitted by these arcs is blinding, and several cases of severe retinal burns are known to have caused blindness. Theoretically, the emitted infrared radiation should be substantial, giving rise to concern about chronic exposures that might possibly enhance cataract formation. Apart from being extremely hazardous in itself the radiation can produce other harmful and toxic substances such as ozone, nitric oxides and particulates.

Considering the number of people that are exposed it seems not only worthwhile but also necessary to know in detail the spectrum emitted by these sources, the parameters that affect the emitted radiation, and the necessary steps to be taken to safely work with such sources. This implies detailed knowledge of the properties and performance of protective devices.

TERMINOLOGY AND TECHNIQUES

The welding terminology in Europe differs from the American terminology. A few comments on this matter will hopefully lessen the confusion.

In Europe, the term Manual Metal Arc (MMA) welding refers to the type of welding where the arc is produced between a consumable electrode ("rod"), which is inserted in a handle, and the work-piece. Each time the electrode has burned up a new electrode must be inserted into the handle. Apart from a transformer to supply the power to the arc no extra welding equipment is required. In the USA this type of welding is referred to as Shielded Metal Arc Welding (SMAW). Although this welding technique is old it is still used frequently, typically when welding thick plates, say more than 3-4 mm. It is therefore used extensively in shipyards.

The european designations MIG (Metal Inert Gas) and MAG (Metal Active Gas) welding are both referred to as GMAW (Gas Metal Arc Welding) in the USA. Argon is the most frequently used inert gas with helium as an alternative. By active gas is usually meant CO_2 or mixtures of Ar and CO_2 or/and O_2 but with the "definition" that the Ar content is less than 95%. In MIG/MAG welding the consumable electrode ("wire") is fed continously through a tube from the welding machine to the torch together with a coaxial flow of shielding gas.

It is therefore a fast welding technique. However, there's a limit to the penetration depth of the weld, it is used to weld relatively thin materials, up to about 5 mm in thickness.

TIG (Tungsten Inert Gas) welding in Europe is called GTAW (Gas Tungsten Arc Welding) in the USA. Tungsten refers to the material of the non-consumable electrode in the torch held by the operator, inert gas refers to argon (or helium). The "filler" metal, if any is needed, is brought to the weld pool with a separate wire. A shield gas flows through the torch to protect the weld from oxidation. The TIG welding technique resembles soldering and should not be confused with MMA welding. This type of welding is for thin plates where a very fine surface finish is required.

A new type of MIG/MAG welding has appeared lately which provides the possibility of welding thick plates and this technique has therefore great potential in shipyards. It is called Synergic Pulsed MIG welding in Europe (Amin, 1981) but covers active shielding gases as well. The current is pulsed with frequencies up to about 100 Hz with peak currents up to 500 Amps. The term synergic refers to the fact that the wire-feed-speed, current and voltage are not set independently but are "working together" (synergic) in a fashion that permits single metal droplet transfers in the arc.

Plasma arcs are used in cutting and welding processes, and in spraying processes. In plasma spraying a powder material is blown into a hot arc and directed toward a surface such that the surface will obtain a very hard and durable coating. It is used to reduce wear of machine parts. The arc is usually completely exposed resulting in high levels of radiation. In plasma cutting and welding the gas which will create the plasma is led through the torch together with a coaxial flow of shielding gas. The arc is often buried in the work-piece and the resulting radiation levels are therefore only moderately high.

MEASUREMENT METHODS AND HAZARD EVALUATION

Measurement methods

To properly assess the hazards of welding and plasma arc radiation it is necessary at least to measure the spectral irradiance at a point which corresponds to a typical operator position. It is advisable to also measure the radiation with broad-band meters e.g., a photometer, a UV meter or any other type of broad-band meter in order the check the integrity of the spectral measurements. It is also advisable to measure the arc luminance with a luminance meter with a 1/2 or 1/3 degree field-of-view (or less) when possible in order to assess the necessary eye protection. The geometrical emission pattern should be known also, at least to a degree that will reveal any positions that will cause an exposure larger than the one corresponding to a typical operator position. Remember also that the emitted and any reflected radiation will depend strongly upon the structure of the work-piece, i.e. is it a flat plate, a v-groove shape, a triangular corner or a tank ? The instrumentation is not nearly as critical as the instrumentation required for solar UV-B measurements. A good quality spectroradiometer will do the job. The angular response (cosine) is usually not critical since most arcs will be confined into a small angle, they will be point sources. However, the arcs will vary in time, and sometimes substantially (Mariutti and Matzeu, 1987) so integration time has to be considered. In this respect spectroradiometers based on diode-array detectors are excellent since they will average out any rapid emission changes and measurements may be performed in, say, 10 second exposures (averaged measurements) (Eriksen, 1986). Repeated measurements of the same arc process will give statistics to indicate the uncertainty in the measurements caused by arc fluctuations. This uncertainty is much larger than the uncertainty that can be achieved with ordinary light sources in the laboratory.

Hazard evaluation

The optical radiation hazards are assessed from the current ACGIH threshold limit values or TLV's (ACGIH, 1988).

UV radiation. The ultraviolet radiation is assessed from the value of the effective (ultraviolet) irradiance, E_{eff}. E_{eff} is calculated from the measured spectral irradiance, E_λ, and the ACGIH UV action spectrum, S_λ,

$$E_{eff} = \Sigma\, E_\lambda\, S_\lambda\, \Delta\lambda \qquad (1)$$

where $\Delta\lambda$ is the bandwidth in nm. For constant E_{eff} during the exposure time t [s], the "effective" dose is $H_{eff} = E_{eff}\, t$, from which the maximum daily exposure time t_{max} [s] can be calculated when the maximum H_{eff} 30 J/m^2,

$$t_{max} = (\ 30\ J/m^2\ )\ /\ E_{eff} \qquad (2)$$

with E_{eff} in W/m^2. The UV-A irradiance should not exceed 10 W/m^2 for periods greater than 1000 seconds and for exposure times less than 1000 s the UV-A dose (radiant exposure) should not exceed 10^4 J/m^2. The UV-A irradiance and dose are unweighted quantities.

Light. Welding light (visible radiation) is assessed from the arc luminance, L_v, the arc blue-light-radiance, L_B, or -irradiance, E_B, and the retinal-burn-radiance, L_R. These quantities are calculated as

$$L_v = \Sigma\, L_\lambda\, V_\lambda\, \Delta\lambda \qquad (3)$$

$$L_B = \Sigma\, L_\lambda\, B_\lambda\, \Delta\lambda \qquad (4)$$

$$E_B = \Sigma\, E_\lambda\, B_\lambda\, \Delta\lambda \qquad (5)$$

$$L_R = \Sigma\, L_\lambda\, R_\lambda\, \Delta\lambda \qquad (6)$$

where L_λ is the arc spectral radiance, V_λ is the photopic eye response (color vision), B_λ is the blue-light hazard weight spectrum, R_λ is the retinal-burn-hazard weight spectrum and $\Delta\lambda$ is the bandwidth (ACGIH, 1988).

The arc luminance is important for selecting the proper eyewear, it is selected such that the operator obtains suitable or comfortable viewing conditions. For welding arcs this luminance L_C (C for comfort) is somewhere between 1 and 10 kcd/m^2 (Sliney et al., 1982; Hoikkala, 1986) depending upon the operator's eye performance. Luminance values above 10 kcd/m^2 will generally cause discomfort glare. However, welders usually observe that the smaller the arc the higher is the arc luminance that yields comfortable viewing conditions. This observation is important, especially for low current MAG welding with CO_2 as shielding gas because it affects the choice of welding filter selection.

The value of L_B and L_R will set the most stringent requirements to the spectral transmittance of the eyewear. Behind the welding filter the product of the exposure time [s] and L_B should be limited to 10^6 $J/m^2/sr$, for exposure times less than 10^4s (2.8 hrs). For longer exposure times, and not common to welding, L_B should be limited to 100 $W/m^2/sr$. Also, L_R [$W/m^2/sr$] should be limited to $10^4/(\alpha t^{1/2})$ where t is the exposure time in seconds, limited to between 1µs and 10 s, and where α is the (longest possible) angular subtense of the source in radians.

For sources (welding arcs) subtending an angle less than 11 mrad (0.63°) the above limits on L_B are relaxed: the product of the exposure time [s] and the blue light irradiance E_B should be limited to 100 J/m^2. This relaxation is important to some welding operations, one of them is MAG welding with CO_2 as shielding gas and with low currents because the arc is very small.

Because of fluctuations in a welding arc direct measurements of spectral radiance are very difficult, if not impossible, since the arc must be imaged onto the entrance slit of the spectroradiometer (monochromator/spectrograph). One way to circumvent this difficulty is to calculate L_λ from measurements of E_λ and the solid angle subtended by the arc at the spectroradiometer input optics (diffuser), Ω, i.e. $L_\lambda = E_\lambda/\Omega$. The solid angle Ω may be obtained from a photograph of the arc and subsequent analysis of the picture (negative) with a microdensitometer. Ω equals the arc area divided by the square of the distance between the arc and the position where E_λ was measured. The area is taken as an effective area: it is the area within the L_v(peak)/e -isoline, where e=2.7183 (Marshall et al., 1976). The validity of this procedure may be checked by comparing calculated and directly measured arc luminance values.

IR radiation. Finally, the infrared radiation is assessed from the IR irradiance E_{IR} (λ > 760nm) which should be limited to 100 W/m^2. This quantity may be measured with a simple radiometer and cut-off filters since it is an unweighted quantity.

CHARACTERISTICS OF WELDING AND PLASMA ARC RADIATION

Characteristics

There are several process-parameters that affect the emission from welding and plasma arcs. Among the most important process-parameters are the operating current, the electrode (material) type, the shielding gas (type and flow rate), the exhaust/ventilation (if any). Other less important parameters are arc voltage, electrode (wire) dimension, wire-feed speed (MIG/MAG) and melt down rate. It is extremely important to be aware of all the variables involved when analyzing and assessing welding arc radiation, especially when comparing measurements from different experiments, a slight difference in parameters may easily give very different spectra. An example of such a situation is given in table I for MIG welding on aluminum with two different electrodes. Also, it takes a well trained operator to run a welding arc properly.

The following applies to welding arcs in particular but many features are similar with plasma arcs. Representative examples of welding arc spectra are shown in figs.1, 2 and 3, others may be found in the references.

Arc current and voltage. Most welding arc currents are DC, although many TIG arcs are operated on AC current. The current "determines" the overall output level, the arc voltage plays only a minor role, but affects the length of the arc. The experienced operator will know how to set the current and voltage correctly for a given welding job. The irradiance at any given distance scales approximately as the square of the current.

Electrode. The material composition of the electrode determines primarily the spectral output in the central parts of the arc, especially in the UV and visible. In MIG/MAG welding electrodes containing iron (Fe) results in complex spectra whereas electrodes containing aluminum (Al) and magnesium (Mg) results in simpler spectra with "few" spectral lines dominating the output. In TIG welding the filler material will produce spectral lines. In MMA welding with cored electrodes the result is very complex spectra that can only be resolved with very high resolution spectrometers. The presence of neutral, single and double ionized atoms in the arc is easily detected with 1nm resolution.

Shielding gas. The shielding gas is mainly responsible for spectral lines in the infrared and near infrared and are mostly confined to the peripheral parts of the arc. Only when Helium is used does it reveal itself in the visible and UV-A.

Exhaust/ventilation. If a local fume exhaust is used it has the effect of removing scattering particles in the optical path, it therefore increases the irradiance in typical operator positions, in the UV by as much as 50 percent.

Welding arc ignition

The ignition phase of MIG/MAG welding arcs lasts approximately 25 milliseconds. During this period the emission increases rapidly until a sufficient plasma has been created to sustain the arc and then falls off to a "stable" level characteristic of the particular process. In electronics jargon, there is an "overshoot" of radiation output during the first 25 ms (Eriksen, 1985). The peak output during the overshoot has been observed to be about 10 times the output of a steady burning arc. The overshoot is highest in the UV, possibly because the density of ionized atoms in the arc has to increase dramatically during ignition in order to establish an arc. With a diode-array spectrometer full spectra (200-800 nm) were recorded during the ignition of a 200 ampere MIG (alu) arc every 5 milliseconds (Eriksen, 1985). Fig. 4 shows the emission during ignition at particular wavelengths characteristic of this type of welding, i.e. a spectrum similar to the one shown in fig.1 except for the scale in spectral irradiance.

Assessment of optical radiation

Table I shows typical data for several welding and plasma processes (Eriksen, 1986). The radiation is assessed as described in the section on Hazard evaluation.

Ultraviolet radiation. From the data presented in table I it's evident that the ultraviolet output is indeed very high, it is always of concern. Although most UV injuries are reversible (self-healing) the problem of possible chronic effects remain, and there is not much known about chronic effects at the present time. It follows that the maximum exposure time (seconds) allowed *per day* ranges from fractions of a second to several tens of seconds (for low currents). The risk of having skin burn (erythema) and/or arc-eyes or welders' flash (photokeratitis) is therefore very high. In a recent survey among 200 welders and cutters approximately 50% had suffered from at least one welders' flash within the last year (Eriksen, 1986), also within the last year of questionning 7% had taken eye-drops to relieve eye-pain before commencing work. It appears that the welder's education in radiation protection is inadequate because it is not difficult to protect against UV radiation, protective eyewear and clothing (Sliney et al, 1987) is readily available.

Light. The most severe injury from welding arc radiation is partial loss of sight on one or both eyes resulting from retinal burns, whether by a thermal effect or a photochemical (blue-light) effect. Fortunately, such cases are rare. We have recently had a case of a female who suffered a 60% decrease in visual acuity on both eyes because of exposure to welding light. She suffered the injury while she was on job training. She had been working in a service job (office) before and had no previous experience with welding arc radiation. The lack of welding experience partly explains the injury.

To assess the risk of retinal injury L_R, L_B, or E_B are needed, eq. (4)-(6). E_B is only used to assess the blue-light hazard for small arcs that subtend an angle of less than 11mrad, for larger angles L_B is used. L_R and L_B were calculated from E_R and E_B (where E_R is an auxillary quantity that may be defined like E_B in eq. (4) with B_λ substituted by R_λ). Because R_λ is approximately 10 times B_λ in the wavelength range where the dominant weight in eqs. (4)-(6) occurs (ACGIH, 1989) E_R or L_R will be approximately 10 times E_B or L_B. For the welding arcs in table I, the longest arc dimensions (lengths) range from about 3mm to about 10mm.

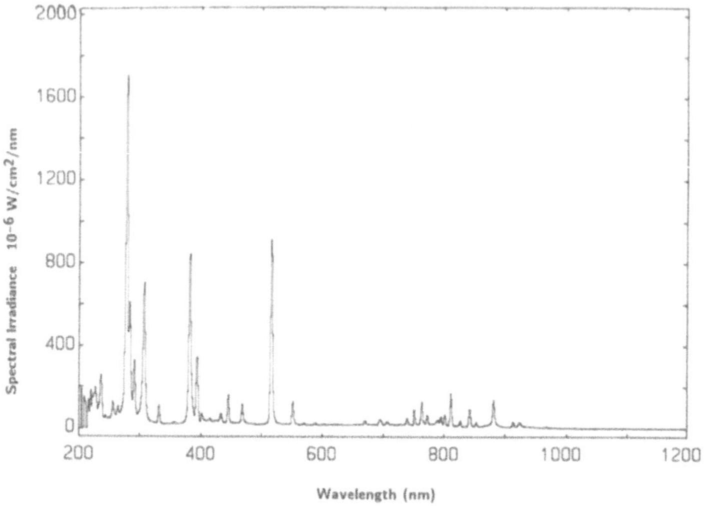

Fig.1. Measured spectral irradiance at 0.5m from a 300 ampere MIG welding arc. Base metal: aluminum, shield gas: Ar, electrode: AlMg5 with ESAB designation OK 18.15. Peaks at 280, 382 and 518 nm are caused by Mg, at 309 caused by Al.

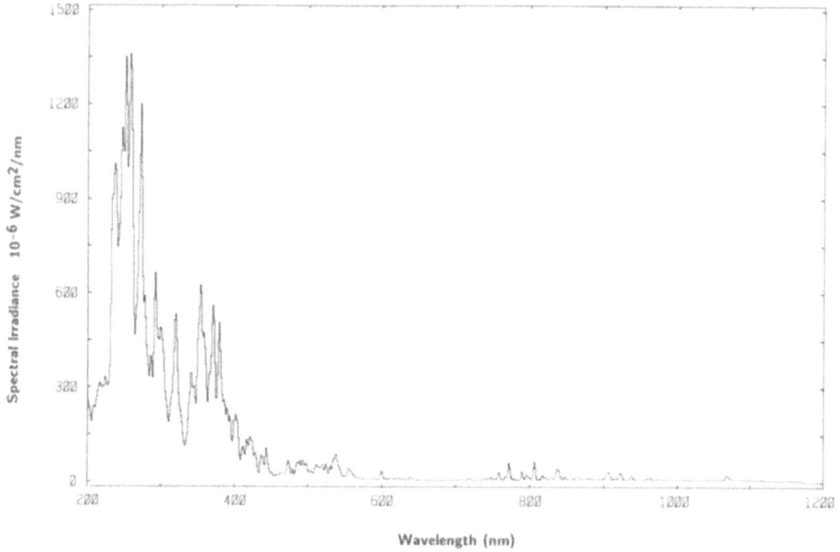

Fig.2. Measured spectral irradiance at 0.5m from a 250 ampere Synergic Pulse MIG welding arc. Base metal: mild steel, shield gas: Ar/O_2, electrode: ESAB designation OK 12.51 (Fe).

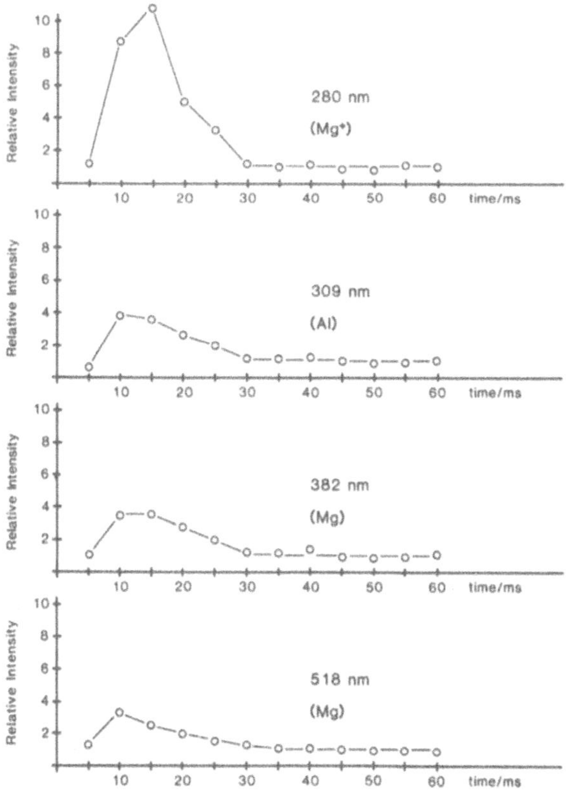

Fig.3. Measured spectral irradiance at 0.5m from a 170 ampere MMA welding arc. Base metal: stainless steel, electrode: AVESTA designation 832.

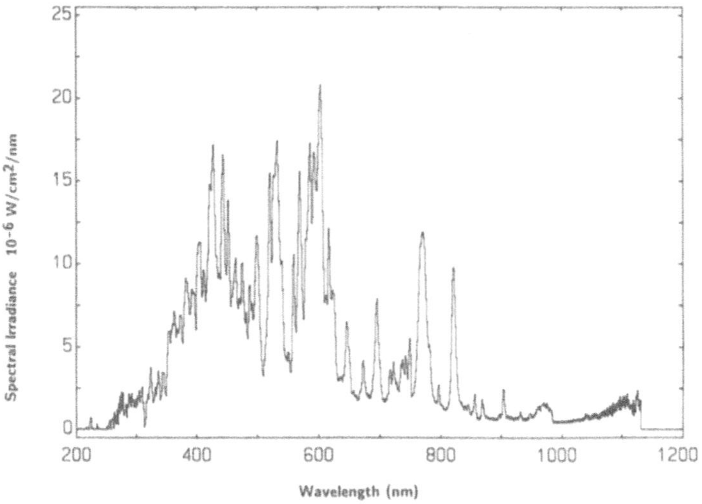

Fig.4. From time resolved spectral measurements of a 200 ampere MIG (aluminum) welding arc ignition (fig.1) the time evolution of the major peaks in the steady state spectrum is shown here during the ignition phase. By permisssion of Am.Ind.Hyg.Assoc.J. 46;101:1985.

TABLE I. Typical data for some welding and plasma arcs at a distance of 0.5m.

	shield gas	base metal	elec-trode	current amps	E_{eff} W/m^2	E_{UVA} W/m^2	E_B W/m^2	E_v klux	E_{IR} W/m^2
1) SP-MIG	Ar/CO_2	ms	OK12.51	200	200	150	30	150	35
2) MAG	Ar/CO_2	ms	OK12.51	200	40	25	20	100	25
3) MAG	CO_2	ms	OK12.51	150	1	10	2	2	10
4) MIG	Ar	alu	OK18.15	200	100	70	25	300	35
5) MIG	Ar	alu	OK18.04	200	20	20	10	100	10
6) TIG	Ar	alu	HAlSi5	100	3	3	6	4	15
7) TIG	Ar	ss	HR316L	100	3	3	3	15	20
8) MMA	NA	ms	OK33.80	200	3	10	20	25	40
Cut	Ar/H_2	ms	N/A	200	5	7	4	2	200
Spray	Ar/He	N/A	N/A	400	20	30	25	20	80
SUN	N/A	N/A	N/A	N/A	0.01	10	130	120	600

NOTES to TABLE I:
1) Cut and Spray refers to plasma arcs, entries 1)-8) to welding arcs.
2) SP-MIG is an abbreviation of Synergic-Pulse-MIG welding.
3) Base metal abbreviated: ms= mild steel; alu= aluminum; ss= stainless steel.
4) Electrode abbreviations: HR316L and HAlSi5 are AGA designations, others are ESAB designations.
5) N/A= Not applicable.
6) Data for the sun refers to summer-noon in the mediterranean with clear skie

TABLE II. Further data for the welding arcs in table I.

	process parameters	arc area mm^2	visual angle mrad	L_R $/10^5$ $W/m^2/sr$	t_{max} (burn) s	L_B $/10^5$ $W/m^2/sr$	t_{max} (blue) s	L_v $/10^8$ cd/m^2
1) SP-MIG	1) table I	18	12	40	0.05	4	2.5	20
2) MAG	2) table I	18	12	30	0.08	3	3.3	14
3) MAG	3) table I	4	4	10	6	1	10	1
4) MIG	4) table I	28	12	20	0.1	2	5	30
5) MIG	5) table I	28	12	10	0.4	1	10	9
6) TIG	6) table I	4	6	40	0.2	4	2.5	3
7) TIG	7) table I	6	6	10	3	1	10	6
8) MMA	8) table I	60	18	10	0.3	1	10	1
SUN	table I	N/A	9	250	0.002	20	0.5	20

NOTES to TABLE II: same as 2) thru 6) in table I.

Let us immediately notice the results for the sun in table II. The retinal-burn radiance L_R is so high that the unprotected eye should limit the exposure time to 2 ms. Clearly this is a very conservative result. It is well known that looking momentarily at the sun does not cause permanent damage to the retina although it is usually very unpleasant. Thus the limiting exposure time of 0.1 - 0.2 s controlled by the eyes' aversion response does protect the retina. Apparently the hazard assessment from L_R is too conservative, or too simplistic; the problem is the source size dependence of damage threshold and the R_λ values in the 400-500 nm region. A more detailed analysis of the thermal hazard to the retina may be required (Sliney & Freasier 1973) if a more accurate hazard assessment is needed. The same conclusion holds for the welding arcs listed in table II, they will generally not produce a retinal burn within the limited exposure time afforded by the aversion response.

Table II shows that E_B rather than L_B should be used to assess the blue-light hazard for the welding arcs of entries 3), 6) and 7). This also affects the requirements to welding filters in eyewear as discussed below.

Infrared radiation. Most welding arcs emit relatively little in the IR part of the spectrum and the welder does usually not need substantial protection against IR radaition. This is evident from table I and the TLV for the infrared irradiance $E_{IR}(TLV) = 100\ W/m^2$.

PROTECTIVE FILTERS

Welding filters are used in eyewear to protect the eyes against harmful radiation, light as well as ultraviolet and infrared radiation. This puts requirements to the transmittance of the filters. Welding filters serve another important purpose namely to reduce the arc luminance to a level such that the operator obtains comfortable viewing conditions. The selection of an appropriate welding filter depends upon the welding process, the welding current and the operators perception of "comfortable" viewing conditions.

UV and IR transmittance

It is not difficult to achieve sufficiently low tranmittance in the UV. Most welding filters are made from mineral glasses and have very low transmittance below 350nm (Eriksen, 1986) and it is not difficult to achieve spectral transmittance lower than 0.0003 % below 313nm as required by the standards (ISO 4850, 1979; prEN 169, 1986). Even when the welding filter is made from plastic the demands to UV transmittance can be met by adding UV absorbers to the plastic material. Since it is easy to achieve the low UV transmittance, and because this effectively eliminates UV hazards behind the filter, the current UV transmittance requirements are adequate although overly restrictive for low welding currents.

The requirements of the international standards (ISO 4850, 1979; prEN 169, 1986) to IR radiation are overly restrictive since most welding arcs, even at high currents, emit relatively little in the IR. The requirements to the IR transmittance are met by welding filters made from mineral glass, but are hard to meet for filters made from plastics, unless a reflective coating is applied on the front surface of the filter (hot mirror type).

Light transmittance

Welding filters are characterized by their shade number, or scale number, N, which is determined from the filter's luminous transmittance, τ_v

$$N = 1 - \log_e\tau_v = 1 - (7/3)\log_{10}\tau_v = 1 + (7/3)D_v \qquad (7)$$

where D_v is the luminous optical density. The shade numbers that are actually used are 1.2, 1.4, 1.7, 2.0, 2.5, and 3 to 16 in integer increments. The luminous transmittance is measured with a CIE-A illuminant and a photometer with V_λ response. Shade number 10 is most common in welding and is probably used by about 75% of the welders according to sales information from suppliers of welding filters. Shade numbers lower than about 8 are used for brazing, gas welding and cutting etc.

For any given welding operation there is a shade number, N_C, that will provide comfortable viewing conditions. Most standards on requirements to welding filter transmittances contain tables that to a given welding operation recommends which shade number should be chosen to obtain safe and comfortable viewing conditions (ISO 4850, 1979; prEN 169, 1986). Let N_{REC} denote a recommended shade number to a given welding. Supposing that the UV and IR hazard is already eliminated by using a welding filter the blue-light hazard is left to analyze and will determine the final demands to the filter.

The question is whether N_C and N_{REC} will provide sufficient protection against the blue-light hazard from long term observation of the arc through the filter. Let N_B denote the lowest shade number that provides sufficient protection against the blue-light hazard for a given welding i.e., any shade number equal to or above N_B will provide sufficient protection.

To analyze the blue-light protecting properties of a welding filter we must know the arc's L_λ, or E_λ in a typical operator position, and the spectral transmittance of the filter, τ_λ, and determine L_B or E_B behind the filter from eqs. (4) and (5). Let L'_B and E'_B denote the values of L_B and E_B behind the filter. They are calculated from

$$L'_B = \Sigma L_\lambda \tau_\lambda B_\lambda \Delta\lambda \qquad (8)$$

and a similar equation for E'_B because, behind the filter, the arc's spectral irradiance is observed as $L_\lambda \tau_\lambda$. For the time factor involved in the analysis a maximum daily exposure time of 10^4 s = 2.8 hrs is assumed. This is exceeded only in extreme situations according to people in the welding community. Since we cannot measure τ_λ for every welding filter we have to make approximations to solve eq. (8). One approximation is to assume τ_λ to be constant, then we can put it outside the summation in eq.(8). One way to do this is to assume that $\tau_\lambda = \tau_v(N)$ i.e., the spectral transmittance equals the luminous transmittance for a given shade number N. Another approximation can be made to solve eq.(8) if τ_λ can be approximated mathematically for a given shade number N. This happens to be the case with filters made from mineral glass (Eriksen, 1986): the function $k(N) / \exp((\lambda-530)/65)^2$, where k(N) is a constant depending on shade number N, and where λ is the wavelength in nm, makes a very good fit to measured spectral transmittances in the wavelength region 400-600 nm. For example $k(9) = 6\ 10^{-4}$ and $k(10) = 2\ 10^{-6}$ gives very good fits. However, both approximations turn out to give almost the same results so the first mentioned approximation can be used in confidence despite it's simplicity (Eriksen, 1986).

The luminance that yields comfortable viewing conditions, N_C, is based upon a "comfortable" arc luminance (behind the filter). In most work on standardization, and thereby on recommendations, the value that is and has been used so far is the "magic" 730 cd/m^2 (Sutter et al., 1972). The origin and justification of this value is unknown to the author despite several searches into the litterature, and I do not know for which field of view (viewing angle) it is defined. However, for welding situations it is certainly much too low a value (depending on field of view, though) and a comfortable luminance for welding is between 10^3 cd/m^2 and 10^4 cd/m^2 (Sliney and Wolbarsht, 1981; Hoikkala, 1988) for a half-degree field of view appropriate for welding arcs (table II).

The results from a big survey of MIG, MAG and TIG welding processes on welding filter selection and protection properties along the lines described above can be summarized as follows: N_C is 1-2 shade numbers lower than N_{REC} and N_B is 2-5 shade numbers lower than N_{REC}, with N_{REC} according to the standards (ISO 4850, 1979; prEN 169, 1986). In other words: let the operator choose a welding filter that gives comfortable viewing conditions - it will provide the necessary protection against blue-light, UV and IR.

Electro-optic filters

A new class of "automatic" welding filters based on liquid crystals has recently been introduced on the market. They are characterized by a "light" state in which the operator has sufficient visibility through the filter to see where to start the welding, and a "dark" state corresponding to a normal welding filter. When the arc is struck a photodiode in the filter senses the light emitted during ignition and the filter switches from the light state to the dark state in a few milliseconds (for the better products) (Buhr and Sutter, 1989). They can be recommended, especially for thack welding.

REFERENCES

Amin, M., 1981. Synergic Pulse MIG Welding. Metal Construction, 13: 349-353.

Buhr, E., Sutter, E., 1989. Dynamic filters for protective devices, In: "Dosimetry of laser radiation in medicine and biology," Müller, G.J., Sliney, D.H. (eds.), SPIE vol. IS 5, Bellingham, WA.

Eriksen, P., 1985. "Time resolved optical spectra from MIG welding arc ignitions", Am. Ind. Hyg. Assoc. J., 46: 101-104.

Eriksen, P., 1986. "Optical radiation from MIG and TIG welding arcs," Danish Work Environment Fund, Copenhagen. (in Danish, English summary).

Eriksen, P., 1987. Occupational applications of ultraviolet radiation: risk evaluation and protection techniques, In: "Human exposure to ultraviolet radiation. Risks and regulations," Passchier, W.F., Bosnjakovic, B.F.M. (eds.), Elsevier, Amsterdam.

ACGIH, 1988. "Threshold Limit Values and Biological Indices for 1988-89," American Conference of Governmental Industrial Hygienists, Cincinnati, OH.

Mariutti, G., Matzeu, M., 1987. Measurement of ultraviolet radiation in welding precesses and hazard evaluation. In: "Human exposure to ultraviolet radiation. Risks and regulations," Passchier, W.F., Bosnjakovic, B.F.M. (eds.), Elsevier, Amsterdam.

Sliney, D.H., Wolbarsht, M.L., 1981. "Safety with lasers and other optical sources," Plenum Press, New York.

Marshall, W.J. et al., 1977. "Non-ionizing radiation protection special study no. 42-0312-77. Evaluation of the potential retinal hazards from optical radiation generated by electric welding and cutting arcs," US Army Environmental Hygiene Agency, Aberdeen Proving Ground, MD, USA.

Lyon, T.L. et al., 1977. "Non-ionizing radiation protection special study no. 42-0053-77. Evaluation of the potential hazards from actinic ultraviolet radiation generated by electric welding and cutting arcs," US Army Environmental Hygiene Agency, Aberdeen Proving Ground, MD, USA.

Sliney, D.H. et al, 1987. "Transmission of potentially hazardous actinic ultraviolet radiation through fabrics," Appl. Ind. Hyg. 2:36-44.

Sliney, D.H., Freasier, B.C., 1973. "The evaluation of optical radiation hazards," Appl. Opt. 12: 1-22.

Sutter, E., et al., 1972. Strahlungsmessungen an verschiedenen Lichtbogen-Schweisserverfahren, Bericht Optik 2/72, Physikalisch-Technische Bundesanstalt, Braunschweig.

ISO 4850, 1979. "Personal eye-protectors for welding and related techniques - Filters - Utilisation and transmittance requirements," International Organization for Standardization, Geneva.

prEN 169, 1986. "Personal eye-protectors for welding and related techniques. Transmittance requirements," Commission Europeen de Normalisation, Paris.

ULTRAVIOLET EXPOSURE IN THE OUTDOOR ENVIRONMENT: MEASUREMENTS OF AMBIENT ULTRAVIOLET EXPOSURE LEVELS AT LARGE ZENITH ANGLES

David H. Sliney, Rodney L. Wood, Jr.
Patrick M. Moscato, and Wesley J. Marshall

Laser Microwave Division
US Army Environmental Hygiene Agency
Aberdeen Proving Ground, MD 21010-5422 USA

Mr. Paul Eriksen

Danish National Institute of Occupational Health
Lerso Parkallee 105
DK-2100 Copenhagen
Denmark

BACKGROUND

The ambient levels of ultraviolet radiation (UVR) incident upon the eye and skin vary dramatically with latitude, time of day (solar zenith angle), and geometry of exposure [1-10]. The first two factors have been extensively studied for meteorological purposes and to quantify UVR exposures for epidemiological studies of geographical incidence of skin cancer [11-12], but the last factor--geometry of exposure--has often been neglected in studies of chronic eye exposure to sunlight and in related epidemiological studies of skin cancer and cataract. For this reason, measurements of UVR irradiance in the 295-315 nm spectral band were performed in several geographical locations at several elevations with varying ground reflectance factors to provide a better quantitative estimate of UVR exposure in different climates.

The purpose of this paper is to better quantify the ambient UVR exposure to exposed vertical skin areas and the eye to permit more accurate epidemiological studies of the relative causal factor (i.e., the etiological factor) of UVR exposure in the development of skin cancer and changes in the anterior segment of the eye, such as cataract. It will be shown that the reflectance of the terrain plays the greatest role in

Light, Lasers, and Synchrotron Radiation
Edited by M. Grandolfo *et al.*, Plenum Press, New York, 1990

varying UVR exposure to the eye and to shaded areas of the skin, and that cloud cover plays a lesser role than most would expect.

The UVR incident upon the eye from ground reflections and from the sky near the horizon is of primary importance to any epidemiological study of UVR induced corneal changes, cataractogenesis, pterigium, or potential retinal changes. Apart from some relative dose studies by Diffey, et al., this data has been unavailable. Quite obviously, the ocular exposure varies significantly from skin exposure, since photokeratitis is seldom experienced during sunbathing, yet the threshold for UV photokeratitis is less than that for erythema of the skin. During the past 20 years, concerns of UVR induced skin cancer and depletion of the ozone layer in the stratosphere led to a worldwide program of outdoor, daily UVR measurements. Maps of annual UVR exposure were compiled for epidemiological studies of skin cancer, etc. [13]. However, these measurements were all of direct and/or "global" UVR, i.e., the UVR falling upon a horizontal surface from all directions (both direct and scattered radiation). While such measurements would be of interest in calculating the UVR exposure dose to a prone individual, they are of very limited value in estimating exposure of the eye and shaded skin surfaces.

Several epidemiological and related studies have suggested a relationship between outdoor ambient UVR levels and the incidence of corneal degeneration and senile cataract [14-22]. Since the ambient UVR exposure rate from overhead increases with elevation, some data which show less incidence of cataract in the Hymalayan mountains appeared to argue against this hypothesis of UVR cataractogenesis. However, Brilliant, et al argued that one must be careful to consider the shading effect of mountains in most Nepalese mountain villages which would explain this apparent contradiction [21]. Unfortunately, while mountains produce extensive shady areas, this shade is minimal during the midday periods when UVR ground irradiances are the highest. Thus a more refined inspection of the physical factors was to be warranted. Sliney argued that variation in ground reflectance would explain the contradiction [23].

Prior to a detailed presentation of UVR exposure measurements it is useful to review the daily variations in the terrestrial solar spectrum. Although visible light falling upon the ground may vary by only 20% from noon to 3:00 p.m. (local solar time) in the summer, the UVR spectrum undergoes a dramatic change during this period. At a wavelength of 300 nm the spectral irradiance decreases by 90% [from approximately 1.0 to 0.1 $\mu W/(cm^2\text{-}nm)$]!

Although we visually perceive some changes in ambient daylight, the preferential Rayleigh scattering of shorter wavelength radiation (i.e., UVR and blue-violet visible wavelengths), we are perceptually unaware of the dramatic variations in our UVR environment. Although most visible sunlight falling on the earth's surface is direct and causes distinct shadows when intercepted by a tree or parasol, this is not true for UVR. The majority of 300-nm UVR incident upon the body is indirect, scattered radiation. Hence, one can actually be sunburned while sitting in the shade of a beach umbrella even with little ground reflection. The principal point to be made here is that the UVR exposure of the cornea and crystalline lens is from scattered radiation. Direct solar radiation normally does not strike the cornea during the hours of peak sunlight when UVR irrradiances are highest (approx. 9:00 a.m. to 3:00 p.m. local time) [24].

UVR MEASUREMENTS

Instrumentation and Methods

Measurements of both directional and global UVR during daytime hours were performed in various environments: snowfields, desert, mountain, and grassy terrains. Since laboratory studies show that severe sunburn, skin carcinogenesis and acute cataractogenesis followed exposure to the 295-320 nm wavelength band, this measurement effort concentrated on this spectral region (which is frequently referred to as part of the "UV-B" spectral band by photobiologists).

An International Light Model 730 UV Radiometer, which had a spectral response closely following the ACGIH action spectrum for UVR health hazard evaluations was employed to measure the UVR "effective irradiance. Figure 1 shows the action spectrum and approximate spectral response of this instrument.

To calibrate the UV radiometer against the solar spectrum, an EG & G Model 585 Spectroradiometer system was modified to employ a double monochromator to obtain low-stray-light spectral irradiance measurements at Edgewood, MD. Figure 2 shows the spectral irradiance measured for a zenith angle of 15°. Measurements of near-UV (UV-A) wavelengths between 320 nm and 400 nm were made, but the variation of measured values was small compared to the 295-320 nm band; and since UV-A is approximately three orders of magnitude less effective in producing most known biological effects of UVR, these results are omitted from this paper.

The sun's position was recorded in azimuth zone (N, S, E, & W) and by zenith angle. Figure 3 clarifies the goemetry of measurement and the zenith angle of the sun [the angle to the sun's position from a vertical line to the nadir (zenith) overhead].

The effective irradiance, E-eff, was measured with a cosine-receptor detector with three angular fields-of-view (FOV): hemispherical, quarter-spherical, and a 40° cone. With the hemispherical response, global UV was measured by directing the detector toward zenith; direct irradiance was obtained by directing the detector axis at the sun; and horizontal irradiance was obtained by directing the axis horizontally either north (N), south (S), east (E), or west (W). The quarter-spherical readings were taken with the detector axis oriented horizontally with a hood over the detector to limit incident radiation to elevation angles below 20° as might simulate facial exposure when an individual wears a broad-brimmed hat, or when an individual is walking with a downward gaze. The 40° conical FOV (i.e., 0.42 steradian) was used to obtain measurements of sky radiance and localized UVR.

To measure terrain reflectance, the detector (with a cosine response and hemispherical FOV) was directed downward along a vertical axis (i.e., Z = 180°), then vertically upward toward zenith (i.e., Z = 0°). The ratio of the two readings was then recorded as the diffuse reflectance. All measurements were performed with the cosine-corrected hemispherical UV-B detector head of IL 730 radiometer. Reflectance is the ratio of ground "down"/zenith measurement.

Results

Table 1 lists some representative UVR values measured in various locales. Table 2 provides typical ground surface reflectance values.

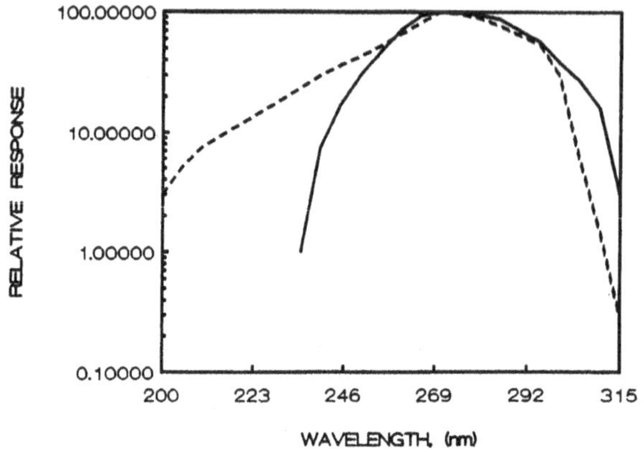

FIGURE 1. The Action Spectrum for UV Hazard Assessment Instrument and the Solar Spectrum. The ACGIH action spectrum (---) is closely fit by the UVR radiometer (-) used in this study. However, the terrestrial solar spectrum overlaps to such a small extent that problems of stray light must be dealt with by constant checks with a filter which blocks wavelengths less than 320 nm.

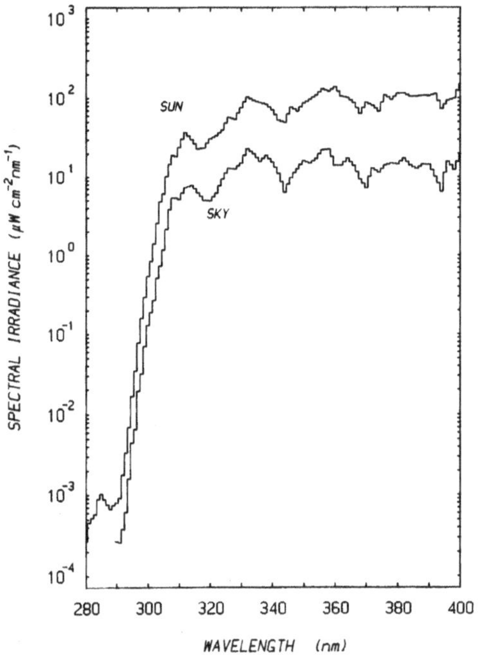

FIGURE 2. The solar spectrum from the spectroradiometer compares favorably with previous studies (12) for a zenith angle of 15°.

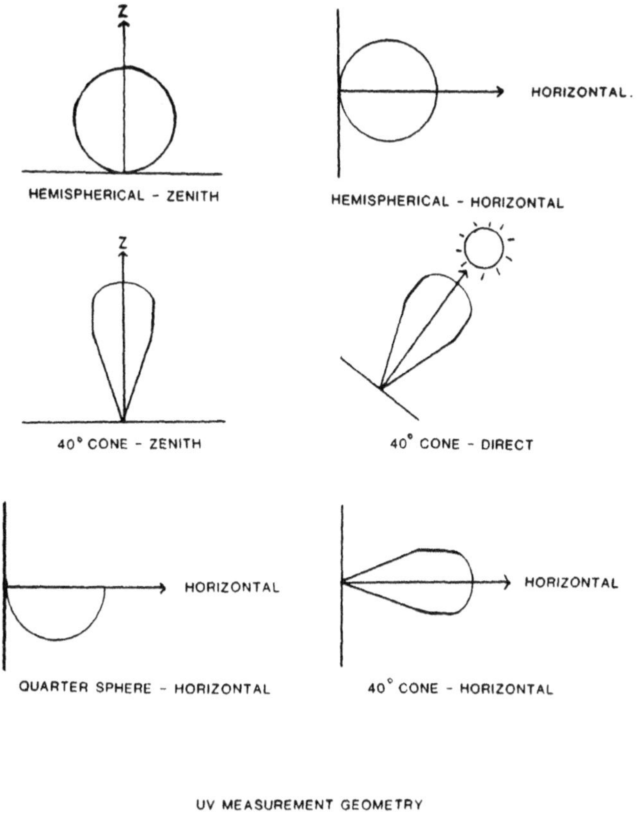

FIGURE 3. UV Measurement Geometries. The zenith angle is illustrated along with the angular response of the three types of detector fields-of-view used in this study: a cosine detector with either an hemispherical, a quarter-sphere, or a 40°-cone response was employed. The upper left panel shows the geometry for measuring "global" radiation as most often used in previous environmental UVR measurements. The cosine-response characteristics of the detector is illustrated by circular coordinate plots.

Table 1. Representative UVR Measurements at Different Locales for Several Zenith Angles (Z).

Geographical Location	Elevation ASL (feet) & (m) Lattitude	Sky	Zenith Angle	Measured UVR Effective Irradiance (uW/cm²) Global Diffuse	Direct (hemi.)	North (hemi.)	North (quart.)	South (hemi.)	South (quart.)	East (hemi.)	East (quart.)	West (hemi.)	West (quart.)
Wildwood, New Jersey	1 0.3 38Δ59'N	CB	25Δ-S	1.8	1.8	0.7	0.3	0.75	0.3	0.75	0.3	0.75	0.3
Wildwood, New Jersey	1 0.3 38Δ59'N	CL	49Δ-SW	2.0	2.5	0.56	0.20	2.0	0.39	0.76	0.29	1.5	0.31
Edgewood, Maryland	50 15 39Δ24'N	PC	16Δ-S	4.2	4.3	1.0	0.9	2.6	1.8	2.6	1.1	2.2	1.0
Edgewood, Maryland	50 15 39Δ24'N	UC	34Δ-S	2.8	3.2	0.52	0.04	0.44	0.08	0.52	0.05	0.81	0.06
Fallston, Maryland	500 150 39Δ30'N	PC	73Δ-W	0.57	0.70	0.21	0.07	0.21	0.06	0.21	0.09	0.7	0.2
Fallston, Maryland	500 150 39Δ30'N	PC	32Δ-S	1.0	1.5	---	0.12	---	0.12	---	0.11	---	0.13
Salt Lake City, UT	4800 1460 40Δ45'N	PC	18Δ-S	4.2	4.0	0.8	---	1.2	---	0.8	---	1.4	---
Vail Mtn. Colorado	9000 2750 39Δ45'N	PC	25Δ-S	4.5	5.0	0.7	0.3	3.3	0.5	1.3	0.4	1.0	0.25
Vail, Colorado	8150 2500 39Δ45'N	CL	45Δ-SE	2.7	3.6	0.38	0.08	1.5	0.06	1.8	0.2	---	---
Loveland Pass, CO	12000 3660 39Δ45'N	PC	56Δ-W	0.65	1.3	---	0.12	---	0.11	---	0.09	---	---
San Francisco, CA	50 15 37Δ40'N	UC	40Δ-S	2.6	3.1	0.46	0.2	2.0	0.3	---	---	---	---

All measurements recorded above were made with an IL-730 Actinic UV Radiometer with a spectral response which approached that of the ACGIH action spectrum for UV hazard evaluations (wavelengths less than 315 nm). All data for summer foliage. Sky characterics: UC = unusually clear day with very low humidity; CL = clear (less than 5 % clouds); PC = partly cloudy (less than 50 % cloud cover); CB = cloudy, bright (mostly cloudy or hazy); OC = overcast. --- = obstruction in line-of-sight.

Note that with the exception of snow, natural ground reflectance values in the 295-315 nm region are very low. It is clearly for this reason that photokeratitis from outdoor daylight is so rare. Figures 4 and 5 show typical angular variations of E-eff across the sky as a function of zenith angle for a sunny day and an overcast day. Note that on a hazy day the irradiance at zenith is greatest, and actually exceeds the irradiance from the sun's position behind the clouds. Furthermore, the irradiance values near the horizon (70° to 90° zenith angles) do not vary dramatically with clouds: the cornea is exposed to an irradiance of the order of 0.1 $\mu W/cm^2$ during daylight hours where the sun is at elevation angles above 20° (i.e., Z = 0° to 70°) when an individual is in an open field. However, when hills, trees, or buildings loom in one's field of view, this level is greatly reduced--often by an order of magnitude. When the sky is overcast, the lids open more widely; therefore, a slightly overcast sky can lead to a greater ocular exposure dose to a worker in an open field than when the sun is brightest with no clouds present!

TABLE 2. Representative Terrain Reflectance Factors for Horizontal Surfaces Measured with a UV-B UVR Radiometer and Midday Sunlight (290-315 nm).

Material	Percent Reflectance
Lawn Grass, Summer, Maryland, California and Utah	2.0 - 3.7
Lawn Grass, Winter, Maryland	3.0 - 5.0
Wild Grasslands, Vail Mountain, Colorado	0.8 - 1.6
Lawn Grass, Vail, Colorado	1.0 - 1.6
Flower Garden, Pansies	1.6
Soil, clay/humus	4.0 - 6.0
Sidewalk, Light Concrete	10 - 12
Sidewalk, Aged Concrete	7.0 - 8.2
Asphalt Roadway, Freshly Laid (black)	4.1 - 5.0
Asphalt Roadway, Two-years Old (grey)	5.0 - 8.9
Housepaint, White, Metal Oxide	22
Boat Dock, Weathered Wood	6.4
Aluminum, Dull, Weathered	13
Boat Deck, Wood, Urethane Coating	6.6
Boat Deck, White Fiberglass	9.1
Boat Canvas, Weathered, Plasticised	6.1
Chesapeake Bay, Open Water	3.3
Chesapeake Bay, Specular Component of Reflection at Z = 45	13
Atlantic Ocean, New Jersey Coastline	8.0
Sea Surf, White Foam	25 - 30
Atlantic Beach Sand, Wet, Barely Submerged	7.1
Atlantic Beach Sand, Dry, Light	15 - 18
Snow, Fresh (Two Days Old)	88(50)

All measurements performed with cosine-corrected hemispherical UV-B detector head of IL 730 radiometer. Reflectance is ratio of "down"/zenith measurement.

Other interesting findings resulted from the measurements of the angular distribution of UVR relative to solar position and cloud distributions. A cloud obscuring the sun had no effect upon the UVR radiance of the open blue sky or the horizon sky. However, when the sun

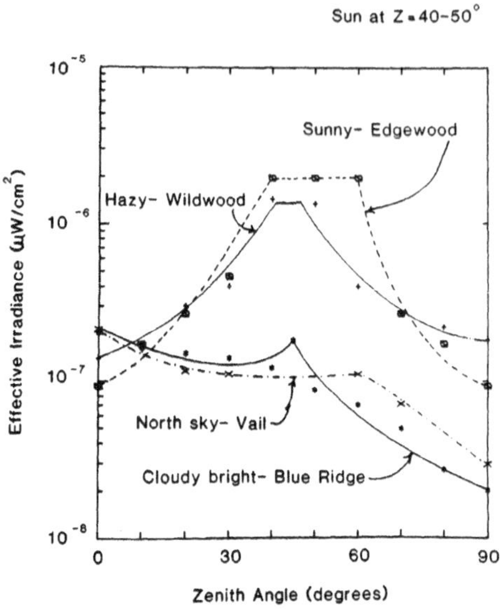

FIGURE 4. Semilogarithmic Plots of the Angular Distribution of 290-315 nm UVR for a Solar Zenith Angle of 45°. Upper most curve shows that direct UVR from the sun is more than two times greater than scattered UVR normally incident upon the eye at near-horizon angles where the zenith angle Z = 70° - 90°. Most surprising is the similarity of blue sky and cloudy sky UVR irradiances at zenith or near the horizon.

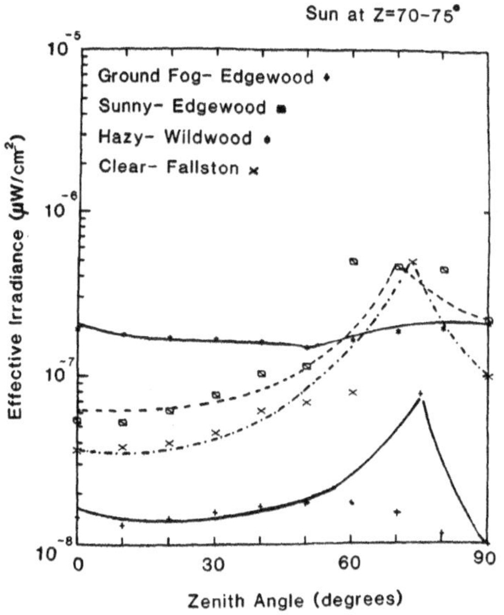

FIGURE 5. Semilogarithmic Plots of the Angular Distribution of 290-315 nm UVR for a Solar Zenith Angle of 70-75°. Except for the ground-fog condition, the greatest UV-B irradiance was near the horizon at greater zenith angles. Even at azimuths away from the sun, the irradiance values were similar. The north-sky UVR irradiances were typically close to the values at zenith. Note that under hazy conditions the irradiance was almost uniform at all angles. Severe ground fog acted as a total diffuser and also attenuated the UVR significantly in the measurement series shown. The UVR distribution under heavy overcast also appeared similar to the ground fog condition.

was "out" (i.e., in an open sky), clouds near the horizon opposite the sun apparently reflected more UVR than would otherwise be present from the blue sky. This is not really surprising when one studies photographs of the sky taken through a narrow-band filter at 320 nm [24]. Such photographs reveal that the sky looks almost uniformly bright even when clouds are present; the clouds disappear into a uniformly hazy sky. Only the sun stands out, as would be expected from the plots of Figures 4 and 5. When the sun is near the horizon and can be looked at without great discomfort (i.e., at Z = 75°-90°), the UVR effective irradiance is again of the order of 0.1 $\mu W/cm^2$, e.g., about 0.3 to 0.4 $\mu W/cm^2$ at an elevation angle of 12° to 15°. When the sun was at the horizon, the greatest UV exposure was obtained not by aiming the detector directly at the sun, but by directing the detector to zenith.

The greatest ocular exposures result from ground reflections from snow, followed by sand, concrete pavements, wooden walkways, boat decks, and water, and not from the horizon sky or from green grass or foliage. In fact, the reflected levels of UVR from light sand should be sufficient to cause a threshold photokeratitis within exposure periods of six to eight hours.

Ground reflectance measurements in the 295-320 nm band, with the sun at high elevation angles, revealed a surprisingly large range of values. While snow has the highest reflectance values of any natural surface (50-85%), sand reflectance values were in the general range of 8-15%, whereas grass and humus were in the 1-2% range. Hence, the ground reflectance in the UV plays a major, if not dominant, role in determining the ocular exposure to UVR. Table 2 summarizes the measured reflectance values. The presence of a tree-line or hill would also reduce the horiontal sky measurements to values reflected from the ground.

Spectacle Lenses

In addition to the outdoor skylight measurements, the UV-B transmittance of several spectacle lenses were made using the IL-730 radiometer. Measurements performed at noon in Edgewood (Z = 25°) showed transmittances of 4.4 % for a 2-mm thickness of crown glass, 1.1 % for a clear, 2-mm thick CR-39 plastic lens, 2.9 % for a 2-mm thickness of clear polycarbonate and < 1.0 % for a darkened 2-mm thick Corning Photogray-Extra lens. These figures can be considered only approximate because of the lack of close agreement between the IL 30 response curve and the ACGIH TLV.

CONCLUSIONS

The use of global UVR measurements for ophthalmic epidemiological studies can lead to serious errors. Ground reflectance plays the greatest role in determining UVR exposure of the eye and shaded skin surfaces (e.g., beneath the chin). Representative grasses and soils were only 1-2 % reflective in the UV-B spectrum, whereas snow was 50 - 80 % reflective in this spectral region. With the sun near zenith, the reflectance from open water was about 3.3 - 8 %, and from sand and concrete, about 10 %. Hence, while grassy terrain reflections account for less ocular exposure to UV-B radiation than the horizon sky, sand and concrete are by far the greatest contributors to ocular exposure. In most east-coast settings, tree-lines limited horizon-sky exposure significantly. In the Rocky Mountains, the mountain-side covered much of an individual's field-of-view and limited exposure. Only in the plains, atop a mountain ridgeline, in desert regions, or out on the sea does one typically have a significant exposure from an horizon sky. Although overhead UVR measurements at high mountain elevations were significantly above those found in sea-level environments, the ocular exposure would often be less due to low ground reflectance. Most people in mountainous regions live in a valley or aside a mountain--not on the peak where the horizon is in the center of view. Residents of non-mountainous arid regions where grass is not present apparently experience the greatest UV-B doses. Thus one is left with the conclusion that local UVR measurements are necessary along with a study of the geometry of exposure for any careful epidemiological study. Some general exposure factors are given in Table 3.

TABLE 3. UVR Exposure Factors Useful in Estimating Ocular Exposure.

1. Using the global UVR measurements presented in previous studies [39-43], multiply those values by the following exposure values to extimate ocular exposure.

2. Because of the wide variation in atmospheric characteristics and ground reflectances, these factors should be considered approximate. The use of ocular UVR dosimeters should provide more accurate values for future epidemiological studies.

Condition	Exposure Factor	
	No Hat	Brimmed Hat
<u>A. Environmental</u>		
Snowfield	0.8	0.45
Desert (sand only)	0.40	0.16
Desert (sage, cacti, etc)	0.25	0.12
Grasslands	0.16	0.05
Boating	0.25	0.10
Built-Up Urban Areas, No trees	0.20	0.08
Forest	0.001	0.001
<u>B. Use Factors</u>		
Wearing sunglasses		0.01 - 0.1
Exposure limited to \ 2 hours of local noon		0.9
Exposure in morning/afternoon; indoors \ 2 hours of noon		0.1
Always looking away from the sun		0.7
Always looking toward the sun with squint		0.2*

* Very difficult to estimate. Needs further research.

NOTE: To use this table, multiply use factor by environmental factor. This product is then multiplied by global irradiance or exposure for final estimate of ocular irradiance or ocular exposure dose.

Since ocular exposure to most of the cornea and lens is strongly affected by the degree of lid opening, it is worthwhile to point out that the lids are wide open only when the sky is overcast or when the head is turned away from the sun over a grassy plain. Hence, the values in Figure 3 are particularly important. The eye is actually exposed to an irradiance E-eff ranging from 0.03 to 0.5 $\mu W/cm^2$ when the lids are open and the sky is overcast or the sun is not in the field-of-view during midday hours when a hat is not worn. These values agree with the known threshold exposure doses for human photokeratitis when integrated over a daily exposure. Remember that most UV-B exposure occurs in the middle of the day, and the hours near sunrise and sunset do not contribute to the total exposure, unless one faces into the sun above the horizon.

An understanding of the UVR absorption properties of the lens as a function of wavelength should provide important clues to the etiological role of UVR in different forms of cataract. If a cataract were limited to changes near the anterior capsule, then a physical factor such as radiation absorption or temperature elevation which was highly localized there could reasonably be studied as a potential etiological factor. For example, Pitts found that acute UVR induced cataract appeared in the anterior subcapsular region where the effective wavelengths were highly absorbed [10].

The UV-B measurements in this paper permit one to estimate the relative exposure of the most exposed skin surfaces (nose, tops of the ears and forehead) to the lesser exposed areas such as the bottom of the chin. By comparing the incidence of skin cancer in these two areas, one can estimate more accurately the "threshold" radiant exposure for human skin cancer. A further insight could be gleaned from the study of skin cancer incidence of persons spending most of their life outdoors on sand/concrete/asphalt surafaces, vs grassy surfaces.

ACKNOWLEDGEMENTS

The measurements of UVR in many locations would not have been possible without the aid of several colleagues at the U.S. Army Environmental Hygiene Agency who aided me in performing measurements at Edgewood and in various locations around the US while travelling on other assignments. These other colleagues were: P. Conner, W. P. van De Merwe, and T. L. Lyon. Gladys Harrison made helpful suggestions and corrections in editing and typing the manuscript.

REFERENCES

1. Henderson, ST: Daylight and Its Spectrum. American Elsevier Publishing Co., New York, 1970.
2. Kostkowski HJ, Saunders RD, Ward JF, Popenoe CH, and Green ASE: Measurements of Solar Terrestrial Spectral Irradiance in the Ozone Cut-Off Region, Self-Study Manual on Optical Radiation Measurements, Part III, - Applications Chapter 1, NBS Technical Note 910-5, National Bureau of Standards, Washington, DC, 1982.
3. Scotto J, Fears TR, and Gori GB: Measurements of Ultraviolet Radiations in the United States and Comparisons with Skin Cancer Data. U. S. Department of Health, Education and Welfare Publication No. (NIH)80-2154, 1980.
4. Mercherikunnel AT, and Richmond JC: Spectral Distribution of Solar

Radiation. NASA Technical Memorandum 82021, Greenbelt, Goddard Space Flight Center, Sept. 1980.
5. Garrison LM, Murray LE, Doda DD, and Green AES: Diffuse--direct ultraviolet ratios with a compact double monochromator. Appl Opt. 17(5):827-835, 1978.
6. Doda DD and Green AES: Surface reflectance measurements in the UV from an airborne platform. Part 1. Appl. Optics 19(13): 2140-2145, 1980.
7. Diffey BL: A comparison of dosimeters used for solar ultraviolet radiometry. Photochem Photobio 46: 55-60, 1987.
8. Giese AC: Living with Our Sun's Ultraviolet Rays. New York, Plenum Press, 1976.
9. Urbach F, and Gange RW: The Biological Effects of UVA Radiation. New York, Praeger Publishers, 1986.
10. Parrish JA, Anderson RR, Urbach F, and Pitts DG: UV-A, Biological Effects of Ultraviolet Radiation with Emphasis on Human Responses to Longwave Ultraviolet, New York, Plenum Press, 1978.
11. Pirie A: Photo-oxidation of proteins and comparison of photo-oxidized proteins with those of cataractous human lenses. Isr Med Sci 8:1567-1573, 1972.
12. Lerman S: Human ultraviolet radiation cataracts. Ophthalmic Res 12:303- 314, 1980.
13. Taylor HR, West SK, Rosenthal FS, Munoz B, Newland HS, Abbey H, and Emmett EA: Effect of Ultraviolet Radiation on Cataract Formation. New England Journal of Medicine 319: 1429-1433, 1988.
14. Hiller R, Giacometti L, Yuen K: Sunlight and Cataract: An epidemiological investigation. Am J Epidemiol 105:450-459, 1977.
15. Wright RE: The possible influence of solar radiation on the production of cataract in certain districts of southern India: a preliminary investigation: Indian J Med Res 24:917-930, 1936.
16. Young JDH and Finley RD: Primary spheroidal degeneration of the cornea in Labrador and Northern Newfoundland. Am J. Ophthalmol 71(1):129-134, 1973.
17. Hollows F and Moran D: Cataract--the ultraviolet risk factor. Lancet 2:1249-1250, 1981.
18. Goldsmith R, Rothhammer F, and Schull W: The multinational Andean genetic and health program. III. Ophthalmic disease and disability among the Aymara. Bull Pan Amer Health Organiz 13:58-65, 1980.
19. Vines AP: An epidemiological sample survey of the highlands, mainland and island regions of the territory of Papua and New Guinea. Port Moresby, New Guinea, Government Printer, 1967.
20. Seung WS: A survey of senile cataracts among high altitude living Tibetans in Chang-du district, Tibet. Chinese J Ophthalmol 15:100-104, 1979.
21. Brilliant LB, Grasset NC, Pokhrel RS, Kolstad A, Lepkowski JM, Brilliant GE, Hawks WN, and Pararajasegaram R: Associations among cataract prevalence, sunlight hours, and altitude in the Himalayas. Am J Epidemiol 118(2):250-264, 1983.
22. Sliney DH: Physical Factors in Cataractogenesis: Ambient Ultraviolet Radiation and Temperature. Investigative Ophthalmology & Visual Science 27(5):781-790, 1986.
23. Zigman S, Datiles M, and Torczynski E: Sunlight and human cataracts. Invest Ophthalmol Vis Sci 18:462-467, 1979.
24. Sliney DH: Eye protective techniques for bright light. Ophthalmology 90(8):937-944, 1983.
25. Livingston W: The landscape as viewed in the 320 nm ultraviolet. J Opt Soc Am 73(12):1653-1658, 1983.

ANATOMY AND PHYSIOLOGY OF THE EYE

Alessandro Checcucci

CNR - Istituto Ricerca Onde Elettromagnetiche
Via Panciatichi 64
50127 Florence, Italy

The eye is a sensory organ which allows the reconstruction in our brain of the images of the surrounding reality. For this purpose the eye is structured as a dioptrical apparatus (cornea and lens) which focalize the light becoming from external objects on the retina, the sensitive portion which in turn transforms the incident light in nervous pulses, successively transmitted to the visual cortex in the brain.

THE STRUCTURE OF THE EYE

The eyeballs are lodged in the *periorbital cavities*, immediately underneath the frontal bone. An orbital fat pad surrounds the eyeball protecting it from the shocks. Seven muscles allow to orient the eye in almost all directions. The brows and eyelids, with their lashes and mucosa conjunctiva, shelter externally the eye.

The eye is essentially a quasy-spherical sac, whose walls are opaque except in the front (*cornea*). The eye, filled with transparent media, is divided by iris in two parts: the *anterior chamber* immediately behind the cornea, and the ring-shaped *posterior chamber*.

Rays of light coming from the exterior are collected and refracted by the cornea. This is the anterior portion of the *sclera*, the outer coat of the eye. The sclera, surrounded by the capsule of Tenon, is formed of collagenous and elastic fibres, arranged to minimize the effects of changes in intraocular pressure and external forces. The thickness of the sclera varies from 0.5 mm at the equator to 1 mm at the back where the optic nerve enters, and in front where the sclera blends in the cornea. The stiffness of the sclera, together with the internal pressure, maintains the spherical shape of the eye. The inner surface of the sclera has a brownish colour due to pigment cells.

The *limbus*, the transition zone between the clear, smooth and transparent cornea and the opaque sclera, blends also with the *conjunctiva*, a membrane covering the internal surface of the eyelids and the anterior portion of the sclera. The limbus is well supplied of blood and lymphatic vessel. The cornea is formed by a proliferating epithelium, separated from a fibrous connective tissue, the stroma, by the Bowman's membrane. The regular arrangement of the stromal fibres is very important to keep the corneal transparency. The Descemet' membrane divides the stroma from the innermost layer of the cornea, the endothelium, wetted by the *aqueous humour* a liquid contained in the anterior chamber. This space is limited anteriorly by the cornea and posteriorly by the lens and by a complex diaphragm composed of the iris and the ciliary body.

The corneal epithelium has a very high metabolism also in order to assure a likewise elevated turnover of reproduction, due to the short cell lifetime (about 48 hours). The corneal surface is protected by a film of tears which are rich in mucin. The continuous *blinking reflex* distributes this protein on the corneal surface forming an hydrophilic coat upon which tears can easily glide. The *lacrimal ducts* shed over the cornea the tears provided by the *lacrimal gland*. The surplus tears are eliminated by evaporation and by the pumping action of the lacrimal passages (*lacrimal sac, canaliculi, nasal duct, nasal cavity*). The rates of tear production and elimination are normally so well balanced that the eye is kept moist without any overflowing of tears or dryness.

The *iris* is the most anterior portion of the *uveal tunic*. This second coat of the eye is divided in two segments. The posterior *choroid membrane* is a thinner and larger portion containing numerous minute blood vessels and pigment cells for supplying the photosensitive cells of the retina, and forming a dark lining on the inside of the eyeball. Toward the front end of the eyeball the uveal tunic is thickened into a special structure called the *ciliary body*. This is a series of minute, radially arranged, ridgelike processes containing smooth muscle fibres and blood vessels. Some muscle fibres are arranged radially and other are circular. Their contraction or relaxation varies the refractive power of the lens. Two inner pigmented layers secrete the aqueous humour. The most anterior portion of the uveal tunic is the iris, the pigmented diaphragm whose contraction regulate the amount of ligth entering the eye trough the *pupil*. The colour of the iris varies among individuals from dark brown to light blue or green, depending on the number and distribution of pigmented cells. The iris contains two minute sets of smooth muscle fbres: the circular, or sphincter, fibres surrounding the pupillary margin and the radiating, or dilator, fibres. The two sets are involuntary and antagonistic in function. When the environment is strongly brigth the sphincter fibres contract reducing the pupillary aperture (*myosis*) and thus the amount of light entering the eye; on the contrary, in the dark the contraction of the dilator fibres widens the pupil (*midryasis*), thus admitting more ligth in the eye. The anterior layer of the iris is a continuation of the endothelium of the cornea, while the pigmented epithelium of the iris continues as the epithelium of the ciliary body.

The *lens*, located immediately behind the pupil, is a virtually transparent elastic body of biconvex shape. The perfectly smooth surfaces of the lens are curved, the posterior surface more so than the anterior. The rind is softer than the core and this relative softness allows the lens to change shape under slight forces. The lens consists of fibrous sheets, like the layers of an onion, enclosed in a clear, elastic membrane, the *capsule*. A number of delicate tendinous threads, the *suspensory ligament* or Zinn's zonule, suspend the lens to the ciliary body. The contraction of the muscle fibres of the ciliary body relaxes the zonule and the curvature of the lens surfaces increases. Changes of lens shape enable the eye to accomodate visually to near and distant objects.

The *vitreous body* is the transparent, semisolid gel lying behind the lens and in front of the retina. The vitrous body, composed of a collagenlike protein, occupies two-thirds of the eye volume and it is firmly attached only at level anteriorly of the ciliary epithelium and posteriorly of the optic disk.

In its early development the retina is a thin outpocketing of the brain around which the eye forms. The *retina*, the innermost coat of the eye, lines most of its posterior part. The so called *ora serrata* is the anterior limit of the retina; beyond it a simple epithelium continues on the inner side of the ciliary body and the iris. In microscopic retina sections, perpendicular to its surface, ten layers customarily may be distinguished:

a. single-layered *pigment epithelium* closest to the choroid and farthest from the vitreous body;
b. *layer of rodes and cones* which are the outer parts of the visual cells, whitin which the first step of vision take place;
c. *outer limiting membrane*, a thin line formed by the supporting radial fibres of Mueller;
d. *outer nuclear layer* consisting of the bodies of the rodes and cones;
e. *outer plexiformis layer* made up of the terminal fibres of the visual cells and of the dendrites of cells that form the sixth layer;
f. *inner nuclear layer* containing the bodies of the horizontal cells, bipolar cells, and amacrine cells;

g. *inner plexiform layer* consisting of processes both of preceding cells and of the ganglion cells;
h. *ganglion cell layer*
i. *layer of axons of ganglion cells*;
l. *inner limitinng membrane* formed by the joining of the inner ends of the supporting radial fibres of Mueller.

The histological image of the retina is the result of a tight packing of the elements of the neural chain: receptors, ganglion cells and connection cells. The sensory path is formed by the *visual receptors* (rods and cones), linked with the *ganglion cells* which are in turn afferent to the brain cortex. *Bipolar, horizontal* and *amacrine* cells connect different group of sensory paths. Then, retina layers express the compacting of alternate layers of cells and fibres.

Retinal pigment epithelium with its protrusions (*microvilli*) surrounds the outer segments of rods and cones and play an essential part in the retinal metabolism as it is shown by the atrophy following the retina detachment. The histological structure of the retina varies considerably in its different regions. The most important region is the one around the visual axis. In fresh specimens the point where the visual axis meets the retina shows a yellow colouring, the *macula lutea.* Here the retina is much thicker because of the greater number of cells, except that in its centre where an excavation is formed. This central part of the macula lutea, which is called *fovea,* shows the sharpest acuity of the vision. On the contrary, the spot at which the optic nerve fibres leaves the eye, *optic papilla* or *disc*, is a blind zone because it cannot contain receptor cells. From the centre of the fovea the structure of the retina becomes cruder toward the periphery, especially close to the ora serrata.

The ratio between cones and rods is 1:20-25. The rods are elongated cylindrical cells with a narrow central part separating an outer and an inner segment. In the outer segment are found several hundred flat membrane-bound sacs which contain the visual pigment. The inner segment contains the nucleus and other citoplasmic components as the mitochondria. The far end of the rod is the connection region with the ganglion cell. The cones differ from rods because their outer segment is conical and secondly because its membrane do not form well separate flattened sacs but only a series of invaginations containing the visual pigment. Cones are more concentrated in the central part ofthe retina while rods are more abundant in the periphery. Cones are specialized for color vision and rods for vision in weak light. Finally, cones are less sensitive than rods but since the most part of cones are separately connected with the ganglion cells, they make for high acuity.

THE PHYSIOLOGY OF THE EYE

The physiology of the eye can be divided in two main part: the optical physiology which studies the formation of images upon the retina, and the vision mechanism which deals with the complex chain starting with the visual receptors excitation and ending with the reconstruction of the external images in our brain.

1. Optical physiology

The refractive power of the eye is measured in diopters, which are the reciprocal of the effective focal length expressed in meters. The total refractive power of the relaxed normal eye is approximately 59 D, the 76% of which is provided by the cornea. In fact, the refractive index of the interface air-to-cornea is 1.376.

The contraction of the ciliary muscles releases the tension of the suspensory ligaments so that the front surface of the lens becomes more curved than in the relaxed state of the eye. Consequently,the refractive power of the lens increases from 20 D to 32 D (*accomodation*). The ability to accomodate is progressively lost with in aging persons which must uses glasses for reading (*presbyopia*). When the distance between the lens and the retina exceeds the accomodation power of the lens because is too much long (*hyperopia*) or short (*myopia*), additional lenses are necessary to allow the image focalization on the reti- na. Some times a normal eye (emmetropic) can have a different radius of curvature in one axis than in another in the cornea or in the lens (*astigmatism*); also in this case correction lenses are necessary.

The visual field is the maximum angle viewed by the eye in different planes. Some anatomical features limit the extent of the visual field so that it is 50 degrees upward, 60 downward, 80 nasally and more than 90 temporally.

At least 1% of the radiation over the entire range from 400 nm to 1400 nm is transmitted troughout the ocular media, but the retinal absorption is practically limited in the spectral range between 400 nm and 900 nm. Furtherly, the so-called *aphakic* eye, that is when the lens is removed after a cataract operation, is able to respond to ultraviolet radiation between 320 and 400 nm.

2. Vision mechanism

The visual pigment *rhodopsin*, is a complex protein composed of a proteic part, *opsin*, and a chromophore which is responsible for the absorption of light in the visible spectral region. The chromophore is the *retinal*, a derivate of the vitamin A. The retinal is a chain of coniugated double bonds attached to an ionine ring at one end and an aldehyde group at the other. The chromophore can assume two stereoisomeric configurations: 11-cis-retinene and all-trans-retinene. The primary molecular event of the vision, after the absorption of one photon, is the isomerization from 11-cis to all-trans-retinene. The shape of the retinal becomes incompatible with the opsin which, after a series of conformational changes, separates from the chromophore. The all-trans-retinene diffuses in the pigmented epithelium and, after the reduction to all-trans-vitamin A, is reconverted by light and enzimatically into the 11-cis form. About ten milliseconds after the lighting onset, an iperpolarization of the outer segment membrane occurs and the Na^+ permeability increases. A sequence of elettrophysiological steps, yet partially unknown, causes generator potentials which then give rise to the firing of action potential in optic nerve fibres.

The visual cortex is the region of the brain containing the central system which rebuilds the images by processing the nerve pulses originated in the visual receptor of the retina. However, there are other intermediate processing centers, also in the retina itself, represented by the connection cells.

The absorption spectrum of the rhodopsin shows a peak at about 500 nm which is the most effective wavelength for human vision in a dark-adapted eye exposed to weak light (*scotopic*, or nightime-rod vision). In strong light (*photopic*, or daylight-color vision) this peak is shifted to about 550 nm (Purkinjie shift).

It is well known that the eye is able to dark adapt, that is the sensitivity threshold decreases so that the person can discern some objects, invisible at first. The dark adaptation process goes trough two times: a first shorter period corresponding to the cone adaptation for the photopic vision and a second longer period during the which the rod adaptation lowers the threshold of the scotopic vision.

LITERATURE

Peyman,G.A., Sanders,D.R. and Goldberg,M.F., 1981, "Ophtalmology" V.B.Saunders, Philadelphia

Stieve,H., 1984, "The molecular mechanism of photoreception Springer, Berlin

ANATOMY AND PHYSIOLOGY OF THE SKIN

Alessandro Checcucci

CNR - Istituto Ricerca Onde Elettromagnetiche
Via Panciatichi 64
50127 Florence, Italy

INTRODUCTION

The human body is protected by the skin, a continuous covering which at level of the natural orifices (eyelids, nostrils, lips, urinary meatus, vulva, anus) gradually changes its structure to assume the histological characteristics of the internal mucosa. The functions of the skin are essential for the maintenance of life so that it can be considered a true and complex organ. In an adult standard man (70 kg) it comprises 4% of the body weight, while its surface extends for 1.6-2 m^2. The skin thickness varies from 10 μm on the eardrum, to 1 mm on the eyelids, to 3-4 mm on the nape. Generally, it is more thick in some particular regions as back, palm, sole. Skin is elastic, and its smooth, soft and pliable surface is broken by intersetting lines, folds, ridges and pores. It is well known the uniqueness of the palmar ridge patterns for each individual, which allows the personal identification by means of the fingerprint patterns. The lost of elasticity, and an increase of folds and ridges are characteristic of the skin aging, together with the appearance of brownish stains. The colour of the skin is largely variable according to the race, sex, age, vasodilatation, effort, emotional state, natural and artificial pigment changes, health state and all possible combinations of these different conditions.

In all the vertebrates the skin consists of three major layers: *epidermis*, the thin outer layer and *dermis* or *corium*, the thick inner part. A third layer, *subcutaneous*, lies under the dermis. Several cutaneous appendages are lodged in the skin: *pilosebaceous units* (*hair follicles* and *sebaceous glands*), *nails*, *sweat glands*. Skin contains hairs and down everywhere, except in foot sole, hand palm, prepuce, labia minora and other small areas.

EPIDERMIS

The epidermis is an extended epithelial layer lodging a number of other cell populations of different embrionic origin. The average thickness of the epidermis is only about 0.1 mm, although it may be 10 to 15 times thicker on the palms and soles, or where subjected to rubbing and pressure.

The epidermis consists of two major layers: the outermost *stratum corneum* (or *horny layer*) and the inner *malpighian layer*. Histologically this last one can be farther subdivided in three portions: a. *germinal basal layer*; b. *stratum spinosum*; c. *stratum granulosum*. Actually, the epidermis is the result of a dynamic equilibrium between cell proliferation in the germinal layer and cell loss of the stratum corneum. Layer differences mainly express the modifications undergone by *keratinocytes*, the epithelian cells continuously proliferating in the monostratified basal layer.

Light, Lasers, and Synchrotron Radiation
Edited by M. Grandolfo *et al.*, Plenum Press, New York, 1990

In human skin about 10 % of keratinocytes of the basal layer is self reproducing daily. The topography of mitosis, however, is not casual because a well-defined vertical organization of keratinocytes in column has been demonstrated, more evident in upper layer. This stacking has led to the concept of an epidermal proliferative unit. About ten basal cells maintain each column and it has been estimated that one cell per day enters each column, while at the same time a cell is lost from the surface. A continuing growth-promoting influence is exerted by the dermis as by providing a substratum of suitable texture, as well by producing factors which promote keratinocyte proliferation.

As keratinocytes move upwards, their shape changes drastically. The cylindrical cells of the basal layer become polyhedral in the stratum spinosum, progressively flatten towards the granular layer and, finally, end as anucleate cornified lamellae in the horny layer. Significant variations of citoplasmic constituents go together with these shape changes. In the germinal layer, keratinocytes produce filamentous proteins (*tonofilaments*) which are loosely arranged in bundles (*tonofibrils*) surrounding the nucleus parallely to the longitudinal axis of the cell cylinder. Going upwards, tonofibrils become more numerous and compact until form in the stratum corneum a mesh of fibrillar structures embedded in an amorphous electron-dense matrix. By different approaches it can be demonstrated that keratin macromolecules of the stratum corneum is directly related to the constituents of the tonofilaments of the viable epidermal cells. Together with tonofilaments, keratinocytes form other two cellular consituents. *Keratohyalin*, a histidin-rich protein, is deposited in the cells of the stratum granulosum, within and around tonofilaments bundles, as an amorphous, highly electron-dense substance. The granular aspect and the basophilia of the granulosum stratum are given by the keratohyaline granules, which appear associated with ribosomes. The chemical structure and the function of the keratohyalin are not yet well known, eccept the probable contribution to the formation of the amorphous and electron-dense matrix of the cornified cells. *Keratinosomes* or *Odland bodies* are the third product of keratinocytes synthesis. They are formed in the upper spinosum and granulosum layer and they appear as membrane-delimited lamellated granules. Their components are phospholipids, nonpolar lipids and proteins and a number of hydrolitic enzymes. Keratosomes disharge their lamellar contents into the intercellular space.

Horny cells lack of nucleus and other cell components and appear as an amorphous matrix containing packed fibrils. A reinforced membrane surround the cell. This final aspect is the result of two different processes: synthesis and degradation. During the first phase of differentiation, keratinocytes form tonofilaments, keratoyalin, Odland bodies; during the successive phase of degradation the cytoplasmic components are dissociated whereas a fibrous-amorphous complex is formed. Some cell components as the ribosomal system, the endoplasmic reticulum and Golgi apparatus as well an effective phagolysomal system can promote, maintain and regulate cell evolution.

Contacts between cells are kept by means of: a. *thight junctions*, prevalently located in the upper levels of the stratum granolusum; b. *gap junctions* of ionic and metabolic nature; c. *desmosomes* which are areas of enhanced adhesiveness. The last two kind of junction are present at all levels. These contacts are formed across the intercellular space, which extend throughout all layers surrounding the keratinocytes. About 20 nm is the rather constant width of the intercellular space. The intercellular substance is a material rich in glicosaminoglycans with an high capacity to bind water so that its permeability for water-soluble compounds is equally elevated. However, in the upper layers of the stratum granulosum the lamellar masses (see above Odland bodies) occlude the intercellur gap. It is evident that the transformation of viable into cornified cells go together with a complete sealing of the intercellular space to control the passage of water and water-soluble substances across epidermis.

In the epidermis are contained other cells different from keratinocytes: melanocytes, Langherans cells, Merkel cells. *Melanocytes* are confined to the basal layer. They sinthetize *melanin* and transfer it as pigmented granules (*melanosomes*) to the keratinocytes throughout dendritic processes. Melanosomes, following the cell migration, are distributed in the whole epidermis. Melanogenesis is stimulated by UV absorption in the skin the browning of which is a protection against sun over exposure. *Langerhans cells* are dendritic cells located in the basal layer. They form a self-maintaining population both by intraepidermal mitosis and migration from connective tissue. Some author thinks that they could have a role in the control of keratinocyte proliferation and their organization in vertical columns, but it seems now established that they are involved in immunologic reactions of the skin. Langerhans cells would be a special type of macrophage acting as the initial receptor for external antigens and their vector towards limphocytes to initiate the cutaneous immune response. *Merkel cells* are mainly present in the distal glabrous skin of the fingers, in the lips, gingiva and palatal rugae. They are thought to function as a specific, slowly adapting, mechanoreceptor.

DERMIS

The dermis, which is interposed between the epidermis and the subcutaneous fat, is a connective tissue composed of fibrous protein embedded in an amorphous substance. Its tickness can vary following the mechanical stresses undergone by the skin. The dermis contains the epidermal appendages, several kinds of cells, and is traversed by the blood vessels, lymphatics and nerves. Two major strata can be distinguished in the dermis: the *papillary layer*, which is strictly touching the epidermis being an exact replica of its ridges, and the underlying *reticular layer*. In all the two layers there is a three dimensional network of *collagen* fibres. The network is much denser in the reticular layer, the fibres of which are also thicker. Between the meshworks there are the *elastic* and the *reticulin* fibres. *Fibroblasts, mast cells* and istiocytes are the typical cells of the dermis, but other several kinds of free cells are present: *Schwann cells, limphocytes, pericytes* and *endothelial cells.* Fibroblasts are mainly generators of fibres. Mast cells produce all the substances (histamin, eparin and probably serotonin) responsible of dermal defense reactions during flogistic or allergic processes. Istiocytes are phagocitic cells.

A continuous tridimensional network of blood vessels, connected with the subcutaneous blood system, pervades the dermis. Following the skin characteristics, the form, number, distribution of the blood vessels are strongly different in order to assure in every situation the blood circulation. Particularly important are the arteriovenous connections which act as an openshut stopcocks and allow the dermal vasculature to adapt the blood circulation to the requirements of the thermoregulation. Together with the blood vessels a sistem of lymphatics is developped in the dermis.

Two main nervous components are present in the skin: somatic sensory nerves and autonomic motor nerves. Sensory fibres, which are the peripheral extension of the nervous cells located in the spinal ganglia, transfer to the central nervous system the impulses originated from a variety of peripheral receptors. Autonomic nerves anatomically elong to the sympathetic section, but some of their fibres must be classified physiologically as parasympathetic, since their neurohumoral trasmitter is the acetylcholine. Autonomic nerves regulate the activity of blood vessels, glands and arrectores pilorum muscles.

A very intrigued question concerns the nature of the peripheral receptors and the mechanisms of the cutaneous sensations. According to the Max von Frey's theory of specific nerves energies, there are different kinds of end organs which convey different sensory impulses throughout their afferent fibres. In other words a specific receptor would exist for every kind of sensation. On the contrary, other authors believe in pattern responses, that is the specificity is due to the tissues containing the nervous terminations. The elementary sensations of the skin are touch, cold, warmth, pain, but other sensations exist which are a combination of the first ones. For instance, itch, roughness, and viscidity are surely combined sensations. An excess of stimulation can transform a sensation in an other one as in the case of high temperature which can evoque pain. Furtherly, it has been demonstrated that the cutaneous sensibility is different following the body parts or the sex. Finally, it is a fact that thermal sensations can be perceived without any esternal variation of the temperature: cold during fear and fever, warmth in moment of passion. Histological studies showed a number of nervous endings, some of which are, probably, only artifacts. Surely, we can recognize a superficial dermal nerve network and the end organs. The *dermal nerve network* is located between the papillary and the reticular layers of the dermis and it is formed by a feltwork of fine, wavy, horizontally oriented, nerve fibres. Among the end organs are described the *mucocutaneous bodies*, the *Meissner, Vater-Pacini, Golgi, Krause* and *Ruffini corpuscles.* All the receptors are constituted by a nervous terminal, generally in form of rolled ball, more or less strictly encapsulated.

SUBCUTANEOUS

The subcutaneous is the most deep skin layer interposed between the dermis and the fascia covering skeletal muscles. Its thikness and distribution vary following the sex, age and body regions. It is more developped in the woman especially at the limb roots, breast, buttoks, umbilical and pubic regions. Subcutaneous is more abundant during the infancy, while it is generally reduced in old pearsons and this reduction, together with the lost of dermis elasticity, is responsible of the wrinkle appearance. *Ipoderm*, the superficial layer of the subcutaneous, is a network of elastic fibres, the meshes of which contain richly vascularized lobes of fat. A gathering of elastic fibres, *fascia superficialis*, divide the ipoderm from the a second deeper layer where the fat lobes are less densely packed. In the thickness of the fascia superficialis are contained the *cutaneous muscles* which allow the skin to move on the underlying planes. In the human beings they are present only in the face to which let assume different expressions.

CUTANEOUS APPENDAGES

The *nails* are platelike, translucent, keratinous structures growing over the ends of the fingers and toes. Three portions are distinguished in the nail: the distal free edge, the body and the root. The last part is covered by a particular kind of epidermis, *eponichius*, in front of which is visible a curvilinear, lighter zone, the *lunula*. The free edge grows from 40 to 60 μm daily. Nails consist of densely compact, higly cornified, dead epithelial cells containing only few residues of degenerated nuclei. Nails rest on several layers of proliferating keratinocytes, the so-called *matrix*.

More than two millions of *sweat glands* occur everywhere eccept in the skin of the lips, prepuce and labia minora. They developp an estimated total tubular length of 13 km. Each glands is an independent, slender and blind duct (0.02 mm in diameter). Its proximal portion is coiled to form a ball (*adenomer*), lying in the lower dermis, while the distal portion is straight and open in a tiny sweat pore at the external skin surface (*escretory duct*). Each gland is formed of two layers: an inner epithelial lining and an outer mioepithelial coating. In the axillary cavity and around the anus, the sweat glands are different because their secretion is not the product of the cell machinery, but the removal of the apical portion of the cell. Also the earwax glands, the ciliary glands of Moll and the mammary gland can be considered of this kind although especially adapted.

Sebaceous glands are multilobated structures lodged in the dermis surrounding the hair follicle. These glands escrete a mixture of lipids, called *sebum*, which impregnates the hair and finally forms the greasy skin surface film. Only in few regions of the body (nipples, labia minora) sebaceous glands open on the outside independently from the hairs. The amount of sebum produced by the sebaceous glands varies with the age, the sex, the race and the skin type.

Pilosebaceous units are densely distibuted in almost all parts of the skin. Each unit includes, further the hair, a sebaceous gland and an erector muscle. In the armpits, perineal and pubic regions, also sweat glands are part of the unit. Hairs are an invagination of the epidermis (*hair canal*) the inner part of which (*root*) shows a strong cell proliferation and keratinization. A hair *stem*, formed of highly keratinized epithelial cells, grows progressively in the canal and finally emerges at the external skin surface. Its free extremity, *apex*, is shapened, while the portion lodged in the dermis, *bulb*, is swollen and wraps the *dermal papilla*. The dermis around the hair canal is organized to form, together with the epidermis invagination, the hair *follicle*. Small muscle bundles (*arrectores pilorum muscles*) are connected with the hair follicles. Their involuntary contraction pushes the follicles outside while the dermis forms a small wrinkel around the hair exit hole, determining the hair erection and the so- called goose-flesh. The length, thickness, shape of the hairs vary following the localization in the different regions of the body. A down is distributed practically everywhere, but in some areas (scalp, male face, pubis, axilla) hairs become more long and thick. In some individuals hairs are particularly developped also in other regions (torax, limbs). There are great individual and regional variations in types and distribution of the hairs which have alternating cycles of growth, rest and differentiation.

FUNCTIONS OF THE SKIN

The skin offers a remarkable protection against a wide range of injurious phisycal, chemical, and biological factors:

a. being nearly waterproof, the skin keeps unchanged the body fluid composition also in dry air or in water;
b. since the skin is elastic, gliding on the underneath layers, and provided of a continuous fat pad, it is a very good shock-absorber;
c. the keratinized layers of the skin, which are poor conductors of heat and electricity, serve with melanin pigment as a screen for the ultraviolet radiation;
d. the skin acts a barrier against chemicals by means of different mechanisms. Alkaline and acid substances can be neutralized by means of the free groups of cutaneous aminoacids or free fatty acids of the skin surface. Substances dissolved in aqueous solutions are repelled by hydrophobic characteristics of the skin surface;
e. the physical integrity, the relatively low pH (4.5-5), the saprophytic flora, and some enzymes are important defenses against bacterial infections of the skin. Moreover, immune defense mechanism of the skin are a well developped section of the immunitary system.

The skin is a very important component of the thermoregulatory system. Body temperature regulation is achieved by a combined action of the vascular plexus and sweat glands. A vasoconstriction stops the heat lost when the environment is cold. In the case of body overheating, a vasodilatation increases the mass of circulating blood while sweat glands take effect. Consequently, the radiative, conductive, and evaporative cooling of the skin are intensified.

Finally, it must be remembered that the vitamin D is produced in the epidermis when the skin is exposed to the sun light and that human body is in relation with the external world also by means of the sensory apparatus of the skin.

LITERATURE

Fitz-Patrick,T.B., Eisen,A.Z., Wolff,K., Freedberg,I.M. and Austen,K.F., 1987, "Dermatology in General Medicine", McGraw-Hill Co., N.York,

Rook,A., Willkinson,D.J., Ebling,F.J:G:, Champion,R.H. and Burton,J.L., 1986, "Textbook of Dermatology", Blackwell Scientific Publications, Oxford

Sliney,D. and Wolbarsht,M., 1980, "Safety with Lasers and Other Optical Sources", Plenum Press, N.York & London

BIOLOGICAL EFFECTS OF COHERENT AND NONCOHERENT IR RADIATION

Louis. A. Court and Daniel Courant

Centre de Recherches du Service de Sante des Armees
Commissariat a l' Energie Atomique
Grenoble and Fontenay aux Roses France

The knowledge of immediate and long-term biological, physiological effects of IR non coherent and coherent (laser and laser devices) light is only the way to establish permissible exposure levels and to define a good approach of medical lasers useful. The research of various biological, physiological functional behavioral criteria, which accurately reflect the changes with IR non coherent and coherent light provide generally valuable information on current, proposed safety standards.

The purpose of this paper is to establish the general approach of evaluation of the IR light effects on biological tissues; however the eye and the skin are the two biological systems most frequently affected by IR non coherent and coherent light. For all cases, the definition and the description of effects are variable with multiple parameters

- physical parameters, physical exposure conditions
- biological and physiological parameters, with response criteria, date of observation, accuracy and sensitivity of methodological approach
- biological data analysis and statistical evaluation.

I - GENERAL CONSIDERATIONS

I-1 Definition

The International Commission on Illumination (C.I.E) divides the optical spectrum into seven photobiologically significant spectral bands and infra red into three bands :

- IR-A Wavelength $760 \leqslant \lambda \leqslant 1400$ nm
- IR-B Wavelength $1400 < \lambda \leqslant 3000$ nm
- IR-C Wavelength $3000 < \lambda \leqslant 10^6$nm

NON COHERENT IR SOURCE	COHERENT IR SOURCE
In all conventional sources of light the atoms of the source emit light	In a laser, the atoms are stimulated to emit light quanta which have nearly

Light, Lasers, and Synchrotron Radiation
Edited by M. Grandolfo *et al.*, Plenum Press, New York, 1990

independently of one another.	the same frequency, the same direction and propagation and the same phase.
The light is emitted : . at all wavelength of the spectrum . with generally wavelength of visible spectrum . in all possible phase directions	The light is emitted : . with great selectivity and with same wavelength monochromatic . with a well defined phase relationship through space and time

I-2 Sources

The IR sources are multiple and of two types, natural and artificial. Sun and many biological structures or organisms are IR natural sources and generally IR spectral irradiances are characteristic of biological tissues or living organisms and their normal metabolism or conditions. These sources are emmitting non coherent light and their spectral irradiance are variable. All coherent sources are artificial. The table I give some examples of IR sources.

TABLE I

WAVELENGTH nm	COHERENT SOURCE	NON COHERENT SOURCE	EYE	SKIN
NEAR INFRA-RED				
IR-A (700-1400)		VARIABLE WITH SPECTRAL IRRADIANCE		
700-820	ALEXANDRITE			
850-912	DYES LASER	. SUN		
		. TUNGSTEN HALOGEN LAMP		
850	GALLIUM ALUMINIUM ARSENIDE	. XENON ARC LAMP		
		. XENON HIGH PRESSURE	CORNEA	
905	GALLIUM ARSENIDE	. QUARTZ HALOGEN	LENS	
		. SUN LAMP MERCURY	RETINA	
1064	NEODYME:YAG	. SODIUM HIGH PRESSURE		
1318-1338	NEODYME:YAG	. CARBON ARC		
		. PHOTOCOAGULATOR		
		. HEAT IR LAMP		
FAR INFRA-RED		. INCANDESCENT MATERIALS		
- IR-B(1400-3000)				
1540	ERBIUM			
2060	HOLMIUM			
2600-2900	MULTILINE HYDROGEN FLUORIDE (HF)		CORNEA	
- IR-C $\geq$ 3000				
3600-3900	MULTILINE DEUTERIDE FLUORIDE (DF)			
5000-7000	CO			
10600	CO_2			

I-3 Physical exposure conditions

1. The light source

The output of the light source is quantified in terms of ENERGY

RADIANT ENERGY - Q -	RADIANT POWER ϕ [W]
with units of Joules J	with units of watt W
or calories C	for continuous source
for pulsed source	

2 The irradiated area

The concentration of energy or power incident on irradiated surface is:

RADIANT EXPOSURE	IRRADIANCE
with units of $J.cm^{-2}$	$W. cm^{-2}$
or $cal.cm^{-2}$	
$dH = \frac{d\phi}{dA}$	$dE = \frac{d\phi}{dA}$

3 Nature of illumination

The notion of direct, reflected, diffuse illumination must be introduced. The effects differ for point source (direct viewing of collimated beam) and extended source (diffuse reflexion) or diverging beam (beam divergence) and with the absorption, transmission, scattering coefficient of irradiated area.

4 Spectrum of light

This is the notion of continuous spectrum or wavelength bandwidths

5 Aperture

The notion of aperture is function of the nature of irradiated area and the interaction mechanism. The limiting aperture define the diameter corresponding to the circular area over which the beam irradiance should be averaged as :

WAVELENGTH (nm)	SKIN (mm)	CORNEA (mm)
760 - 10000	1	1
> 10000	11	11

Pupil size 3mm for diurnal conditions
7mm for nocturnal conditions

6 Time

The parameter time is the distribution of energy over time :

.the duration of exposure pulsed or continuous illumination, if $t \geq 150ms$

.for pulsed illumination (single or multiple pulses) the pulsewidth, the pulse repetition frequency (nb pulses in time unit) PRF

.the time between two exposures, for prolonged and repetitive exposure with the notion of total exposure time (chronical or long term exposure)

.the evolution of source power during exposure, with the particular references for thermonuclear exposure and the evolution of total radiant energy, color temperature source, fireball area and spectrum source during exposure time

7 Coherence

The specific problem of the influence of coherence is the difficult to analyse globaly and in a first appreciation, only physical exposures conditions. Radiant energy or radiant power and radiant exposure or irradiance are the most important paremeters. (cf figure 1, Ham et al.)

For the law levels exposure, it is possible that the coherence of light has an influence but the demonstration of mechanisms is not clear. (long term exposure of monkey's retine and low level exposure of culture cells).

I-4 Biological parameters

The biological parameters differ with the nature of the biological materials or tissue and with the irradiated area.

1. Eye

. Pulse size
. Eye movements - blink reflex
. Adaptation of retina
. Transmittance and absorption of cornea, lens, ocular media, vitreus and retina
. Pigmentation of iris

2. Skin

. Pigmentation of dermis.

3. Choice of animal models (rabbit, monkey) or man

4. Generally, transmittance, reflectance and absorption of tissue, but as well as mitotic index and cell metabolism.
The effects of light on retinal photoreceptors without real possibility of repair, and regeneration are not equivalent to the effects on dermis with high level mitotic index and intense repair process.

I-5 Response criteria

To evaluate the biological effects of illumination it is necessary to use several criteria and to have the same or comparable delay of observation (immediately, one hour, five hours, 24 hours or a long time, for example 1 year after last exposure). Different techniques are required to appreciate with accuracy the threshold of damage on the physiological function.

1 Visible changes

The visible criterion is the clinical aspect which will be described:
. for eye, by a clinical examination of cornea, lens and retina by (ophthalmoscopy with or without fluorescein use fluoresceinic angiography for eye fundus). The damage is characterized by presence of a very light depigmentation, an opacity or an haemorhage.
. for skin : by a minimal erythema or different burn's degrees observed after several hours and 4 to 24 hours after the exposure.

. for culture cell : by expression of the cell morphology, growth rate, chromosomic aberrations (fragments or recombination) oncogen expression, metabolism changes. In all cases, the damage and the notion of threshold vary with the nature of lesions or functional alterations.

2 Histological changes

There are apprécied with classical optical microscopic examination, transmission electron microscopy, histochemical and immunological tests.

3 Functional changes

There are apprecied without morphological changes, but rather by the effect of the exposure on physiological functions.
For eye, there are :

. measure of visual acuity
. appreciation of spectral photopic and scotopic sensitivity
. campimetry or visual field evaluation
. electrophysiological investigations (ERG-electroretinogram and visual evoked potentials)

For skin, there are :

. evaluation of pain threshold after long term repetitive low level exposure

I-6 Definition of threshold

From dose response data obtained for many different exposure conditions a probit analysis must be performed. The ED_n is the effective dose required to produce an observable response on n per cent of population. Confidence intervals about the dose response curve are calculated. From the curve data the ED_{50} (with the greatest statistical significance) is generally refered to as the damage threshold.

I-7 Maximum permissible exposure - MPE -

From biological data and after statistical analysis, the MPE for humans are derived. The MPE is obtained by dividing the ED_{50} for the worst case condition using more sensitive significant criteria by a variable factor of 2 to 100.

II - DAMAGE MECHANISMS AND RADIATION PENETRATION INTO TISSUE

IR-A wavelengths are the most penetrating in all tissues and tissue is more strongly scattering in the IR-A than in the visible. In IR-B the penetration decreases and in the IR-C is negligible (from the CO_2 laser 10,600nm the degree of penetration in most tissues is approximatively 10 nm)

The general attenuation of light in tissue is the complex combination both of absorption of light by molecules and of scattering. But tissues, with layers variable vascularisation, inhomogent distribution of pigments are not optically homogeneous and the absorption law is complex.

The depth at which 63 per cent has been absorbed (the 1/e irradiance depth) is called penetration depth; provide a relative indication of the depth of penetration by light and the effects of different wavelength; the greatest energy would be absorbed near the surface with less and less absorbed in deep.

Essentially two types of damage mechanisms appears with the IR-action: there are thermal and non thermal effects.

II-1 Thermal effects

The light energy is absorbed and for the most part converted in heat and the temperature time history determines the extent of damage: deactivation of enzyme systems, denaturation of protein, coagulation, thermal vaporization with mechanical disruption. For IR exposure, durations from 10ms to 5000 ms, a maximum temperature rise of 20 to 35°C is required to produce an observable lesion, but temperature change of less than 1°C may be involved in some functional changes.

Between 45 and 65°C the injury is due to thermal coagulation and above 65°C, near and than 100°C the process is thermal vaporization with mechanical disruption. The rate, of heat transfer, the area of irradiated tissue, the degree of cooling with vascularisation, play an important role : with a short duration of exposure and small irradiated area the cooling time may be sufficiently short to prevent any coagulation at all.

The mathematical models to appreciate the heat dissipation in several tissues (retina, cornea, lens, skin) are generally precise and an example will be described on the cornea, in ordinary light absorption, the temperature of irradiated tissue is proportionnal to radiant exposure, in in approximately a linear process.
Generally the thermal effects are predominent for the exposures ranging from a few s to a few ns (nanoseconde).

II-2 Non thermal effects

1 Mechanical effects Mechanisms for ultrashort pulses

The exposures above ns, in the ps (picosecond) range, create photo-mechanical damage with acoustic transients and shock waves. If the irradiance is extremly high (laser devices) in the range of giga (10^9) to tera (10^{12}) watts cm^{-2}, several stages appear :

. plasma formation with,
  - preionisation and production of the starting electrons after simultaneously several photo absorption
  - ionisation growth and very high increasing of population of free electrons and ions absorption phase when free electrons population exceeds a critical density.

. plasma expansion creating shock and acoustic pressure waves that mechanically disrupt tissue adjacent to the disintegrated area. The peak pressure in the shock wave associated with plasma formation may be exceed 1000 atm. The amount of irradiance required to produce plasma in a substance or tissue is known as its optical breakdown threshold : the threshold varies with several parameters, pulse duration, focal spot area, wavelength. The avalanche ionisation in the growth stage uses inverse Bremstrahlung and the absorption coefficient for inverse Bremstrahlung increases with the square of wavelength. Tissues adjacent are ruptured when stresses produced by these mechanical waves exceed the elastic limits and properties of the structures.

2 Phototoxic damage

The long term exposure at sub-thermal visible or near IR irradiation produces permanent damage to levels below those thermally hazardous. Generally the damage is:

. homogeneous over the irradiated image
. the same on all the size irradiated area
. very variable with the spectrum of wavelength.

III OCULAR EXPOSURE

Retina, cornea and lens are most frequently affected by IR.

III-1 Retina

Wavelengths between 700 and 1400 nm only are transmitted through the ocular media. The spectral transmission of human ocular media and the relative spectral absorption of human retinal pigment epithelium is characteristic (table II) (data obtained by Geeraets, W. et al. 1)

TABLE II

WAVELENGTH (nm)		% TRANSMISSION	% REDUCTION	% ABSORPTION
Visible	400	4	8	88
	500	8	9	83
	600	14	13	73
	700	18	18	64
	800	25	27	48
	900	30	40	30
	1000	36	42	22

(Absorption is total incident light minus the sum of transmission and reflection)

Within the spectral range 700 - 1400 nm, there are at least three types of retinal damage;

. at power levels too low to produce appreciable temperature rise and at long term exposure (from short wavelength light), photochemical and phototoxical damage
. from µs to s and after long term exposure, thermal damage
. above ns exposure time, mechanical damage.

The type of lesions depends of the wavelength of spectrum, power level exposure time for exposure pulse ranging ns to s, but mechanical damage are generally independant of wavelength and depend only upon sufficient energy absorption to produce sonic transient phenomena.

1 Exposure time ≧ 1s

The threshold values varie with the criterion, generally the lesion very discret opacity without haemorrhage detectable by ophthalmoscopy 1 or 5 hours after exposure. But it is important to consider the difference between lesion, real damage and functional disorders. A permanent lesion in the fovea (with the maximum AV acuity) constitutes always a severe visual decrement. The localisation of lesion elsewhere in the paramacular area will be a severe handicap, but the same lesion in the peripheral visual field will not even be noticed.

Ham W.T and al (2-3) have obtained retinal threshold in the rhesus monkey at wavelengths beyond 600 nm using a 2500 W xenon lamp with interference filters, the threshold radiant exposures on J.cm^{-2} were determined for wavelengths 820 ± 5nm and 910 ± 25 nm. Exposure times ranged from 1 to 1000 s and image diameters on the retina were 500 um. The figure 1 (radiant exposures in J.cm^{-2} plotted logarithmically against exposure times en s) compare similar data previously obtained for laser wavelengths (Nd:YAG) 1064nm (Ar-Kr) 647nm and (He-Ne)632.8nm:

- no significant difference in threshold is noted for 820, 860, 910 nm
- all three lines are straight and approximatively parallel, indicating similar type or mechanism of injury.
- the retinal radiant exposure increases with the time exposure histological examination at 24 and 48 hours post exposure show
- structural damage in the retinal pigment epithelium (RPE)and numerous pyknotic nuclei in the outer nuclear layer.
- minimal thermal lesions always smaller than the image diameter on the retina because temperature is maximal at the center of the irradiated image
- damage maximal at the center of the lesion, tapering off towards the periphery.

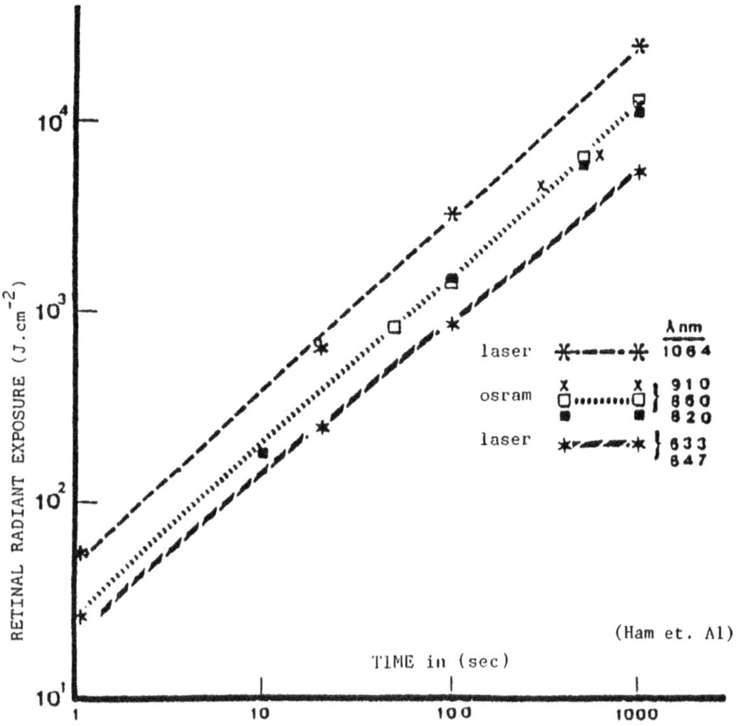

Figure 1 . Comparison of near infrared thresholds at 820, 860 and 910 wavelengths (xenon) with previous laser data at 1064, 647 and 633 nm.

In sharp contrast, the minimal photochemical lesions is fairly uniform with a definite border between normal and injured RPE cells at periphery with only a few pyknotic nuclei in the outer nuclear layer.

Generally

- the corneal power required to produce a lesion decrases by three orders of magnitude in going from 1064 nm to 441nm

- exposure to a thermal lesion at elevated retinal temperatures whereas a long time exposure to visible light produces a photochemical lesion at power levels too low to raise the retinal temperature by an appreciable amount (0,1°C) cf table III.

TABLE III

WAVELENGTH	TRANSMITTANCE	TIME EXPOSURE (s) 1s		16s		100s		1000s	
λ nm	T	HO	K	HO	K	HO	K	HO	K
514.5	0.87	14.5	25	165	18	220	4	320	10
632.8	0.93	29.9	49	243	15.2	840	8.4	5400	9
1064	0.76	56.1	55	600	37.5	3.250	32	24000	23

Sensitivity of the retina to photic damage as a function of wavelength through the monkey ocular media.
T transmittance
HO maximum radiant exposure in $J.cm^{-2}$
K maximum temperature above ambiant on the retina during irradiation in °C (as determined from the mathematical model of Clarke)

2 Long term laser exposure

Many investigations have indicated that there is a nonthermal damage mechanism associated with a production of retinal lesions from long term exposure to visible radiation. This type of damage has been produced using both laser and non laser sources The exposures of macula of rhesus monkey for 120s to Nd:YAG and argon laser radiation at wavelengths of 1060 and 514.5nm by Gibbons et al (4) show that non visible radiation produce observable retinal thermal damage.

Exposures to visible radiation produce distinguishable different lesions at power levels damage. At near-thresholds levels after exposure to 1060nm IR radiation, only pigment epithelial damage was noted, without photoreceptor acter segment involvement seen after the visible 514.5 nm near threshold exposure.

3 Exposure time $10^{-6} \leqq t \leqq 1s$

The average measured threshold irradiance $\bar{H}r$ graphed versus exposure time for several retinal image diameter present approximatively the same nature of graphs. An experimental study of Allen and al.(5) give these data:

For rabbit with the range of exposure time from 165 ns to 100s and the range of image retinal diameter from 0,053 to 1,08 mm (fig 2) and for primates respectively from 4 ms to 250 ms and 0,11 mm to 1,3 mm (fig 3) the theoretical retinal irradiance is :

$$\bar{H}r = At^B + C$$

with t = time A, B and C constant

Several conclusions can be drawn :

- no reciprocity relationship between $\bar{H}r$ and t at exposure times greater than one second
- trend toward reciprocity for $\bar{H}r$ and t at exposure durations less than one second
- definite asymptote associated with each curve and constant C represent this asymptote only for large image size C with respect to retinal image diameter is approximatively proportional to burn area.

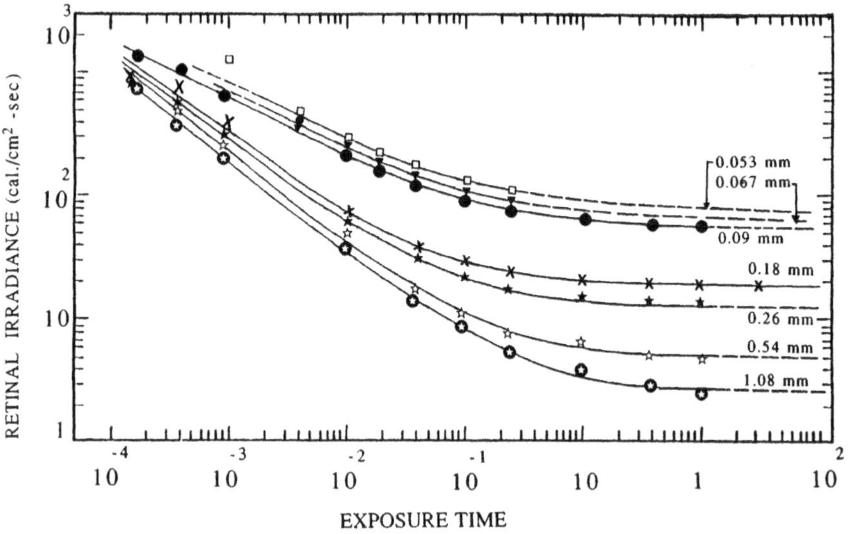

Figure 2 . Average threshold irradiance ($\bar{H}r$) versus time for seven image sizes

4 Ultrashort exposure time .. 10^{-9}s (laser)

The mechanisms of damage is different. Retinal damage thresholds after exposition ultrashort 30 ps pulses were measured by Goldman et al (6) on the monkey for 1064 nm mode-locked Nd:YAG and 532 nn with several retinal exposure diameters 25, 430, 630, 725 µm.

The table IV show the experimental data

From the data it is possible to infer for ultrashort exposure time exposure :

- the change in radiant exposure is not linearly dependant upon exposure diameter, as it would be in the thermal model.

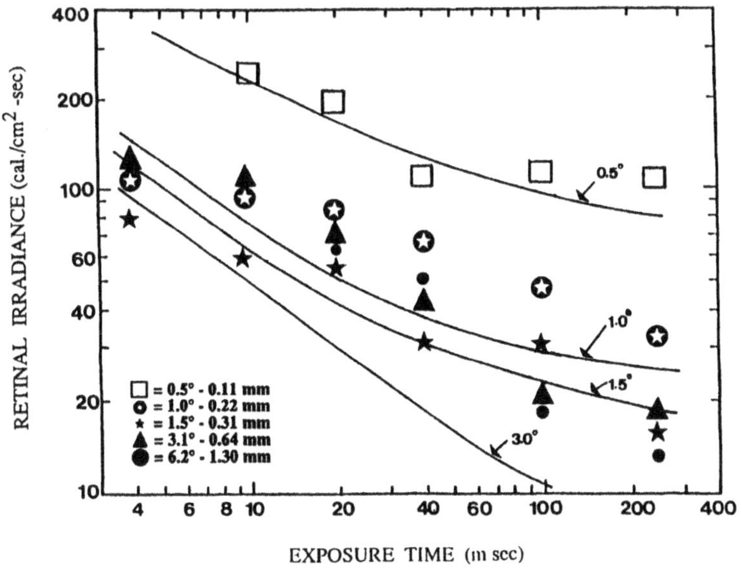

Figure 3 . Average threshold irradiance (Hr) versus time for four image sizes.

TABLE IV

SPOT SIZE µm	WAVELENGTH nm	INTRAOCULAR energy µm	RETINAL RADIANT Exposure $J.cm^{-2}$
25	532	18.2 ± 8.3	6.5
25	1064	8.7 ± 4.8	2.7
430	1064	1521 ± 330	1.6
630	1064	1017 ± 700	0.5
725	1064	1184 ± 334	0.44

(Criterion 24 hr post exposure ophthalmoscopic lesions)

- the damage are independant of image size. When the retinal image diameters varie by a factor of 30, the threshold radiant exposure level, by a factor of only 6. The general pattern of inner segment damage is the same, when the retinal diameter exposure differ by a factor of 20.
- damage is postulated to result from shock waves produced by transient heating of melanine granules. The damage threshold at 532nm appears to be slightly higher for 1064 nm. At 532 nm photochemical damage to the outer segment membranes is surimposed upon photomechanical damage from shock waves
- the other data points represent thresholds for a Q-switched Nd:YAD laser exposure with undetected mode-locked output pulses (10^{-6}–10^{-7}s) or multiple pulse exposures to trains of ~83 ps mode-locked pulses and not thresholds for single pulse exposures at the indicated pulsewidths.

A study of retinal damage thresholds by Allen R et al (7) in the rhesus monkey eye with 1064 N:YAG radiation at four pulsewidth 4, 30, 200 ns and 10 us show a flat trend of threshold with pulsewidth. (fig 4)

5 Retinal effects of multiple pulse

Many retinal damage thresholds data from multiple pulse lasers have been determined. It has postulated by Gibbon and Skeen (8,9,10) with ruby laser that cumulative effects will occur as a function of the rate of deposition of energy as determined by the pulse repetition frequency (PRF)

Thresholds determination for multiple pulses gallium arsenide (GaAs) laser in the 850 to 900 nm wavelength range in the temporal paramacular area of monkey (macaca mulatta) are shown by Ebbers (11) The ED-50 is expressed as total exposure duration rather than energy per pulse. The ED-50 vulve of 0,72 s for the 1,000 pps exposure is significantly lower than the 23.3 s for the 40 pps exposure and this results support the convention that a cumulative effect occurs as the pulse rate increases.

But generally the data for retinal damage thresholds from multiple pulse IR lasers (PRF ≧1000 Hz) are not really sure and the law of interaction for very high frequency is not known.

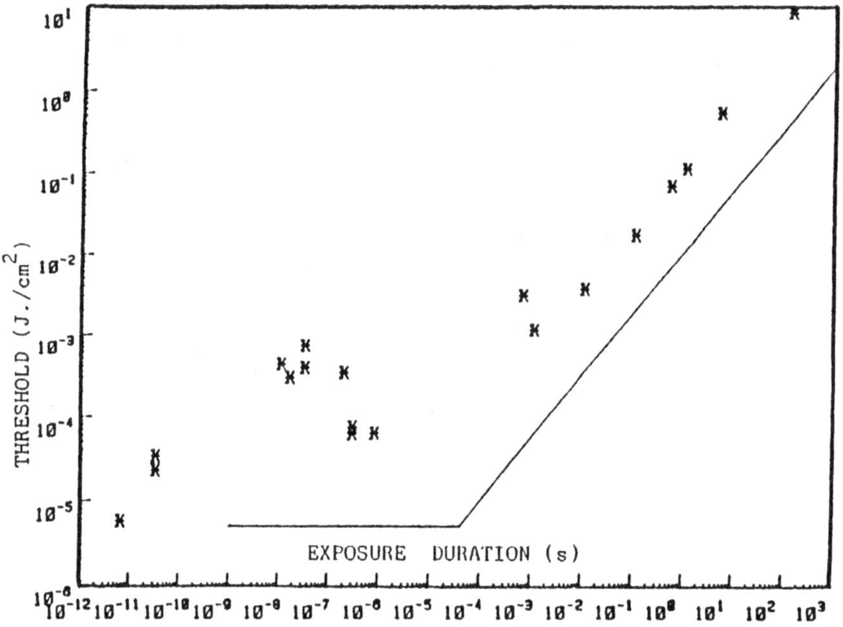

(Allen et. Al)

Figure 4 . Threshold vs pulsewidth plot of retinal damage induced by Nd laser radiation. The solid line is the maximum permissible exposure for near infrared laser radiation

6 Retinal alterations and gallium arsenide radiation

ED-50 FOR RETINAL DAMAGE IN RANGE 900 nm COMPARISON BETWEEN VISIBLE AND ND:YAD IRRADIATION

A study of Lund, Beatrice and Schuschereba (12) precises the ED-50, for single Q-switched exposure determined for 6 laser wavelengths obtained from dye laser. The criterion is the presence of visible extramacular alteration by ophthalmoscopy 1 hour after exposure in rhesus monkey. The table V give the experimental data

TABLE V

WAVELENGTH (nm)	PULSE Duration (λs)	ED 50	95% LIMITS MicroJ	SLOPE OF THE REGRESSION LINE defined as ED84/ED50
850.2	11	9.1	7.8 - 10.7	1.71
859.6	10	6.7	5.6 - 8.0	1.77
867.0	7	5.2	4.4 - 6.0	1.69
880.0	14	6.3	5.2 - 7.7	1.80
899.7	6	4.3	3.4 - 5.3	2.38
912.0	7	5.5	4.6 - 6.7	1.87

The figure 5a show wavelength dependance of the MPE and the ED 50 for retinal damage in the rhesus monkey. A minimum is seen near 900 nm. Histopathology shows at the exposure site the photoreceptors rods and cones damaged throughout their length (specially rods). It is difficult to explain the reduced ED-50 near 900 nm, with different data of optical absorption measurements of rhesus eye.

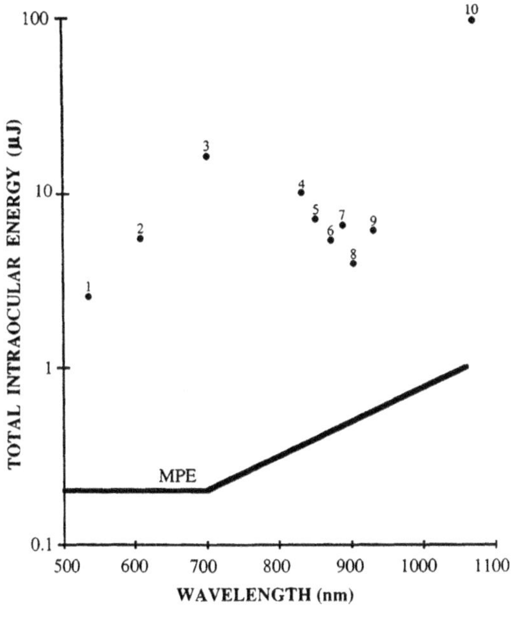

(Lund et. Al)

Figure 5a. Wavelength dependence of the MPE and the ED_{50} for retinal damage in the rhesus monkey

ED_{50} s are for : 1 - 532 nm, 140 ns; 2 - 600 nm, 400 ns; 3 - 694.3 nm, 15 ns; 4 - 850.2 nm, 11 ns; 5 - 859.6 nm, 10 ns; 6 - 867 nm, 7 ns; 7 - 880 nm, 14 ns; 8 - 899.7 nm, 6 ns; 9 - 912 nm, 7 ns; 10 - 1064 nm, 180 ns

D. J. LUND, E. S. BEATRICE and S.T. SCHUSCHEREBA : Bioeffects concerning the safe use of GaAs LASER training devices. In : COMBAT OCULAR PROBLEMS, Sup. (1982).

In a recent paper, D.J. Lund and al. present the threshold for laser chorio-retinal injury in the red end of the visible spectrum and near-infra-red (IR-A) spectral regions and demonstrate an unpredicted wavelength dependence of the injury threshold for single Q-switch pulses. Four lasers were used to determine thresholds at 40 wavelengths between 532 and 1064 nm: a Ruby laser, a Nd-Yag pumped dye laser, an Erbium and an Alexandrite laser. The figure 5b precise the experimental data for solid state lasers (1) and for dye lasers (2). There is no complete or obvious explanation for the significant variation of threshold with small changes in wavelength.

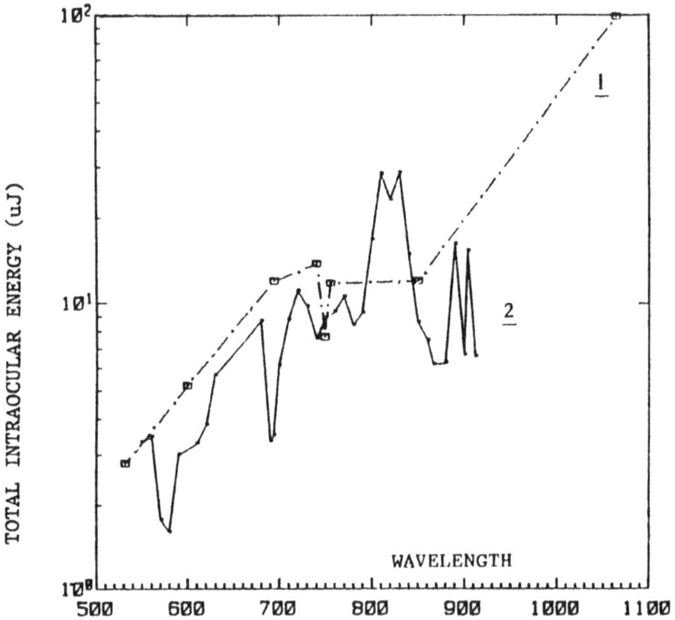

(Lund et. Al)

Figure 5b . Ophtalmoscopically determined retinal injury ED_{50} values for extramacular exposure in the rhesus monkey for the dye laser and the solid state laser. The values are for a one-hour post-exposure examination. When the dye laser data is included, any simple view of the wavelength is destroyed

LOW LEVEL

This retinal alterations are particular and did not produce the type of retinal opacity seen by fundoscopy after laser irradiation, but a pale gray clouding within 10s after initiation of exposure (Beatrice et al 13). The retinal changes are persistent 24 days after exposure, however no fluorescein leckage or histological evidence of retinal alteration appear in any of the exposure sites. The change appears only in the direct observation and none of the technique (angiography, flat preparation, serial microscopy and fundus photography) demonstrate any retinal alteration.

The change may be due to "bleaching" of retinal photopigments or to swelling of photoreceptor or intraretinal neutral layers. The table VI gives the main data.

TABLE VI

LASER Mode PRF (Average repetition rate)	1 WATT Pulses train 100 ns - 1600 Hz	10 WATT
Q p = pulse	.212 erg/p	1.64 erg/p

III-2 Cornea

IR is very hazardous for cornea
The incident energy in the spectral region beyond 1400 nm is absorbed by the outer ocular media (cornea, aquous, lens). The absorption coefficients of the ocular media vary over 3 orders of magnitude in the spectral range from 1400 to 3000 nm.

The criteria are :

- at very low exposure (at a lower exposure dose than the threshold for visible damage), a significant change in corneal curvature with precise keratometry (0,5s 2 W Cm^{-2}.10600 nm).

- generally, a small lesion observed immediatly or at 1 hour after exposure), with subsequent epithelial sloughing or depression exhibiting mild fluorescein staining :
  . lesions near the ED-50 are smaller and less than those produced at 2 times, the ED-50.
  . no lesion is observed at 24 - 48 hours that was not observed at one hour.
  . some corneal lesions observed at 1 hour are not observed at 48 hours.
  . lesions are observed up to 10 months after exposure.

- ED-50 exhibit the wavelength dependance of the damage threshold; it is the wavelength dependance of the ocular media absorption.

There is a direct correlation between the dose at threshold and the penetration depth and the wavelength dependance of the corneal. ED-50 is approximated by the shape of the absorption depth curve. The absorption coefficient of the cornea and outer ocular media approximates that of the water or physiological saline.

The figure 6 (Stuck, Lund, Beatrice (14)) shows the evolution of the ED-50 for the production of corneal lesion as a function of wavelength. The solid curve is the depth at which 95% of incident energy is absorbed in physiological saline.

Doses required to produce an observable corneal lesion in the wavelength range from 1000 nm and 2000 nm are higher than those required at 2800, 3800, 10,600nm. The corneal lesions of a near threshold exposure at the shorter IR wavelengths involve the corneal stroma without rapid repair as reported for the longer wavelength where the threshold lesions at the shorter IR wavelengths can be considered more severe.

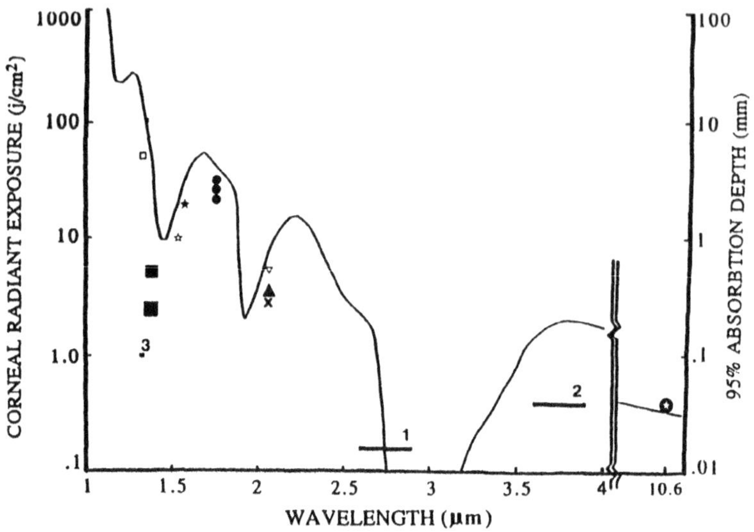

Figure 6 . The ED_{50} for the production of a corneal lesion for exposure conditions given in Table 2 as a function of wavelength. The solid curve is the depth (right hand axis) at which 95% of the incident energy is absorbed in physiological saline (the absorption properties of physiological saline approximate those of the outer ocular media). The data point at 3 is the ED_{50} for the production of a retinal lesion obtained at the 1.33 µm.

	Laser	Wavelength	Exposure duration
□	Nd	1.318 & 1.338 µm	250 µs
■	Nd (Raman)	1.41 "	25 "
★	* Er	1.54 "	50 "
☆	Er	1.54 "	930 "
●	Er	1.732 "	225 "
▽	Ho	2.06 "	42 "
▲	Ho	2.06 "	50 "
X	Ho	2.06 "	100 "
1	HF Multiline	2.6 - 2.9 "	45 "
2	DF Multiline	3.6 - 3.9 "	100 "
3	Nd	1.312 & 1.338 "	625 " (Retinal)
✪	CO_2	10.6 "	100 "

For 1,732 nm laser radiation and for irradiance diameters from 500 to 1000 um, the radiant exposure required to produce a threshold lesion decreased as the beam diameter increased (fig 7).

Generally, the diameter and the depth of the lesions are both dose and wavelength dependant. The corneal lesion after exposure to IR laser radiation is considered to the result of a thermal effect.

A biomathematical model considered to be in good agreement with experimental data has been realized to predict the corneal damage threshold from IR laser exposure. It is an analytical expression, function of absorption coefficient, duration of exposure, beam radius and damage end-point criterion Reed(15), Wray (16) and Vos (17) were the first to develop models to calculate temperature rise after non coherent exposure. The basis is the heat conduction in the ocular fundus.

Clarke et al (18) include Lambert-Beers absorption with the steady-state heat diffusion equation. Mainster et al (19), add or find difference solution for transient temperatures, adapt it to symetrical coherent sources and devises an analytical solution to the problems. Takata et al. (20) generalize the model for retina, cornea and skin.

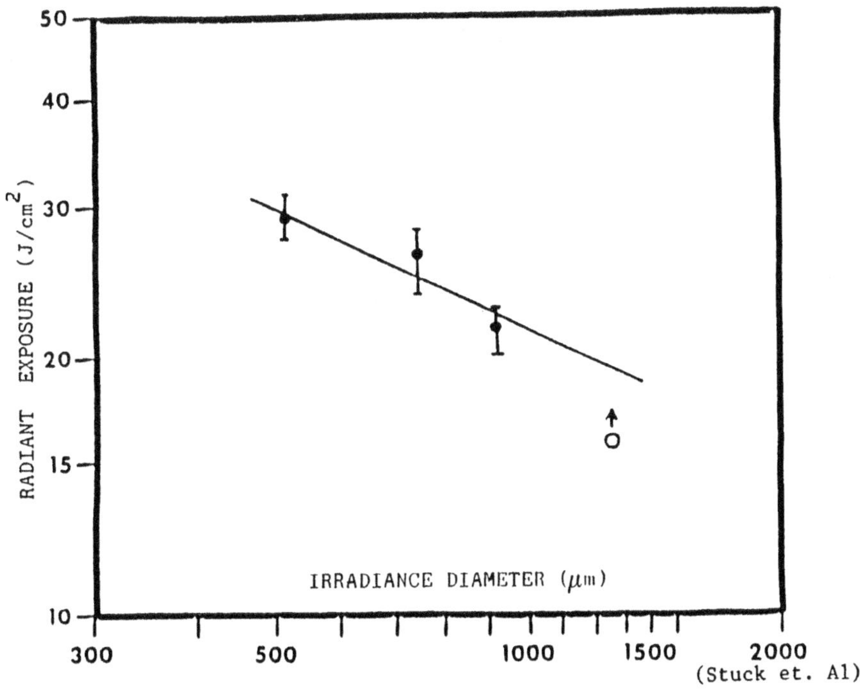

Figure 7 . The ED_{50} and 95% confidence interval about the ED_{50} for the production of a corneal lesion as a function of the irradiance diameter of the incident beam ($d_{1/e}$). No corneal effect was observed for exposures made with the 1200 μm irradiance diameter (open circle with arrow).

The model computer time-varying three-dimensional temperature distributions. The fundamental equation is :

$$\frac{1}{pC} \cdot \frac{dT}{dt} = A + K\nabla^2 T$$

$$\text{with} \quad A = -\frac{d}{dz}[Ho\,(r,t)\,\exp\,(-\,az)]$$

$$= \alpha Ho\ (r,t)\ \exp\ (-az)$$

where P = density
C = specific heat
T = temperature (degrees Kelvin)
t = time
K = thermal conductivity
∇ = Laplacien differential operator in cylindrical coordinates
A = source term
Ho (r,t) = spatial distribution of energy in the beam
z = depth in the beam axes
α = absorption coefficient of Lambert

For known damage-rate constants the model calculate

- damage thresholds directly from temperature-rise histories by a damage-rate integral method
- with critical peak temperature used as the damage criterion

For a Gaussian beam having a 1/e radius of 0.707 mm exposure durations of 10^{-8} to 10^{2}s, H the predicted threshold in $J.cm^{-2}$
$H = (200/alpha + 0,25) + (110/alpha + 2,2)\ t^{2/3} + (35/alpha + 3.3t)$
with $(200/alpha + 0,25) \geqq 0,3$ (alpha in cm^{-1} and t in s).

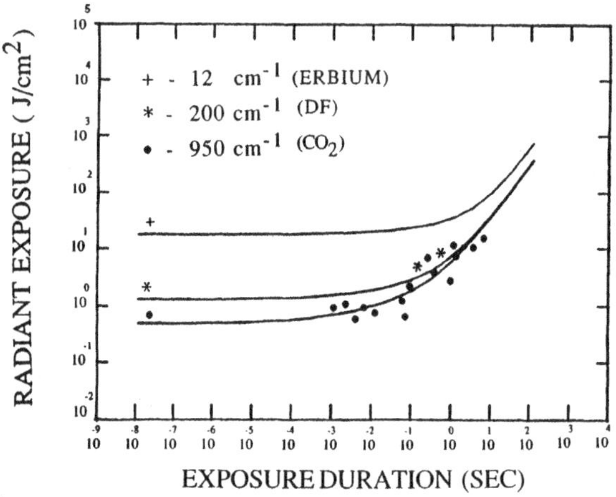

(Egbert et. Al)

Figure 8 . Comparison of experimental and theoretical lesion thresholds for three absorption coefficients. Data points are cited by Egbert and Maher for 10^{8}-10^{2}sec.

The figure 8 represents the comparison of experimental data (Egbert et al) and theorethical (equation) lesion threshold for three specific absorption coefficients and three wavelengths laser erbium, DF and CO_2 : radiant exposure in $J.cm^{-2}$ versus exposure duration in s.

The MPE for corneal injury to wavelengths greater than 1400 nm depend only on the exposure duration. These values have been based on the dose-response reported for 10,600 nm CO_2 radiation : the MPE for exposure to laser radiation at 1732 nm or 2,060 nm is the same as the MPE at 10,600 nm, even through the ED-50 differ by a factor of 10 to 100. For 10,600 nm and 10^{-3}s, the MPE is about a factor of 10 lower than the thresholds, but for Q switched pulses (10^{-9} - 10^{-7}s), the thresholds are equal to or slighty lower than MPE.

III-3 Lens and IR cataract

The figure 9 exhibit absorption of lens (rabbit) : for wavelength range between 1400 and 2000 nm, the % relative absorption is ≃ 80-90%.
Glass blower's cataract results from long time industrial exposure to low levels of IR.

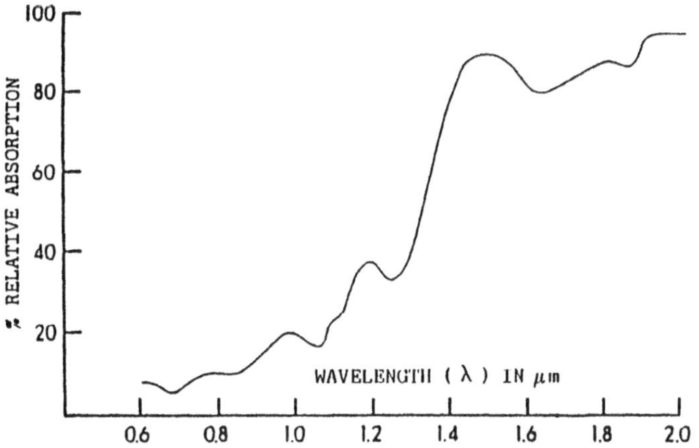

(Egbert et. Al)

Figure 9 . Absorption of rabbit lens. The absorption values are for normally incident light and normal transmitted light and are not corrected for scatter. Adapted from Campbell (1968).

The mechanism and the relation exposure time irradiance and ED-50 of this caractogénesis effect is not clear. Two theories may be formulated, in the purpose to explain the damage :

. absorption by iris and conversion into heat conducted to the adjacent epithelium of the lens Goldmann (23), Langley et al (24).
. direct absorption by the lens, resulting in photochemical degradation of a target molecule Vogt (25) Wolbarsht (26).

The increasing of temperature appears not enough to provoke caractogenesis effect

Wolbarsht exhibits specific changes in the lens proteins following IR exposures near IR with non coherent heat lamp. The experimental data in rabbit confirm in vitro and in vivo an alteration and decreasing of alpha cristallin protein. 1 watt laser irradiation 1,064 nm for 1 mn is enough to provoke a cataract. Low power levels do not produce immediatly cataract but changes in lens proteins can be detected : probably the change of alpha soluble cristallins to an insoluble high molecular weight form and the general alteration of protein conformation.

The difference with UV caractogenesis is only the specific alteration of amino acid and the photooxydation of tryptophan,secondly aromatic aminoacids with coherent 1,060 n ND : YAG Philison cited by Wolbarsht (22) proposes the asymetry growth of cortex and nucleus of lens. The decreased rate of fiber repair and protein synthesis in an enlarged nucleus would contribute to the development of pathological degradation in the cortex.

III-4 Skin

Coherent and non coherent IR radiation effects on the skin have been studied intensively on porcin and human skin. No focusing action is involved and as the cornea the aperture limit is 1 mm. The response criterion most often used to describe the minimal effect is the observation of a minimal erythema from 4 to 24 hours after exposure. The definition of the damage criterion is very important : it is not sure that the erythema threshold is a good and precise evaluation of damage. Many authors prefer the appearance of the second degree of burn with discret oedema and separation of dermis and epidermis. Doses required to induce the erythemic response or second degree burn have been determined for large broadband IR or CO^2 laser radiation (10,600nm) from 10^{-4} to 40s exposure duration.

Threshold doses for skin changes vary with spectrum, wavelength, exposure duration, experimental models and the degree of the skin pigmentation. In general the appearance of IR laser burns ate similar to typical clinical observed burns. Mathematical model as retinal or corneal theorical model are generally in good agreement with experimental data (Laufer et al 27). The table VII show comparison MPE, PAIN and BURN THRESHOLD.

TABLE VII

PULSE RADIATION (s)	MPE	PAIN THRESHOLD	BURN THRESHOLD
			$J.cm^{-2}$
10^{-9} - 10^{-7}	10^{-2}		3.4
8.10^{-3}	0.17	0.45	5.4
4.10^{-2}	0.25	0.52	7.9
1	0.56		27.3

III-4 Biological tissues and culture cell

The large penetration depth of laser radiation into biological time and the redistributive scattering inside the tissue result in heating or mechanical disrupting of volumes. The use of IR laser radiation in surgery will be better if the precise knowledge of absorption of processes. It is a necessary conditions to have a good useful of medical lasers. To obtain such a result it is important to quantify accurately the dimension of the interactive volume as well the derma of necrosis volume in relation with the light parameters.

1 – Tissue model

In two experimental models, rat liver and dog stomach, an homogenous and well perfused tissue and a multilayered tissue with a thin wall, Morgan and al (28) study and evalue the efficiency of two lasers N:YAG (CWL power : 10-40W, short duration : 0,5-2s; HPL power 100-400W, pulse duration : 0,01-0,2s). An IR thermographic camera is used to record the surface temperature of the tissue.

Histologic examination evalue the nature, volume and depth of the necrosis. The surface temperature increases linearly in function of fluence and the results are similar for CWL and HPL in liver (fig.10). The two important mechanisms of heat transfer are conduction and convection. Despite the high content of liquid, no macroscopic convection occurs. The membranes of cells restrict heat convection to a microscopic scale. Thus for thermal calculations, living tissue can considered as a solid. For a single HPL shot the surface temperature increases immediatly (fig.10) if the laser emission is continuous (CWL) the temperature increase is slow (fig 11).

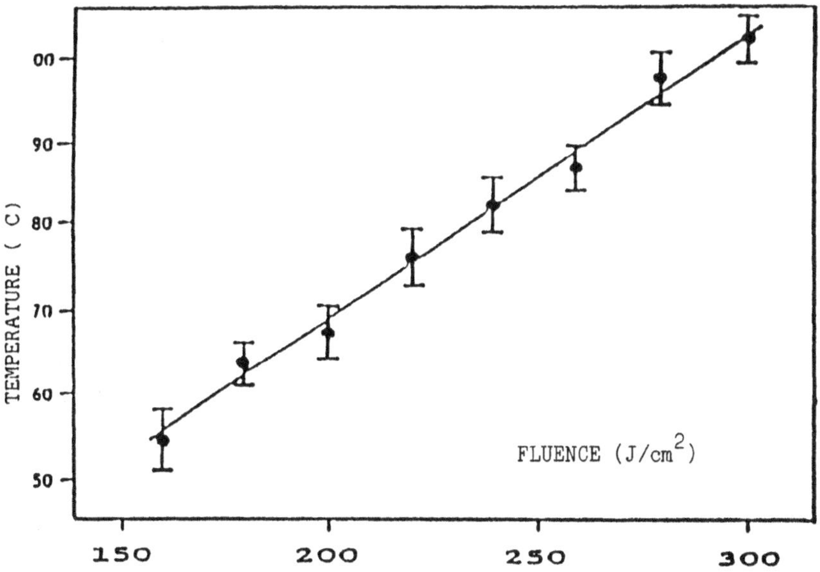

Figure 10 . Surface temperature in function of fluence (J/cm^2). Results are similar for CWL and HPL (in liver)

CWL parameters - Power : 10 to 40 W (Morgan et. Al)
- Exposure time : 5 s to 2 s
- Beam diameter : 4 mm

HPL parameters - Power : 100 to 400 W
- Pulse duration : 0.01 to 0.25 s
- Beam diameter : 4 mm

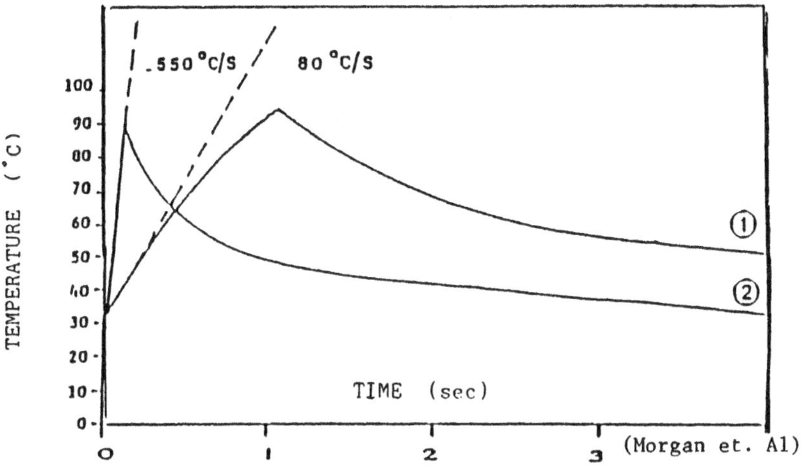

Figure 11 . Comparison of surface temperature evolution for a single CWL shot (1) and a single HPL (2) in liver.

CWL parameters P = 30 W, t = 1s, diam = 4 mm
Coagulation necrosis depth = 1600 µm
Speed of surface temperature increase : 80°C/s

HPL parameters P = 320 W, t = 0.1s, diam = 4 mm
Coagulation necrosis depth = 650 µm
Speed of surface temperature increase : 550°C/s.

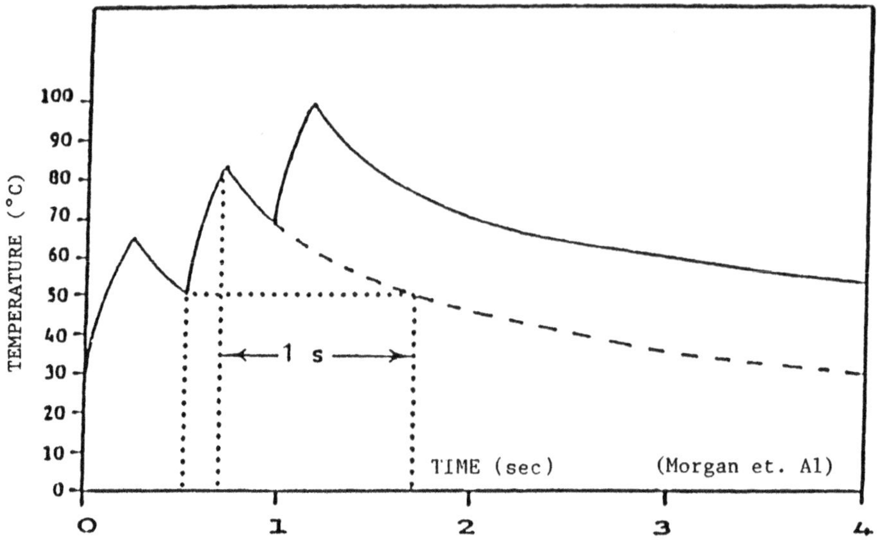

Figure 12 . Surface temperature evolution in rat liver in vivo with HPL. P = 175 W, t = 0.2s and t_2 = 0.3s are maintened constant. This figure actualizes the fact that a pulseto-pulse duration equal (5 t times pulse duration t 1s) is necessary to allow surface temperature decreasing down to 50°C.

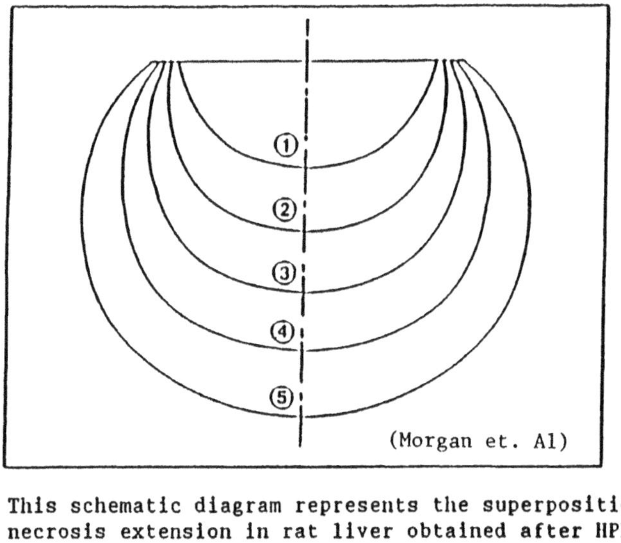

Figure 13 . This schematic diagram represents the superposition of five coagulation necrosis extension in rat liver obtained after HPL irradiation. Different MCS were used: power and spot size were identical for each MCS: power : 175 W and beam diameter : 4 mm. Number of pulses was different and correlatively the duration while surface temperature stepped over 53°C (CD). Necrosis extension depth (ND) is measured by histology.

1. 1 pulse : total energy = 34 J; CD = 1 s; ND = 850 μm
2. 2 pulse : total energy = 46 J; CD = 2.5 s; ND = 1400 μm
3. 4 pulse : total energy = 60 J; CD = 4 s ; ND = 1900 μm
4. 7 pulse : total energy = 96 J; CD = 7 s ; ND = 2350 μm
5. 10 pulse : total energy =130 J; CD =10.5 s; ND = 2750 μm

It is difficult to predict the surface temperature reached at the end of the shot; if the laser emission is divided in several pulses separated by a pause, the heat diffuses into the tissue during the pause. To maintain the surface temperature between 60°C and 80°C,and avoids heat accumulation, it is necessary to have a pause 5 times longer than the shot (fig 12).
For multilayer tissue the thermal diffusivity varies for each layer. The fig 13 give relative diameter of necrosis versus fluence for CWL and HPL laser irradiation.

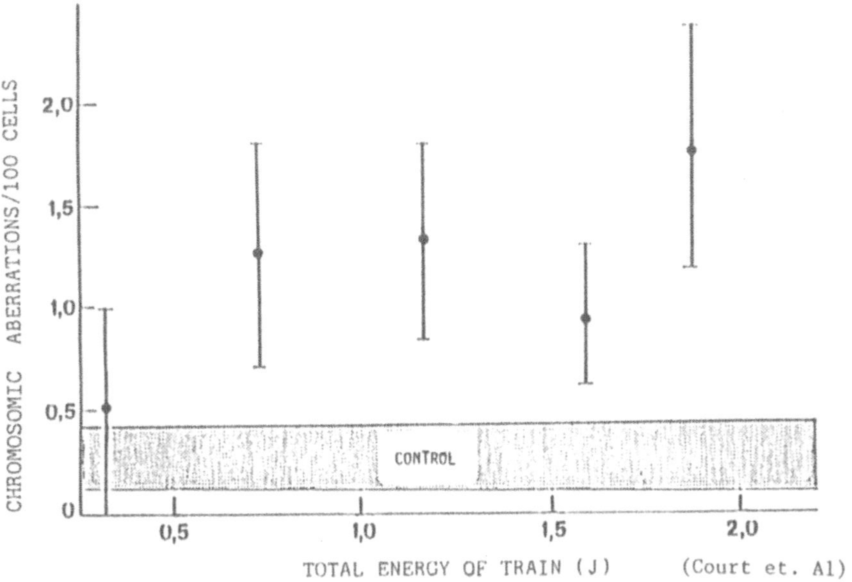

Figure 14 . Expression of chromosomic aberrations/100 cell versus total energy in pulsed Nd : Yag laser.

2 - Effects on culture cell (Court, Courant,Doloy,Dormont et al) (29)

The illumination of lymphocytes by pulsed Nd:YAG laser can induce cytogenetic effects, essentially chromosomic aberrations.
In experimentation using pulsed Nd:YAG (pulse duration 10 ns - NRF: 1000Hz - Total exposure from nb pulses range between 100 and 490 pulses. Total energy range 0,32 and 1,86 J) we have observed on the human lymphocytes culture 48 h after exposure for the first generation, the appearence of chromosomic alterations. These alterations are fragments, without production of dicentrics and the required energy is relatively very low. The table VIII and the fig 14 show the main experimental data.

Only Rounds et al (30) observe with energy of 23 $J.cm^{-2}$ higher than the total energy of our experimental data diameters on rabbit endothelial cells. Nakasija et al (31) found similar data on the V79 hamster, but with very higher total exposure.

The appearance of chromosomic aberrations for very low total energy exposure is probably an important problem in the risk evaluation of medical lasers. The same irradiation of human embryonic fibroblasts show the expression of oncogen and particularly oncogen H-ras.

Although, additional experimental dose-response data are needed for others exposure conditions, wavelength, exposure durations, CW or pulsed irradiations conditions, a reevaluation of technical rules and a restructuration of MPE are probably necessary to define the new medical laser devices and their using, with increased safety margin and without carcinogenic effect.

TABLE VIII

Total energy (mJ)	Control	323	744	1154	1596	1865
NB of pulses		100	200	270	410	490
NB of observed cells	1111	200	400	605	837	400
Chromosomic alterations						
- Dicentrics	0	0	0	0	0	0
- Rings	0	0	1	0	0	0
- Fragments	3	1	4	8	7	7
- Other	0	0	0	à	1	0
. TOTAL /100 cells	0,27 ± 0,16	0,5 ± 0,49	1,25 ±0,56 *	1,32 ±0,47 *	0,96 ±0,34 *	1,75 ±0,66 *

* p : 0,05

IV CONCLUSIONS

The bibliography on biological effects of IR non coherent and coherent radiations is very important. The analytical models for thermals and non thermal effects and for exposure time range from 10^{-11}s to 1000s are generally good and in accordance with the experimental data.

But our knowledge is generally limited to irreversible and immediat effects, but long exposure time with low level irradiation dose and rate have not yet been clearly defined. Future areas of research will involve a concerted effect to establish the relationship between radiation and biological processes.

Many military systems developpers and generally all biomedical laser industry their criticize currently laser safety standards for conservative and complex aspects.

The proliferation in the number of lasers and in the applications of biomedical lasers made to have very sure standards and excellent guidelines in order to minimize or prevent injuries and damage.

Our knowledge must be increased in several ways; the effects of low but repeated and chronic IR laser irradiation and the use of functional criteria for eye, skin and many biological tissues are most important.

The problem of cell stimulation by low level IR coherent irradiation is difficult to observe and to explain.

REFERENCES

1. Geeraets W.J. and Berry E.R.
Ocular spectral characteristics as related to hazards from lasers and other light sources
Am. J. Ophthal. 66:15-20 (1968).
2. Ham W.T. Mueller H.A., Ruffulo J.J. Jr, Cleary S.F., Guerry R.K. and Dupont Guerry
Biological Applications and effects of optical lasers.
Annual/final report contract N° DAMD 17-82-C-2083. U.S Army Medical research and development command fort Detrick, Frederick, MD 21701-5012 (1987).
3. Ham W.T. Jr., Mueller H.A., Ruffolo J.J. Jr and A.M. Clarke
Sensitivity of the retina to radiation damage as a function of wavelength
Photochemistry and photobiology, 29:735-743 51979).
4. Gibbons W. D., and Allen R. G.
Retinal damage from long-term exposure to laser radiation
Invest. Ophthal. Visual Sci., 16,6:521-529 (1977).
5. Allen R.G. Jr., Bruce W.B., Kay K.R., Morrison L.K., Neish R.A., Polaski C.A., and Richards R.A.
Research on ocular effects produced by thermal radiation
report AD 659 146 USAF School of Aerospace Medicine (1967).
6. Goldman A.I., Ham W.T., and Mueller H.A.
Ocular damage thresholds and mechanisms for ultrashort pulses of both visible and infrared laser radiation in the rhesus monkey
Exp. eye Res. 24:45-56 (1977).
7. Allen R.G., Thomas S.J., Harrison R.F., Zuclich J.A., and Blankenstein M.F.
Ocular effects of pulsed Nd laser radiation : variation of threshold with pulsewidth
Health Physics 49,5:685-692 (1985).
8. Gibson G.L.M
Retinal damage from repeated subthreshold exposures using a ruby laser photo coagulator
SAM-TR-70-59 (1970).
9. Skeen C.H.
Ocular effects of repetitive laser pulses
Contract F 41609-71-C-0018, Technology, Inc, San Antonio, Tx, (1972)
10. Skeen C.H.
Ocular effects of near.IR laser radiation for safety criteria
Contract F 41609-71-C-0016, Technology, Inc, San Antonio, Tx, (1972).
11. Ebbers R.W.
Retinal effects of multiple pulse gallium arsenide laser
Report SAM.TR.72.25
Usaf School of Aerospace Medicine (1972).

12a. Lund D.J., Beatrice E.S. and Sliney D.H.
Near infrared laser ocular bioeffects
in Proceedings of First International Symposium on laser biological effects and exposure limits
19, 246-255, L. Court et. Al Eds, 1986

12b. Lund D.J., Beatrice E.S., and Schuschereba S.T.
Bioeffects concerning the safe use of GaAs laser training devices in combat ocular problems. Proceedings of Conference 20-21/10/1980 Division of Ocular Hazard Letterman Army Institute of Research, supp. 15-29 (1982).

13. Beatrice E.S., Lund J.D., and Talsma D.H.
Retinal alterations produced by low level gallium arsenide laser exposure

Letterman Army Institute of Research Report N° 38 (1976)
LAIR Presidio of San Francisco Ca 94129.

14. Stuck B.E., Lund D.J., and Beatrice E.S.
Ocular effects of relatively "eye safe" lasers
Combat ocular problems supp 1-14 (1982). LAIR, San Francisco

15. Reed R.D.
A predictive equation for infrared laser damage to the corneal epithelium
Health Physics 36:73-75 (1979).

16. Wray J.L.
Model for prediction of retinal burns
Headquarters DASA, Washington DC, Tech. Rep. DASA 1282 (1962).

17. Vos J. J.
A theory of retinal burns
Bull. Math. Biophys. 24:115-128 (1962).

18. Clarke A.M., Geeraets W.J., and Ham W.T. Jr.
An equilibrium model for retinal injury from optical sources
Appl. Opt. 8,5:1051-1054 (1969).

19. Mainster M.A., White T.J., Tips J.H., and Wilson P.W.
Spectral dependance of retinal damage produced by intense light sources
J. Opt. Soc. Amer 60,6:848-855 (1970.

20. Takata A.N. et al
General model of laser induced eye damage
II TRI, Techn. Rep. 74-6324 (1974).

21. Egbert D.E., and Maher E.F.
Corneal damage thresholds for IR laser exposure : empirical data, model predictions and safety standards.
USAF School of Aerospace Medicine
Brooks AFB, SAM-TR-77-29 (1977).

22. Wolbarsht M.L.
Damage to the lens from IR
in : Ocular effect of non-ionizine radiation, 121-141 (1980).

23. Goldmann H.
Genesis of heat cataract
Arch. Ophtal. 9:314-316 (1933).

24. Langley R.K., Mortimer C.B., and Mc Cullogh C.
The experimental production of cataracts by exposure to heat and light
Arch. Ophthal. 63/473-488 (1960).

25. Vogt A.
Das experimentelle reine ultrarot Strahlen katarakt des Kaninchen 17 Monaten nach der Bestrahlung der Experimentelle f. des Albinos
Klin. Monats BL Augenh. 89:255-256 (1932).

26. Wolbarsht M.L.
The effects of optical radiation on the anterior structure of the eye
In Current Concepts in Ergophthalmology, pp.24-26, ed by B. Tengroth, P. Epstein, A. Anseth, A. Hedin, A. Keeny, M. Ropper-Hail and D. Sliney.
Societies Ergophtalmologica Internationale, Stockholm (1978).

27. Laufer G., Joachims H.Z., Eliachar I., and Mordechovitz D.
Measurement of laser exposure levels for burn threshold in biological tissue
J. of Biomech. Eng 106:283-284 (1984).

28. Mordon S.R., Cornil A.H., Jensen D.M., Gosselin B.,and Brunetaud J.M.
Nd : Yag laser coagulation comparative study of continuous and high power pulsed lasers in vivo

in : First International Symposium on Laser Biological Effects and Exposure Limits.
L.A. Court et al Ed. 84-101 (1988).

29. Court L., Courant D., Doloy M. T. et Dormont D.
Effets cytogénétiques d'un laser ND:YAG pulse
In Rapport annuel DRET N 86/1027 78-91 (1988).
CRSSA - Grenoble

30. Rounds D.E., Chamberlain E.C., and OKIGAKI T.
Laser radiation on tissue cultures
Ann. N.Y. Acad. Sci, 122:713-727 (1964).

31. Nakajima M., Fukuda M., Kavoki T., and Atsumi K.
Cytogenetic effect of argon laser irradiation on chinese hamster cells
Rad. Res. 93 598-600 (1983).

INFRARED RADIATION AND THE EYE

Per G. Söderberg

Dept. of Ophthalmology (Head: Björn Tengroth) and
Dept. of Medical Biophysics (Head: Rudolf Rigler)
Karolinska Institutet, Box 60 400, S-104 01 Stockholm
Sweden

INTRODUCTION

With the industrial revolution large groups of workers in the steel and glass industry became exposed to high doses of infrared radiation (IRR). In the 18th century this lead to a discussion on whether intense longterm exposures to IRR could induce cataract. The invention of high intensity light sources such as the carbon arc lamp, the xenon lamp and the tungsten lamp created new powerful IRR sources with potential hazards for the eye at moderate exposure times. The recent development of IRR lasers has created a need for detailed knowledge about IRR toxicity to the eye at short to extremely short exposures with high power densities.

TRANSMITTANCE OF IRR IN THE EYE

In the closed eye any IRR directed against the eye is effectively absorbed in the eyelids.

The absorbtion of IRR in the eye (Boettner and Wolter 1962, Maher 1978) is closely related to the absorbtion spectrum of water with absorbtion peaks centered around 980 nm, 1200 nm, 1430 nm and 1950 nm (Fig. 1). At wavelengths longer than 2000 nm only a small fraction is transmitted through the cornea. The aqueous humour limits the transmission of IRR to wavelengths shorter than 2000 nm. The iris completely absorbs all IRR. The lens filters out all the remaining wavelengths longer than 1400 nm. Finally, the vitreous partly absorbs the shortwave IRR between 760 nm and 1400 nm. Energy of IRR absorbed in the eye is converted to heat.

CLINICAL EFFECTS OF IRR

Damage mechanisms

Most clinical effects of exposure to IRR are associated with the heat produced at radiation absorbtion. Thus, heat induced denaturation of biomolecules is the predominant damage mechanism at low power densities. If the temperature rises to 100 ^{o}C water transforms to vapour that

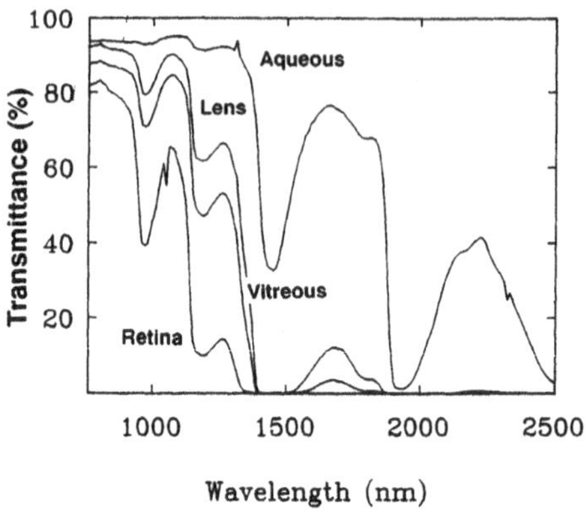

Fig. 1. Cumulated transmittance of the primate eye to infrared radiation reaching the indicated medium. (Plotted from data published by Maher, 1978).

ruptures the tissue by expansion. Moderate power densities in dried tissue causes carbonization. At very high power densities the electromagnetic field induces optical breakdown with a subsequent pressure wave that causes mechanical rupture.

Eye lids

Experimental exposure of the eye lids to moderate doses of IRR induces erythema (Duke-Elder 1972). At high doses the heat may induce blister formation and necrosis (Duke-Elder 1972).

Cornea

In the cornea a medium dose of IRR, delivered at a moderate power density, induces opacification of the epithelium and the stroma due to protein coagulation. However, lid closure is evoked as a pain reaction to the heat prior to formation of damage (Duke-Elder 1972). There are some observations indicating that the posterior layers of the cornea may demonstrate more damage than the anterior layers (Verhoeff & Bell 1915/16). This was believed to depend on better cooling of the anterior than the posterior surface.

Iris

Moderate doses of IRR absorbed in the iris induces aqueous flare, hyperemia and miosis (Duke-Elder 1972). High doses of IRR causes paralytic mydriasis, and congestion of the blood vessels of the iris with secondary hemorrhages, trombosis, and stromal inflammation (Duke-Elder 1972). At days after exposure damaged areas of the iris become atrophic.

Lens

Pathological effects of IRR to the lens has been discussed since an increased incidence of cataract was observed among glass blowers and blacksmiths. These observations have triggered numerous investigations on the potential hazard of IRR to the lens employing experimental and epidemiological techniques.

Experimental studies. In some of the first experimental studies of effects of IRR to the eye it was concluded that a low dose of IRR induces cataract by direct absorbtion in the lens (Vogt 1912, 1932). However, later a series of experiments by Goldman (1930, 1932, 1933a, 1933b, 1933c, 1933d, 1935, 1950) demonstrated that opacities were induced in the anterior subcapsular region just beneath the iris, only if the iris was exposed. Yet, if the free lens was exposed through the pupil no opacification occured. Thus, it was concluded that the IRR induced cataract, at low doses, was an effect of heating secondary to IRR absorbtion in the iris. This has later been supported in more recent investigations (Langley 1960, Pitts 1980).

Other experiments with IRR at different times of exposure have indicated that opacification of the lens after exposures to IRR involves both mechanisms. Absorbtion in the iris with transformation of the energy to heat induces opacification at short exposures to high irradiances (Wolbarsht et al. 1976, Dannheim et al. 1977). However, in addition lens proteins degrade directly when IRR is impinging the lens (Wolbarsht et al. 1976, 1977, 1980) and may play a major role in long-term low-level exposures.

Epidemiological studies. Already 1786 it was mentioned that there is a frequent occurence of cataract among glass blowers and blacksmiths (Wenzel) . During the 19th century and the beginning of the 20th century a number of epidemiological observations of high frequency of cataract among glass workers were reported (Beer 1817, Arlt 1858, Knies 1892, Meyhöfer 1886, Hirschberg 1898, Pröbsting 1899, Robinson 1903, Robinson 1907, Stein 1913, Huber 1914, Wick 1922, Cords 1926, Schläpfer 1930, Fatorelli 1953, Barthelmess & Borneff 1959). These studies are difficult to evaluate because different definitions of cataract and various methods for investigation were used. Furthermore, the design of these investigations did not include matched control groups. In 1950 (Dunn) the causes of industrial cataracts were reviewed among the workers at the Corning Glass Works. It was concluded that the cataracts were caused by other etiological factors than IRR. An examination in 1971 (Wallace) of 1000 steel workers exposed to IRR demonstrated a higher incidence of common cataract at high ages than in the control group.

Glass blowers cataract - a specific entity. Part of the investigation of the possible role of IRR in causing cataract has been focused on attempts to identify a certain entity of cataract, the "Glass blowers cataract". Meyhöfer (1886) concluded that the glass blowers cataract is like any other cataract. However, Robinson (1903) stated that the "Glass blowers cataract" is typically a central posterior subcapsular opacity. This was confirmed by Legge (1907) who related the "Glass blowers cataract" as a blot of ink in the posterior subcapsular region. Goldman (1930) and Vogt (1932) concluded from their experimental studies that the immediate cataract after exposure to high doses of IRR is an anterior subcapsular opacity.

A spontaneous peeling of the lens capsule from the equatorial zone towards the central zone has been stated to be specific for IRR induced lens damage (Elsching 1922). A later investigator (Jaensch, 1955) proclaimed this type of capsular exfoliation to be pathognomonic for IRR damage.

It has been noted that the occupational exposure of glass blowers is assymetrical so that the left eye is allways turned towards the furnace. This has stimulated investigations of if there is a predisposition of side in the "Glass blowers cataract". Some of these studies have found that the left eye is more affected (Meyhöfer 1886, Pröbsting 1899, Huber 1914, Legge 1907) while Robinson (1903) found both eyes equally affected.

Retina and choroid

A moderate power of IRR impinging the anterior segment at low irradiance is potentially hazardous for the retina and the choroid due to the increase of power density evoked by refraction in the optical media of the eye. Thus, in addition to the radiance of the source the transmittance of the eye media, the pupil size, the refractive power of the eye media, and the quality of the retinal image must be known in order to predict the irradiance at the surface of the retina (Sliney 1980).

Damage occurs as a result of the temperature rise if the energy of IRR absorbed in the retina exceeds the energy dissipated. An exposure beyond threshold immediately coagulates the retinal tissue and evokes thrombosis of choroidal vessels leading to choroidal congestion (Duke-Elder 1972). The primary lesion converts into a choroido-retinal fibrotic scarr with proliferation of retinal pigment epithelium. The site of the lesion strictly determines the significance of the lesion for the vision. A wide spread lesion in the periphery of the retina has no inpact on vision while a very limited lesion in the macula of the retina may delete vision almost completely. Some potentially hazardous exposures are the solar disc viewed with magnifying optics, incorrectly projected light during indirect ophthalmoscopy, accidental exposures to laser beams and forced observation of powerful broad band sources.

SWEDISH EPIDEMIOLOGICAL DATA ON IRR CATARACT

In 1984 a controlled epidemiological study of the correlation between occupational cataract in glass and steel industry and exposure to IRR was published (Lydahl). The investigation consisted of measurements of radiation, interviews with 208 iron and steel workers and 208 matched controls, and with 209 IRR exposed glass workers and 298 controls, and clinical examination of the interviewed workers.

Spectral radiance in glass works and steel industry

The heat sources in the glass and steel industry can, for the calculation of spectral radiance, be approximated to be black body radiators. According to Planck's radiation law (Eq. 1) the spectral radiance, L ($W\ m^{-2}\ sr^{-1}$), is a function of the temperature.

$$L(\lambda,T)=\frac{1.191\ x10^{-6}}{\lambda^{5}\ (e^{\frac{1.439\ x\ 10^{-2}}{\lambda\ T}}-1)}\qquad \text{Eq. 1}$$

Here, λ is the wavlelength of the light (nm) and T is the absolute temperature of the black body radiator.

In Fig. 2a it is seen that visible radiation and IRR constitute the main fraction of the energy emitted from black body radiators at temperatures occuring in the steel and glass industry. The amount of ultraviolet radiation (UVR) is quantitatively negligible. The spectral radiance is shifted towards shorter wavelengths at high temperatures. Spectral radiance measurements on a glass works furnace was consistent with the Planck's radiation law (Fig. 2b).

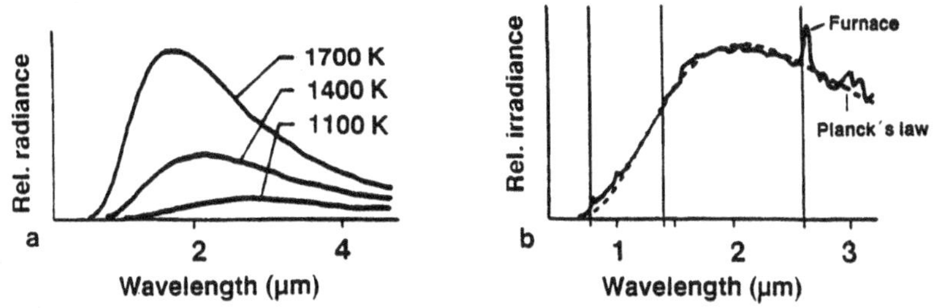

Fig. 2. a) Spectral radiance at three temperatures occuring in the steel and glass industry as predicted by Planck's radiation law. b) Spectral irradiance measured at a fixed distance from a furnace in the glass industry as compared to the spectral irradiance predicted at the actual temperature by the Planck's radiation law. (Redrawn from Lydahl, 1984).

Radiometry

The radiometric measurements were taken with a pyroelectric probe. To determine the dose of radiation received by a certain cathegory of workers the probe was held close to the eye of the worker in activity, and the gaze direction was simulated with the probe. The irradiance settled for a certain activity was the mean of a number of independent measurements and was transformed to dose per eight-hour shift for the steel workers and dose per hour for the glass workers. Application of Planck's radiation law allowed calculation of doses within different wavebands.

Considering maximum irradiances (Fig. 3a) it was found that some occupations in the steel industry recieve higher irradiances than any jobs in the glass industry. When the exposure times were taken into account it was found that the maximal dose per eight hour work shift was about the same for the steel and glass workers but the mean dose was higher for the glass workers (Fig. 3b). This indicates that workers in the steel industry receive higher peak irradiances but glass workers receive peaks more often.

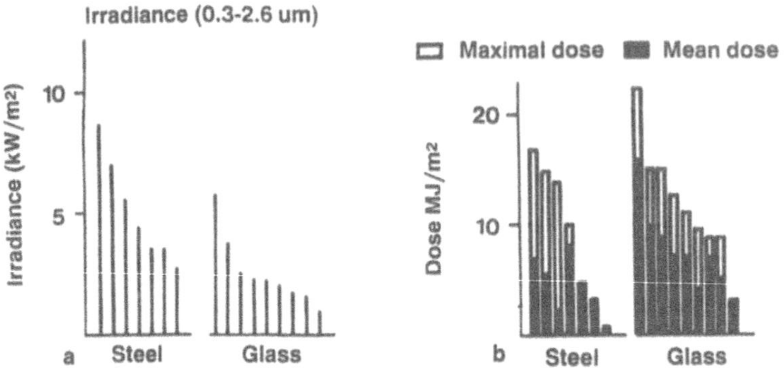

Fig. 3. a) Distribution of occupations in the steel and glass industry for maximum irradiance of optical radiation (0.3-2.6 µm). b) Distribution of occupations in the steel and glass industry for maximum and mean dose of optical radiation (0.3-2.6 µm)/8 hrs work shift (Redrawn from Lydahl, 1984).

In order to establish a dose-effect relationship between IRR and cataract the exposure to the right and the left eye, respectively, was measured differentially for the two eyes with two radiometers mounted in a phantom face. The ratio between the irradiance reaching the left eye and the right eye was recorded, simulating the position of the workers face in several occupations in the glass industry, and in several working environments a higher irradiance of the left than of the right eye was found.

Exposure profiles calculated from the interviews

The age matching between the exposed and the control group was almost complete for the steel workers while for the glass workers the control group was slightly younger than in the exposed group (Fig. 4). Furthermore, because the controls in the group of steel workers were paired to the exposed the exposed and the control group were of the same size for each age group. However, in the examination of the glass workers the exposed and the control groups were not paired and the controls were more numerous.

The life time dose distributions calculated from the interviews demonstrated that most steel workers (Fig. 5a) received lower life lime time doses than the glass workers (Fig. 5b). However, some steel workers received higher life time doses than any of the glass workers.

Eye examinations

Each worker in the steel industry was blindly examined once by each one of the two involved ophthalmologists. The glass workers and their controls were examined by one ophthalmologist and the controls were partly examined uncovered. The examination included visual acuity testing, refraction, determination of intraocular pressure, slit lamp microscopy, and ophthalmoscopy. For comparison of the exposed and the

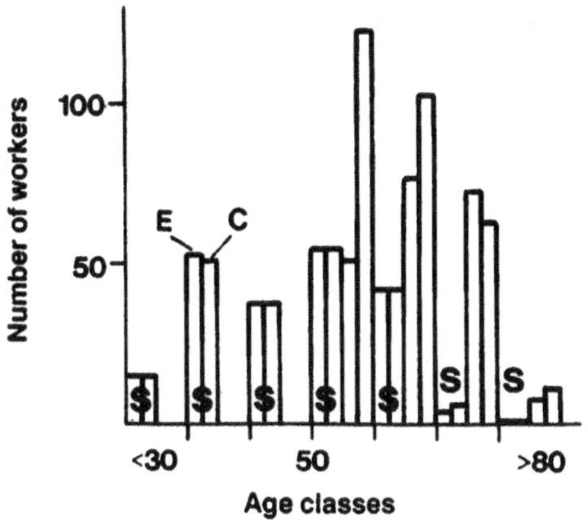

Fig. 4. Age distribution for examined steel workers (S) and glass workers (blank), respectively, devided into the exposed group (E) and the control group (C) (Redrawn from Lydahl, 1984).

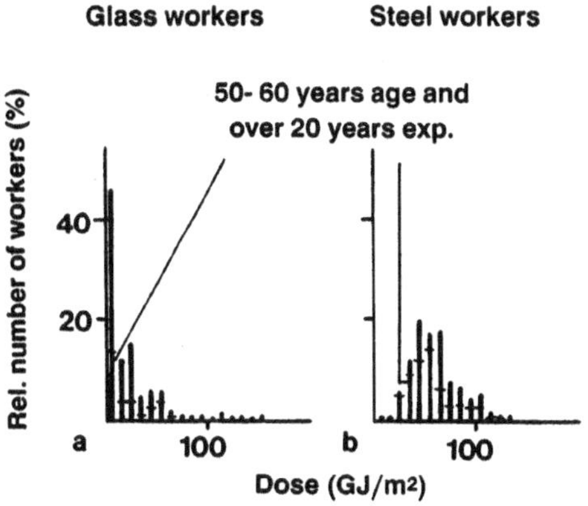

Fig. 5. Distribution of life time dose as calculated for high exposure groups (below bar) and for the complete group (Redrawn from Lydahl, 1984).

control groups the material was stratified in order to select high dose groups. "High exposure" was for the steel workers the 20% largest life time dose, the 20 % highest irradiance, the 20% longest exposure time. For the glass workers "high exposure" was the 25% largest life time dose, the 20 % highest irradiance, and the 25% longest exposure time.

For the steel workers it was found that wedge shaped opacities occured more often in exposed workers compared to controls in the strata > age 60, high exposure, and > age 70 and high exposure. An increased browning of the lens was found in the high exposure groups and one case of capsular exfoliation was identified among the exposed workers.

In the glass workers posterior opacities were found more often in the strata age 60-69, > age 70, high exposure, and > age 70 and high exposure. Subcapsular opacities were equivalently frequent in exposed and controls while wedge shaped and punctate cortical opacities, and increased nuclear turbidity and browning was more common in the exposed workers. Eight cases of capsular exfoliation were found among the exposed workers. The left eye was found to be more affected by cataract than the right eye. A comparison with 700 consecutive non specific cases operated for cataract at the Karolinska Hospital in Stockholm demonstrated that the right eye is more commonly operated as the first eye.

By exclusion of all cases with close heredity for cataract, as detected by the interview, and relative glass worker or blacksmith it was found that advanced cataract is more common if both of the factors heredity and exposure exist than without heredity.

It was concluded that high occupational exposure to IRR increases the risk for development of cataract. Prevalence of cataract in IRR exposed workers increases after age 60. All common types of cataract are more frequent in IRR exposed workers. There is no "Glassworkers cataract entity". The maximally daily dose is about the same in the glass and steel industry but a larger proportion of jobs in the steel industry gives lower IRR exposure. Left eyes of glass workers are more often affected by cataract than right eyes. Heredity for cataract increases the risk for cataract at exposure to IRR. For these reasons all workers occupationally exposed to IRR should wear protective glasses.

ACKNOWLEDGEMENT

I am indebted to Björn Tengroth for reviewing this manuscript and to Lotta Vig for help with the preparation of the manuscript.

REFERENCES

Arlt, F., 1858, Die Krankenheiten des Auges fur praktischen Ärzte, p. 292, Verlag des k.k. Hof-Buch- und Kunsthandlung, Prag.

Barthelmess, G., and Borneff, J., 1959, Uber die gewerbliche Schädigung der Augenlinse durch Wärmestrahlung, Arch. f. Ophth., 160:641.

Beer, G. J., 1817, Lehre von den Augenkrankenheiten, Vol 2, p. 327, Verlage von Heubner und Volke, Wien.

Boettner, E. A., and Wolter, J. R., 1962, Transmission of the ocular media, Invest. Ophthalmol., 1:776.

Cords, R., 1926, Uber den Berufsstar, Zeitschrift f. Augenheilk., 60:251.

Dannheim, F., and Rassow, B, 1977, Lesions of the anterior segment of the eye by lasers of different wavelengths, in: GSF-Bericht-BPT5, pp. 13-1, Neuherberg.

Duke-Elder, S., 1972, Radiation injuries. B. Electro-magnetic Thermal Lesions in: System of Ophthalmology, Vol XIV Injuries, Part 2, Non-mechanical injuries, p. 858, Henry Kimpton, London.

Dunn,K. L., 1950, Cataract from infra-red rays (glass worker's cataract), Arch. Ind. Hyg. Occup. Med., 1:166.
Elsching, A., 1922, Ablösung der Zonulalamelle bei Glasbläsern II, Klin. Mbl. f. Augenheilk., 69:732.
Fatorelli, A., 1953, Catarata dos vidreiros, Rev. Bras. Oftal., 11:219.
Goldman, H., 1930, Kritische und Experimentelle Untersuchungen uber den sogenannten Ultrarotstar der Kaninchen und der Feverstar, Arch. f. Ophth., 125:313.
Goldman, H., 1932, Experimentelle Untersuchungen uber die Genese des Feverstarres; uber Arbeitshyperthermie bei Feverarbeitern, Arch. f. Ophth., 128:648.
Goldman, H., 1933a, Genesis of heat cataract, Arch. Ophthalmol., 9:314.
Goldman, H., 1933b, Experimentelle Untersuchungen uber die Genese des Feverstars, I Teil, Arch. f. Ophth., 130:93.
Goldman, H., 1933c, Experimentelle Untersuchungen uber des Feverstares: die Physik des Feverstars, II Teil, Arch. F. Ophth., 130:131.
Goldman, H., 1933d, Experimentelle Untersuchungen uber die Genese des Feverstares: die Physik desw Feverstares, II Teil, Arch. f. Ophth., 130:140.
Goldman, H., 1935, The genesis of the cataract of the glass blower, Ann. Ocul., 172:13.
Goldman, H., Konig, H., and Mader, F., 1950, Die Durchlässigkeit der Augenlinse fur Infrarot, Ophthalmologica, 120:198.
Hirschberg, J., 1898, Ueber den Star des Glasbläser, Centralbl. f. prakt. Augenheilk., 22:113.
Huber, A., 1914, Berufskrankenheiten und Berufsverletzungen der Glasarbeiter, Das Österreichische Sanitätswesen 26:349.
Jaensch, P. A., 1955, Berufliche Linsentrubungen bei Metallarbeitern, Arch. f. Gewerbepath. Gewerbehyg., 13:743.
Knies, M., 1892, Grundriss der Augenheilkunde, Teil I, p. 289, Verlag von J F Bergmann, Wiesbaden.
Langley, R. K., Mortimer, C. B., and McCulloch, C., 1960, The experimental production of cataracts by exposure to heat and light, Arch. Ophthalmol., 63:473.
Legge, T. M., 1907, Home office report on cataract in glass workers. Report of the medical inspector in factories and workshops, Annual report for 1907, p. 250, London.
Lydahl, E., 1984, Infrared radiation and cataract, Acta Ophthalmol. (Copenh.), Suppl. 166.
Maher, E. F., 1978, Transmission and absorbtion coefficients for ocular media of the rhesus monkey, Report SAM-TR-78-32, USAF
Meyhöfer, 1886, Zur Aetiologie des grauen Staars. Jugendliche Katarakten bei Glasmachern, Klin. Mbl. Augenheilk., 24:49.
Pitts, D. G., Cullen, A. P., and Dayhaw-Barker, P., 1980, Determination of ocular threshold levels for infrared radiation cataractogenesis, DHHS (NIOSH) Publication No. 80-21, NIOSH # 77-0042-7701.
Pröbsting, 1899, Ueber Starbildung bei Feuerarbeiten, Centralbl. f. allg. Gesundheitspflege, 18:425.
Robinson, W., 1903, Bottle-finishers' cataract, Brit. Med. J., 191.
Robinson, W., 1907, On Bottle-makers's cataract, Brit. Med.J., 381.
Schläpfer, H., 1930, Neue Beobachtungen uber Glasmacherkatarakt, Klin. Mbl. Augenheilk., 85:285.
Sliney, D., and Wolbarsht, M., 1985, Effects of optical radiation on the eye, in: Safety with lasers and other optical sources, p. 101, Plenum Press, New York.
Stein, L., 1913, Untersuchungen uber Glasbläserstar, Archiv f. Augenheilk., 74:53.
Verhoeff & Bell, 1915/16, The pathological effects of radiant energy upon the eye, Proc. Am. Acad. Art Sci., 51:629.
Vogt, A., 1912, Einige Messungen der Diäthermasie das menschlichen Augapfels und seiner Medien, sowie des menslichen Oberlides, nebst

Bemerkungen zur biologischen Wirkung des Ultrorot. Arch. f.Ophth., 83:99.
Vogt, A., 1932, Fundamental investigations of the biology of infrared. Klin. Monatsbl. Augenheilk., 89:256.
Wallace, J., Sweetnam, P. M., Warner, C. G., Graham, P. A., and Cochrane, A. L., 1971, An epidemiological study of lens opacities among steel workers, Brit. J. Industr. Med., 3:265.
Wenzel, M., 1786, Traite' de la cataracte avec des observations p. 5, PJ Duplain, Libraire, Cour du Commerce, Paris.
Wick, W. 1922, Zur Frage der Starbildung bei Glasbläsern, Arch. f. Ophth., 109:224.
Wolbarsht, M. L, Yamanashi, B. S., and Orr, M. A., 1976, The origin of cataracts in the lens from infrared laser radiation, Annual Report U S Army Medical Research and Devlopment Command Contract DAMD 17-74-C-4133.
Wolbarsht, M. L, Yamanashi, B. S., and Orr, M. A., 1977, The origin of cataracts in the lens from infrared laser radiation, Report 7-20-1977: 62772A 3E62772A813.00.013, Duke Univ. Eye Center, Durham, NC, Contract DAMD 17-74-C-4133 US Army Med. Rsh. and Dev. Command, Washington, D.C.
Wolbarsht, M. L., 1980, Ocular effects of non-ionizing radiation, Proc. of the Society of Photo-Optical Instrumentation Engineers, 229: 121

INFRARED ENERGY AND SKIN: BIOLOGICAL EFFECTS AND HEALTH HAZARDS

Alessandro Checcucci

CNR - Istituto Ricerca Onde Elettromagnetiche
Via Panciatichi 64
50127, Florence, Italy

INTRODUCTION

The infrared region (IR) has been subdivided (1) into three biologically significant bands:

IR-A between 0.75 and 1.4 μm
IR-B between 1.4 and 3.0 μm
IR-C between 3.0 and 10^3 μm

Both artificial or natural IR sources can be encountered in the working environment. The main natural source is the sun whose total radiant power in the IR region is approximately 0.7 kW/m^2. Artificial sources include incandescent, fluorescent, and high-intensity discharge lamps, flames, heaters and lasers. The following occupations are associated with potential exposures at artificial or natural IR sources: bakers and cooks, braziers, chemists, construction workers, dryers, electricians, farmers, firemen, foundry and furnace workers, glassblowers, heat treaters, IR laser operators, iron workers, kiln operators, solderers, steel mill workers, welders. The number of workers exposed to broad band of IR is increasing consequently to the appearance of new kinds of specialized industrial heat sources, even if other occupations disappear.

The maximum reflectance of the skin occurs between 0.7 and 1.2 μm, which is comparable to the wavelength of maximum intensity for some IR heaters. In the spectral range 0.4 - 2.8 μm, skin reflection follows Lambert's cosine law for all wavelengths. This means that the skin can be considered a source which diffuses the light uniformously in all the directions. Finally,for wavelengts beyond 2 μm the skin reflectance is largely affected by the pigmentation and blood flow (2).

The water (60-70% of total weight) is the main IR absorber of the skin. However, the IR absorption by the skin depends also upon the presence of pigments, blood or other substances, and upon the degree of scattering due to the microstructures of the skin. This scattering is responsible of an unusual situation: the coincidence of the reflection and transmittance maxima. In general, good reflectance and transmittance are inversely related, but in this case a considerable part of reflected IR radiation is actually scattered reflection from deeper skin layers. The skin is essentially opaque to wavelengths beyond 2 μm. The maximum penetration is approximately at 1.2 μm: at least 50% of the radiation penetrates to a depth of about 0.8 mm, thus interacting with nerve endings and capillaries. Consequently, the spectral region of the near-IR is more hazardous than the other IR ranges because of the higher penetration depth, while the region beyon 2 μm causes surface heating (3).

Light, Lasers, and Synchrotron Radiation
Edited by M. Grandolfo *et al.*, Plenum Press, New York, 1990

While the effects on skin and eye of IR radiation are sufficiently known from a symptomatologic point of view, meaningful biological information is lacking as regards the action mechanism, effect thresholds levels, effects on organs other than skin and eye, potential synergistic effects, heat stress. In general, photochemical reactions in tissues are excluded, because of the low energy level of IR photons and their action mechanism is assumed to be essentially thermal in nature. The differences between thermal effects and photochemical effects are important. In fact, thermal effects are highly dependent upon irradiance (W/cm^2) and are not strongly wavelength-dependent. Photochemical effects, which dominate in the UV and short wavelength visible radiation, are dependent upon exposure-dose (J/cm^2) and have a long-wavelength cutoff where the photon energy is insufficient to cause an effect.

THERMAL EFFECTS

When absorbed by the skin, IR radiation produces tissue heating, and the consequent temperature increase depends on the capability of the tissue to dissipate the heat, that is on the thermal characteristics of the skin: conductivity, density and specific heat.

Thermal radiation, differently from other wavelengths, arouse thermal sensations. Ebaugh and Thauer (4) did not find significant changes in threshold of warmth sensation when skin temperature was raised from 15 °C to 41 °C. A severe pain sensation is induced in human skin when tissue temperature is elevated above 45 °C. Contrary to what seen for the warmth sensation, Hardy and coworkers observed (5) that the pain threshold varies linearly with skin temperature and at a skin temperature of 45 °C no further stimulus is required. In other words, the pain threshold represents that amount of radiation necessary to rise the skin temperature to 45 °C. This means that the pain threshold is dependent upon skin temperature alone and not upon the rate of heating of the skin, nor on the rate of change of internal thermal gradients.

It is well known that a visible effect of the skin heating is the reddening due to an increased vasodilatation of the arteriolar system. External heat applied to restricted skin areas increases the blood flow and consequently the heat transported by the blood from the internal tissues. If the skin temperature reaches values above the internal temperature, the gradient in the heated area becomes reversed and the blood distributes now the added heat troughout the body. If the heat intensity is sufficiently high, the vasodilatation can diffuse until to rather remote areas. This consensual reaction is used in heat therapy. The spreading of vasodilatation elicited by local heat application or throughout a consensual reaction is due to true spinal reflexes starting from the cutaneous receptors (6). Other effects are the changes in the skin pigmentation. Kollias and Baqer (7) observed a bluish appearance in the skin of the lower inner arm of voluenteers irradiated with a 0.39 - 1.7 µm light source. The photoproduct formed was characterized by a weak absorption in the blue range. In individual exposed to IR radiation is often visible a gradual increase of the skin pigmentation. This iperpigmentation can persist for a longer period of time. The development of an erythematous-like appearance is frequent among certain occupational groups (glass workers, furnacemen, etc.) exposed to high intensities of infrared radiation.

The sweat is a thermoregulation way which allows the body disssipate the heat. Wyss (8) has shown that in non-acclimatized subjects sweat began to appear after an exposure of 180 sec first on the abdomen and thorax and then on the face. The succession was reversed in acclimatized individuals who, on the other hand, began to sweat in a shorter time (9).

The most severe effects of IR radiation on the skin are the so-called **flash burns**. However, they are rather rare because they occur with high intensity and short duration exposures. The heat sensation would normally limit exposures of this intensity so that such high irradiance levels are encountered in very few working locations in industry, as for instance by hot gases or steam. Usually, burns are classified according to the gravity of the lesions, going from a superficial reddening up to a tissue destruction. In the following table are indicated the symptoms of the burns of different gravity and the effective irradiance values.

BURN DEGREE	SYMPTOMS	MINIMAL IRRADIANCE (W/m^2)
First	Superficial reddening	$1.2x10^5$
Second	Blistering	$2.4x10^5$
Third	Tissue destruction	$3.4x10^5$

POTENTIAL SYNERGISTIC EFFECTS

It is known that the biologic effects of one type of radiation may be modified by other kinds of radiation so that antagonism or true synergy can occur. It has been demonstrated (10, 11) that prior exposure to UV radiation sensitizes microorganism to thermal killing. On the other hand, stronger effects were observed by Hill and Eidenow (12) in human solar erythema and by Freeman and Knox (13) in mice inflammatory reaction when the skin was heated before UV irradiation. Lipson and Baldes (14) found that heat adversely influenced photosensitisation by hematoporphyrin in mice. Kays et al. (15) demonstrated that the prior skin heating suppressed the phototoxic response to methoxsalen as evidenced by an increase in the threshold erythema dose, while no detectable effect were seen on sunburn erythema or phototoxic reaction provoked by anthracene. Other authors (13, 16) reported that heat enhanced the development and incidence of UV radiation-induced tumours in mice.

LITERATURE

1. International Commission on Illumination: "International Lighting Vocabulary", 3rd ed. Paris 1970 [Publication n. 17(E-1-1))]
2. Jacques,J.A. and Kuppenheim,H.F., 1955, Spectral reflectance of the human skin in the region 0.7-2.6 μm, J.Appl.Physiol., 8:305
3. Hardy, J.D., 1956, Spectral reflectance and transmittance of excised human skin, J.Appl.Physiol., 9:257
4. Ebaugh, F.G.J. and Thauer,R., 1950, Influence of various environmental temperature on cold and warmth thresholds, J.Appl.Physiol., 8:173
5. Hardy, J.D., Goodell,H., and Wolff,H.G., 1951, Influence of skin temperature upon pain threshold as evoked by thermal radiation, Science, 114:149,1951
6. Fisher,E., and Solomon,S., 1972, Physiological responses to heat or cold, in: Therapeutic Heat and Cold, S.Licht Ed., E.Licht, New Haven
7. Kollias,N. and Baqer,A., 1984, An experimental study of the chamges in pigmentation in human skin in vivo with visible and near infrared light, Photochem.Photobiol. 39:651
8. Wyss,V., 1965, Sugli effetti del calore radiante su regioni diverse del corpo umano. Nota 2, Med.Lavoro, 56:293
9. Wyss,V., 1966, Sugli effetti del calore radiante su regioni diverse del corpo umano. Nota 3, Med.Lavoro, 57:262
10. Giese,A.C. and Crossman,E.B., 1946, The action spectrum of sensitisation to heat with ultraviolet light, J.Gen.Physiol. 29:79
11. Tyrrel,R.M., 1976, Synergistic lethal action of ultraviolet radiation and mild heat in Escerichia coli, Photochem.Photobiol. 24:345
12. Hill,L. and Eidenow,A., 1923, Biological action of light: the influence of temperature, Proc.Roy.Soc.London, Biol. 95:163
13. Freeman,R.G. and Knox,J.M., Influence of temperature on ultraviolet injury, Arch.Dermatol. 89:858
14. Lipson,R.L. and Baldes,E.J., 1960, Photosensitivity and heat, Arch.Dermatol. 82:517
15. Kaidbey,K.H., Witkowski,T.A., and Kligman,A.M., 1982, The influence of infrared radiation on short-term ultraviolet-radiation-induced injuries, Arch.Dermatol., 118:315
16. Bain,J.A., Rusch,H.P., and Kline,B.E., 1943, Effects of temperature on ultraviolet carcinogenesis with wavelengths of 280-340 nm, Cancer Res. 3:610

POTENTIAL OCULAR AND SKIN HAZARDS FROM LAMPS AND PROJECTOR SYSTEMS*

James Franks

Laser-Microwave Division
US Army Environmental Hygiene Agency
Aberdeen Proving Ground, MD

INTRODUCTION

Few retinal injuries from non-laser sources of optical radiation have been reported in the literature. In fact, those that have been documented have not been thermal injuries but have been photochemical injuries from short wavelength visible light. It may be argued in fact that man-made non-laser sources other than the nuclear fireball are not capable of causing retinal thermal injury except under unusual conditions where large retinal image diameters are produced or where an individual overcomes his natural aversion to bright light and forces himself to stare into a source for a longer than normal time. Bare arc lamps that have no plastic or glass cover have been known to cause photokerititis or erethema from the ultraviolet B and C content in the lamp's output. Recent exposure guidelines such as those contained in the American Conference of Governmental Industrial Hygienists Threshold Limit Values Suggested Changes for 1990 (ACGIH, 1990) suggest that sources such as tungsten halogen lamps can exceed exposure limits for short wavelength visible light. This is commonly referred to as the "blue-light" hazard. Allistair McKinlay describes this standard in detail in another section of this proceedings.

LAMPS AND PROJECTOR SYSTEMS

There are two important properties of lamps and projector systems that make hazard analysis more difficult for them than for laser sources; the image size on the retina is not minimal as it is assumed to be in a laser exposure, and the source emits optical radiation over a wide spectral region. To properly evaluate hazards from a lamp or projection system you must determine the spectral irradiance or radiant exposure over the total range of output wavelengths and the retinal image diameter as a function of distance and viewing angle (Sliney and Wolbarsht, 1980). Retinal image diameters for a bare lamp can be calculated from geometrical optics while image diameters from projection systems are magnified by the reflective or transmissive optics used to collimate the source. The solid angle subtended by a bare lamp decreases with distance while the

*The opinions or assertions contained herein are the private views of the author and are not to be construed as reflecting the official views of the U.S. Department of the Army or the U.S. Department of Defense.

apparent solid angle subtended by a lamp that is collimated in a projection system remains constant with increasing distance near the lamp and starts decreasing with distance only after the so-called flash distance is reached. The radiometric quantity that combines irradiance and the solid angle subtended by the projected solid angle along the direction of observation is radiance. Fortunately this quantity is a property of the source and not the method uses to observe the source so that it is constant for a given lamp source even when collimated by a projection system.

The spectral content of the lamp or projection system determines whether the organ at risk is the eye or the skin and what parts of these two organs are at risk. The total irradiance determines potential thermal skin hazard. The effective ultraviolet (180 nm-400 nm) dose (effective UV irradiance x time) determines the photochemical hazard to the cornea (photokerititis) and the skin (erethema). The thermally spectrally weighted radiance from 400 nm-1400 nm is used to evaluate the potential for thermal injury to the retina. The blue-light weighted irradiance or the blue-light weighted radiance from 400 nm-700 nm is used to evaluate potential phototoxic hazards to the retina.

The table lists common types of lamps and the potential hazards associated with them.

TABLE 1. COMMON LAMPS AND POTENTIAL OCULAR OR SKIN HAZARDS

Lamp Type	Example	Potential Hazard	Note
Incandescent	tungsten-Halogen	UV, blue-light	UV hazard only for lamp with quartz envelope aversion response limits blue-light exposure.
	60 W light bulb	thermal contact	Don't touch hot lamp
Low pressure discharge	mercury/ germicidal	actinic UV	emits 95% of power at 254 nm
Flourescent	40 W cool white	theoretical UV hazard at lamp surface	glass envelope eliminates almost all actinic UV
High-intensity discharge	400 W mercury	retinal thermal and blue-light	can also be a UV hazard if outer glass envelope is damaged
Short-arc	xenon	thermal skin, actinic UV, blue-light, retinal thermal	extremely bright source so aversion response limits eye exposure

REFERENCES

American Conference of Governmental Industrial Hygienists (ACGIH), Threshold Limit Values and Biological Exposure Indices for 1989-1990, ACGIH, Cincinnati, OH, 1989.

Sliney, D.H. and Wolbarsht, M.R., Safety with Lasers and Other Optical Sources, Plenum, NY, 1980.

ULTRAVIOLET RADIATION AND THE EYE

David H. Sliney

Laser Microwave Division
US Army Environmental Hygiene Agency
Aberdeen Proving Ground, MD 21010-5422 USA

BACKGROUND

A number of in-vitro laboratory studies have shown that ultraviolet radiation (UVR) can produce optical changes in lens proteins that could explain at least one causal factor in some forms of cataractogenesis [1-7]. Cataracts have also been produced in rabbits from acute exposure to UVR in the 295-320 nanometer (nm) UV spectral band [8-10]. Additionally, several epidemiological and related studies have suggested a relationship between outdoor ambient UVR levels and the incidence of corneal degeneration and senile cataract [11-22]. Weale estimated that a factor of ten in the incidence rate of cataract was due to sunlight [23]. In addition to these delayed effects, one can experience the painful symptoms of snowblindness or welder's flash (photokeratoconjuctivitis, or simply photokeratitis) from acute UV exposures.

ULTRAVIOLET RADIATION EXPOSURE OF THE EYE

The ambient levels of UVR incident upon the eye vary dramatically with latitude, time of day (solar zenith angle), and geometry of exposure [24-27]. The first two factors have been extensively studied for meteorological purposes and to quantify UVR exposures for epidemiological studies of skin cancer [26-28], but the last factor--geometry of exposure--has often been neglected in epidemiological studies. The majority of 300-nm UVR incident upon the body is indirect, scattered radiation. Hence, one can actually be sunburned while sitting under a beach umbrella. The principal point to be made here is that the UVR exposure of the cornea and crystalline lens is from scattered radiation. Direct solar radiation normally does not strike the cornea during the hours of peak sunlight when UVR irradiances are highest (approx. 9:00 a.m. to 3:00 p.m. local time) [28].

Light, Lasers, and Synchrotron Radiation
Edited by M. Grandolfo *et al.*, Plenum Press, New York, 1990

The UVR incident upon the eye from ground reflections and from the sky near the horizon is of primary importance to any epidemiological study of UVR induced corneal changes, cataractogenesis, pterigia, or potential retinal changes.

There are many misconceptions regarding the risks of exposure to UVR. Few people in the general public would question the health benefits of sunlight. Many flock to the beaches each summer to develop a "healthy" tan. Since the 1920's scientists have recognized that most of the benefits--and risks--of sunlight exposure result from the UVR present in sunlight. Dermatologists warn sunbathers to avoid exposure or protect themselves against the intense midday UVR, or risk skin cancer. A growing number of scientists warn of hazards to the eye if UVR--perhaps even shorter visible wavelengths--are not filtered by lenses. In addition to any intentional exposure for health or cosmetic purposes, many people are exposed to UVR without being aware of it. UVR penetrates clothing, and people indoors are exposed to UVR from artificial lighting.

Despite a substantial literature on the adverse effects of UVR upon the eye, controversy still surrounds the question of what level is safe for chronic exposure and the degree to which natural and man-made light sources pose realistic hazards under most normal living conditions. The uncertainty with regard to ocular exposure is particularly significant. Scientific evidence accumulates to indicate that chronic exposure conditions may accelerate aging processes in the retina, but the question of: "How much is safe?" remains to be conclusively answered. Occupational exposure limits (EL's) exist for acute exposure to UVR [29-32], but these may not be sufficient to assure total protection from delayed effects from chronic, repeated UVR exposure. In the industrial work environment from arc-welding production lines to the laser laboratory, exposure can be controlled, but persons will only follow safety measures if they understand that a real hazard exists.

Occupational Exposure Limits

As with most other chemical and physical hazards in our environment, current scientific knowledge defines levels of UVR exposure that are clearly hazardous and, to a lesser extent, those levels which are clearly beneficial or apparently harmless. An intermediate zone between clearly safe and clearly hazardous also may exist and may be within the normal physiological range, i.e., where the majority of the population is exposed (as may be the case with customary environmental solar ultraviolet radiation in some moderate climates). Occupational exposure limits (EL's) developed for protection against conventional arc sources [29-32] were adopted for occupational exposure to UV laser radiation [33-34]. Studies of skin carcinogenesis clearly show that because of the nature of UVR damage to biological tissue, there is no clear evidence that damage at the molecular level will always be repaired withing every cell [35-36]. The present IRPA EL's which are widely accepted are given in Table 1.

These were recently updated in the less hazardous UV-A (315 - 400 nm) spectral band. Although exposures to optical radiation in the physiological range have generally been accepted as safe, greater knowledge today suggests that the absense of acute effects does not mean that some damage has not occurred at the molecular level. Delayed effects appear to be possible from UVR exposure to the eye as well as the skin. Furthermore, growing evidence suggests that retinal degradation may also occur from

TABLE 1
ULTRAVIOLET RADIATION EXPOSURE LIMITS AND SPECTRAL WEIGHTING FUNCTION
IRPA/INIRC 1988 REVISION

Wavelength* (nm)	EL (J/m)	EL (mJ/cm)	Relative Spectral Effectiveness S
180	2,500	250	0.012
190	1,600	160	0.019
200	1,000	100	0.030
205	590	59	0.051
210	400	40	0.075
215	320	32	0.095
220	250	25	0.120
225	200	20	0.150
230	160	16	0.190
235	130	13	0.240
240	100	20	0.300
245	83	8.3	0.360
250	70	7.0	0.430
254#	60	6.0	0.500
255	58	5.8	0.520
260	46	4.6	0.650
265	37	3.7	0.810
270	30	3.0	1.000
275	31	3.1	0.960
280#	34	3.4	0.880
285	39	3.9	0.770
290	47	4.7	0.640
295	56	5.6	0.540
297#	65	6.5	0.460
300	100	10	0.300
303#	250	25	0.190
305	500	50	0.060
308	1,200	120	0.026
310	2,000	200	0.015
313#	5,000	500	0.006
315	1.0×10^4	1.0×10^3	0.003
316	1.3×10^4	1.3×10^3	0.0024
317	1.5×10^4	1.5×10^3	0.0020
318	1.9×10^4	1.9×10^3	0.0016
319	2.5×10^4	2.5×10^3	0.0012
320	2.9×10^4	2.9×10^3	0.0010
322	4.5×10^4	4.5×10^3	0.00067
323	5.6×10^4	5.6×10^3	0.00054
325	6.0×10^4	6.0×10^3	0.00050
328	6.8×10^4	6.8×10^3	0.00044
330	7.3×10^4	7.3×10^3	0.00041
333	8.1×10^4	8.1×10^3	0.00037
335	8.8×10^4	8.8×10^3	0.00034
340	1.1×10^5	1.1×10^4	0.00028
345	1.3×10^5	1.3×10^4	0.00024
350	1.5×10^5	1.5×10^4	0.00020
355	1.9×10^5	1.9×10^4	0.00016
360	2.3×10^5	2.3×10^4	0.00013
365#	2.7×10^5	2.7×10^4	0.00011
370	3.2×10^5	3.2×10^4	0.000093

TABLE 1 (Continued)
ULTRAVIOLET RADIATION EXPOSURE LIMITS AND SPECTRAL WEIGHTING FUNCTION
IRPA/INIRC 1988 REVISION

Wavelength* (nm)	EL (J/m)	EL (mJ/cm)	Relative Spectral Effectiveness S
375	3.9×10^5	3.9×10^4	0.000077
380	4.7×10^5	4.7×10^4	0.000064
385	5.7×10^5	5.7×10^4	0.000053
390	6.8×10^5	6.8×10^4	0.000044
395	8.3×10^5	8.3×10^4	0.000036
400	1.0×10^6	1.0×10^5	0.000030

Notes:

*Wavelengths chosen are representative; other values should be interpolated at intermediate wavelengths.

#Emission lines of a mercury discharge spectrum.

chronic exposure to short wavelength light [37-40], and possibly from the trace amount of UV-A that reaches the retina as well. Considering the enormous range of genetically determined individual and ethnic susceptibility to both UV induced skin cancer [34-36], and to the accelerated onset of cataract and skin ageing, it is easy to see that realistic guidelines for chronic exposure present a challenge. The present exposure limits probably reduce the risk from all of these delayed effects quite significantly, but one cannot argue that the risk is zero. As with almost all physical and chemical agents, one cannot speak of "zero risk" -- it is scientifically specious [42].

Most UVR effects are photochemical in nature and have the characteristic reciprocity between irradiance and exposure duration. The same injury can be produced by a low irradiance (e.g., reflected UVR from a laser) repeated over a daylong period. By contrast thermal injury is affected by heat conduction and requires a very intense exposure within seconds to cause a burn, otherwise surrounding tissue conducts the heat away.

In addition to the threshold exposure dose, one must also consider the action spectrum. An action spectrum is the relative effectiveness of different wavelengths to elicit the response. Action spectra indicate the relative risk of exposure to different laser wavelengths. Wavelengths in the UV-C band, although representing more energetic photons, hardly penetrate tissue, whereas wavelengths in the UV-B (280-315 nm) are in a band where photons penetrate and also have sufficient energy to cause significant injury.

Biological Injury Mechanisms

Ultraviolet radiation, like other optical radiations, is considered "non-ionizing radiation" because photon energies for UV wavelengths greater than about 200 nm are insufficient to individually ionize atoms found in important biological molecules. Since a threshold appears to exist for each acute biological effect, such as UV erythema and photokeratits, the acute biological effects of ultraviolet radiation are considered non-stochastic. In the ultraviolet and visible regions of the spectrum, photochemical damage mechanisms are demonstrable. Thermal injury mechanisms dominate for most pulsed laser exposures and infrared radiation exposures. However, the concept of subcellular stochastic biological injury mechanisms similar in some respects with those applicable to ionizing radiation (e.g., X rays), probably applies to delayed effects such as skin cancer.

There are at least five separate types of adverse effects to the eye from UVR exposure which have been considered in the development of EL's:

Photokeratoconjunctivitis (acute inflammation of the cornea and conjunctiva as in "welders' flash") has been defined for wavelengths from 200 nm to 400 nm. Cataractogenesis (lens cataract) has been demonstrated principally in the wavelength range from 290 to 320 nm, and perhaps occurs at greater wavelengths [7-10]. The Xe-Cl 308-nm excimer laser should therefore be considered a "cataract machine."

The development of pterygium (a vascular growth over the eye's surface, including the cornea) and spheroidal degeneration (loss of sphere) of the cornea appears to be related to chronic environmental UVR exposure.

"Blue-light" photochemical injury to the retina of the eye has been defined principally from 400 nm to 550 nm in the visible; but, wavelengths from 310 nm to 400 nm in the UV can also produce effects in aphakic eyes (i.e., eyes with the lens surgically removed), and possibly even somewhat in normal eyes [39-40].

Accelerated aging of the macula (age-related macular [central retinal] degeneration) may also be related to chronic retinal exposure to the trace amounts of UV-A that reach the retina (possibly including even wavelengths as short as 300 nm) and short-wavelength visible light [34,39,40, 43]. Figure 1 summarizes these effects as a function of spectral band.

Individual biological molecules are damaged by ambient, low-level UVR all the time, but the natural on-going processes of renewal at the cellular level maintain an equilibrium condition with these damaging processes [12]. Thus deficiencies in an individual's natural repair mechanisms may render the individual more susceptible to injury from optical radiation--especially in the UV spectrum. For accute exposures, such increased sensitivities are referred to as "photosensitive" reactions and are most notable for skin reactions. Other photosensitive reactions can occur as a result of exposure to a photosensitizing chemical. Dealing with the synergistic effects of photosensitizers adds just one more problem in setting realistic safety standards. Some common drugs, such as the common antibiotic, tetracyline, are photosensitizers.

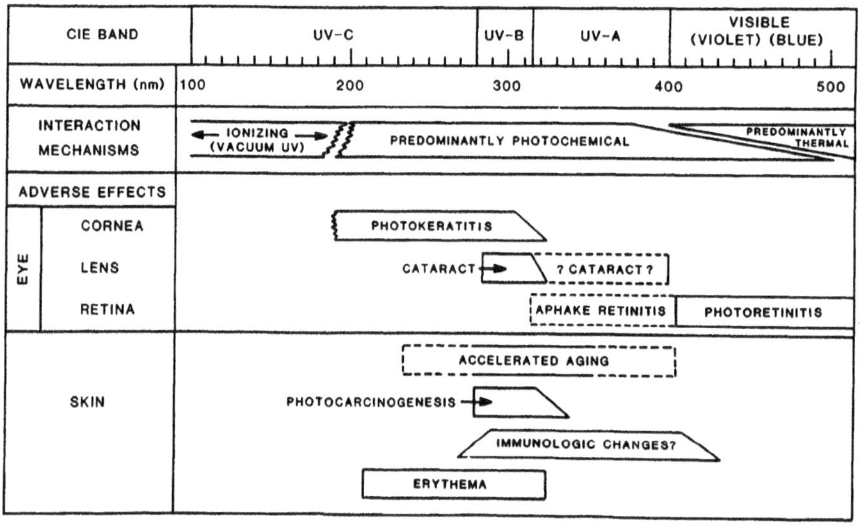

Figure 1. Adverse Biological Effects Attributed to Ultraviolet Radiation and Short-Wavelength Light.

Although individual and racial factors play an important role in determining the susceptiblity of the skin to injury from exposure to UVR, there is no evidence of such variability for photokeratitis. The susceptibility to UVR photochemical injury of the skin varies widely with racial factors, thickness of the stratum corneum and degree of pigmentation (i.e., presence of melanin pigment). These factors simply do not apply with regard to the eye.

The Geometry of Ocular Exposure

The more one studies the biological effects of UVR upon the eyes, one is struck by the enormous importance of the geometry of exposure. The exposure of the eye and skin to direct UVR in the indoor environment differs from outdoor exposure both spectrally and spatially. Mankind appears to be remarkably well adapted to our natural sunlight environment, and customs of dress play an added importance in protecting both the skin and eye in arctic and desert environments [38]. When the sun is near zenith at noontime, the spectral irradiance at 300 nm is ten times greater than three hours prior to or following noontime. For this reason dermatologists warn sun-bathers to remain out of the sun during midday hours. Unfortunately, not too many beach devotees follow this advise. Natives of tropical climes often follow this dictum and go indoors for a siesta during this period, but not tourists from North America.

Although the cornea is more sensitive to UVR injury than the skin, we seldom experience a corneal burn when out in sunlight. Again, the geometry of exposure helps. When the sun is overhead and UVR exposure is most severe, the brow ridge and upper lid shield the cornea, and the more intense scattered UVR from overhead strikes the cornea at a grazing incidence where most is reflected and little is absorbed. Only when the incident UV rays are parallel to the pupillary axis are most rays aborbed (in fact, 98 % are absorbed). When one looks down at the snow the UVR is reflected directly into the eye, hence the eskimos' traditional eye protector, the slit in whalebone, provided geometrical rather than spectral protection aginst the UVR exposure. The strong angular dependence of Fresnel reflection explains the survivability of the cornea in an overhead bath of UVR, but also the glare experienced over water. Only approximately 2-4% of overhead sunlight is reflected from water; but when the sun is low in the sky, much of the incident light is reflected, and the UVR and blue light are then filtered out of the direct rays by the atmosphere--and are therefore harmless.

Protective Measures

Enclosures and baffles, clothing and goggles are all easily employed in the research environment [5,14]. Studies of the filtration of clothing textiles show that most clothing has an optical density of four or greater and that risk is only assumed when wearing open-weave fabrics [5,11].

A key factor to remember is that UVR effects are linearly additive over a period of about a day; hence, exposure to reflected UVR on days when a laser is operated infrequently does not prove that injuries will not result on another day when the laser is operated many more times.

Since the beam of excimer lasers is also not visible, or if visible due to the fluorescence of particulates and target materials will not appear at all bright, one can expect more accidents with this type of laser. Scientists working with excimer lasers must be cautioned against thinking that a lack of brightness indicates a less intense beam.

References

1. Pirie A: Photo-oxidation of proteins and comparison of photo-oxidized proteins with those of cataractous human lenses. Isr Med Sci 8:1567-1573, 1972.
2. Lerman S: Human ultraviolet radiation cataracts. Ophthalmic Res 12:303-314, 1980.
3. Zigman S, Schultz J, Yulo T, and Griess G: The binding of photo-oxidized tryptophan to a lens gamma-crystallin. Exp Eye Res 17:209-215, 1973.
4. Borkman R and Lerman S: Evidence for a free-radical mechanism in aging and UV irradiated ocular lenses. Exp. Eye Res 25:303-309, 1977.
5. Kurzel RB, Wolbarsht ML, and Yamanashi BS: Ultraviolet radiation effects on the human eye. In: Photochemical and Photobiological Reviews, vol 2, Smith KC. editor, New York, Plenum Press, 1977, pp 133-167.
6. Dillon J, Garner MH, Roy D, and Spector A: The photolysis of lens proteins: Molecular changes. Exp Eye Res 34:651-658, 1982.
7. Yamanashi BS, Gacker H, and Klintworth GK: Wavelength dependence and kinetics of UV-induced free radical formation in the human cornea and lens. Photochem Photobiol 30(3):391-395, 1979.
8. Bachem A: Ophthalmic ultraviolet action spectra. Am J Ophthalmol 41:969-975, 1956.
9. Pitts DG, Cullen AP, and Hacker PD: Ocular effects of ultraviolet radiation from 295 to 365 nm. Invest Ophthalmol Vis Sci 16(10):932-939, 1977.
10. Cullen AP: Additive effects of ultraviolet radiation. Am J Optom Physiol Opt 57(11):808-814, 1980.
11. Roger FC, Cuthill JA, Fyvelor PJ, and Lenham AP: Ultraviolet radiation as a possible cause of corneal degenerative changes under certain physiographic conditions. Acta Ophthalmologica 52:777-785, 1974.
12. Taylor HC: The environment and the lens. Brit J Ophthalmol 64:303-310, 1980.
13. Van Heyningen R: The human lens. I. A comparison of cataracts extracted in Oxford (England) and Shikarpur (W Pakistan). Exp Eye Res 13:136-147, 1972.
14. Hiller R, Giacometti L, Yuen K: Sunlight and Cataract: An epidemiological investigation. Am J Epidemiol 105:450-459, 1977.
15. Wright RE: The possible influence of solar radiation on the production of cataract in certain districts of southern India: a preliminary investigation: Indian J Med Res 24:917-930, 1936.
16. Young JDH and Finley RD: Primary spheroidal degeneration of the cornea in Labrador and Northern Newfoundland. Am J. Ophthalmol 71(1):129-134, 1973.
17. Hollows F and Moran D: Cataract--the ultraviolet risk factor. Lancet 2:1249-1250, 1981.
18. Goldsmith R, Rothhammer F, and Schull W: The multinational Andean genetic and health program. III. Ophthalmic disease and disability among the Aymara. Bull Pan Amer Health Organiz 13:58-65, 1980.
19. Vines AP: An epidemiological sample survey of the highlands, mainland and island regions of the territory of Papua and New Guinea. Port Moresby, New Guinea, Government Printer, 1967.
20. Seung WS: A survey of senile cataracts among high altitude living Tibetans in Chang-du district, Tibet. Chinese J Ophthalmol 15:100-104, 1979.
21. Brilliant LB, Grasset NC, Pokhrel RS, Kolstad A, Lepkowski JM, Brilliant GE, Hawks WN, and Pararajasegaram R: Associations among cataract prevalence, sunlight hours, and altitude in the Himalayas. Am J Epidemiol 118(2):250-264, 1983.
22. Zigman S, Datiles M, and Torczynski E: Sunlight and human cataracts. Invest Ophthalmol Vis Sci 18:462-467, 1979.

23. Weale R: Senile cataract, the case against light. Ophthalmology 90:420-423, 1983; see also: Brit J Ophthalmol 66(1):31-34, 1982.
24. Sliney, DH and Wolbarsht ML: Safety with Lasers and Other Optical Sources, New York, Plenum Publishing Corp., 1980, pp. 101-245.
25. Henderson, : Daylight and Its Spectrum. Elsevier, New York check.
26. Garrison LM, Murray LE, Doda DD, and Green, AES: Diffuse--direct ultraviolet ratios with a compact double monochromator. Appl Opt. 17(5):827-835, 1978.
27. Sliney, DH: Eyc protective techniques for bright light. Ophthalmology 90(8):937-944, 1983.
28. Sliney DH (1986) Physical factors in cataractogenesis: Ambient ultraviolet radiation and temperature, Invest Ophthalmol Vis Sci 27(5): 781- 790.
29. American Conference of Governmental Industrial Hygienists: Documentation of the Threshold Limit Values. 4th Edn. pp. 451-455, Cincinnati, ACGIH, 1980.
30. ACGIH (1987) TLV's, Threshold Limit Values and Biological Exposure Indices for 1987-1988, American Conference of Governmental Industrial Hygienists, Cincinnati, OH.
31. Sliney DH (1987) Unintentional exposure to ultraviolet radiation: Risk reduction and exposure limits, In: Human Exposure to Ultraviolet Radiation: Risks and Regulations (Eds: Passchier WF, Bosnjakovic BFM), New York, Excerpta Medica Division, Elsevier Science Publishers, pp 425-437.
32. WHO (1980), Environmental Health Criteria No. 14, Ultraviolet Radiation, joint publication of the United Nations Environmental Program, the International Radiation Protection Association and the World Health Organization, Geneva.
33. ANSI (1986) Safe Use of Lasers, Standard Z-136.1-1986, American National Standards Institute, New York, published by Laser Institute of America, Toledo.
34. IRPA/INIRC (1989) Guidelines for exposure to ultraviolet radiation, Health Physics, Jun 89
35. Forbes PD, Davies PD (1982) Factors that Influence Photocarcinogenesis, in: (J.A. Parrish, M. L. Kripke, and W.L. Morison, Eds.) Chapter 7, "Photoimmunology," Plenum Publishing Corp., New York.
36. Cole CA, Forbes, DF, Davies PD (1986) An action spectrum for UV photocarcinogenesis, Photochm Photobiol 43(3):275-284.
37. Pitts DG, Cullen AP, Hacker PD (1977) Ocular effects of ultraviolet radiation from 295 to 365 nm, Invest Ophthal Vis Sci 16(10):932-939.
38. Ham WT, Jr, Mueller HA, Ruffolo JJ, Jr, Guerry D, III Guerry RK (1982) Action spectrum for retinal injury from near-ultraviolet radiation in the aphakic monkey, Am J Ophthalmol 93(3): 299-306.
39. Mainster MA (1978) Spectral transmission of intraocular lenses and retinal damage from intense light sources, Am J Ophthalmol 85:167-170.
40. Yanuzzi, LA, Fisher YL, Krueger A, Slakter J. (1987) Solar retinopathy, a photobiological and geophysical analysis. Tr Am Ophthalmol Soc, 85:120-158.
41. Sliney DH, Benton RE, Cole HM, Epstein SG, Morin CJ (1987) Transmission of potentially hazardous actinic ultraviolet radiation through fabrics, Appl Ind Hyg 2(1):36-44.
42. Sliney, DH, The Optical Spectrum: Laser and Ultraviolet Sources and Standards, at Lloyd's Conference on Radiation and Energy, Dec 1983.
43. Bos AJJ, DeHaas MP (1987) On the safe use of a high power ultraviolet laser, In: Human Exposure to Ultraviolet Radiation: Risks and Regulations (Eds: Passchier WF, Bosnjakovic BFM), New York, Excerpta Medica Division, Elsevier Science Publishers, pp 377-382.
44. Boettner EA: Spectral Transmission of the Eye. Final Report AF 41-609-2996, USAF School of Aerospace Medicine, Brooks AFB, TX. DTIC No. AD 663-246, July 1967.

ULTRAVIOLET MUTAGENESIS

Ileana Quinto, Massimo Mallardo, Maria R. Ruocco
Alessandro Arcucci, and Giuseppe Scala

Dipartimento di Biochimica e Biotecnologie Mediche, II
Facoltà di Medicina e Chirurgia, Università degli Studi
di Napoli, Via S.Pansini 5, I-80131 Napoli, Italy

SUMMARY

Ultraviolet radiation is a powerful mutagen in eukaryotes and prokaryotes due to its ability to produce premutational lesions of DNA. Of the damage induced by UV-irradiation of DNA at 260 nm, the cyclobutane-type dipyrimidine and the pyrimidine-pyrimidine (6-4) lesions at sites of adjacent pyrimidines are principal cause of mutations. These bulky lesions greatly modify the structure and function of DNA and cause block of DNA replication and cellular death. Prokaryotic and eukaryotic cells are able to repair the DNA lesions as for example the UV-induced pyrimidine dimers. For example, in E.coli, a set of repair enzymes (uvr A,B,C) recognize and eliminate the length of structurally altered DNA, so that resynthesis of DNA can proceed. Thus, the cell survives unmutated to the DNA damage. If the DNA damage is unrepaired, mutations are fixed thanks to a complex interplay of factors e.g. the site of DNA lesion, the replication machinery and, at least in E.coli, a few UV-inducible cell functions.
This paper reviews the mechanisms of UV mutagenesis in E.coli and mammalian cells together with the qualitative and quantitative analysis of UV-induced mutations.

UV MUTAGENESIS IN ESCHERICHIA COLI

Genetic and biochemical studies on the mutagenic effects of UV irradiation led to the discovery of DNA damage-inducible functions in E.coli (for review: Witkin, 1976; Walker, 1984). Weigle first reported evidence of UV-inducible functions involved in DNA repair and mutagenesis (Weigle, 1953). The survival and mutations of UV-irradiated lambda bacteriophage were shown to be increased in UV-preirradiated E.coli (W reactivation and W mutagenesis). Later, Defais et al.(1971, 1976) demonstrated that the W reactivation and W mutagenesis is recA- and lexA-dependent and that it requires de novo protein synthesis. Similar results were obtained for the UV mutagenesis of the bacterial chromosome (Witkin, 1967, 1969, 1976; Miura and Tomizawa, 1968; Kondo, 1969; Mount, 1972).
In the wake of these results, Radman (1974) hypothesized that "SOS" repair functions are temporarily induced in E.coli after UV irradiation, to allow the DNA replication machinery to proceed across the blocking lesions represented by the UV-induced pyrimidine dimers. The term "SOS"

indicated the alarm situation created in the cell by the presence of lesions causing the arrest of DNA replication.
The SOS hypothesis was confirmed by subsequent genetic and biochemical studies from many laboratories (for rewiew: Walker, 1984). Although some aspects of SOS mutagenesis are still unanswered, the present knowledge of the SOS response can be summarized in the following scheme (Fig. 1).

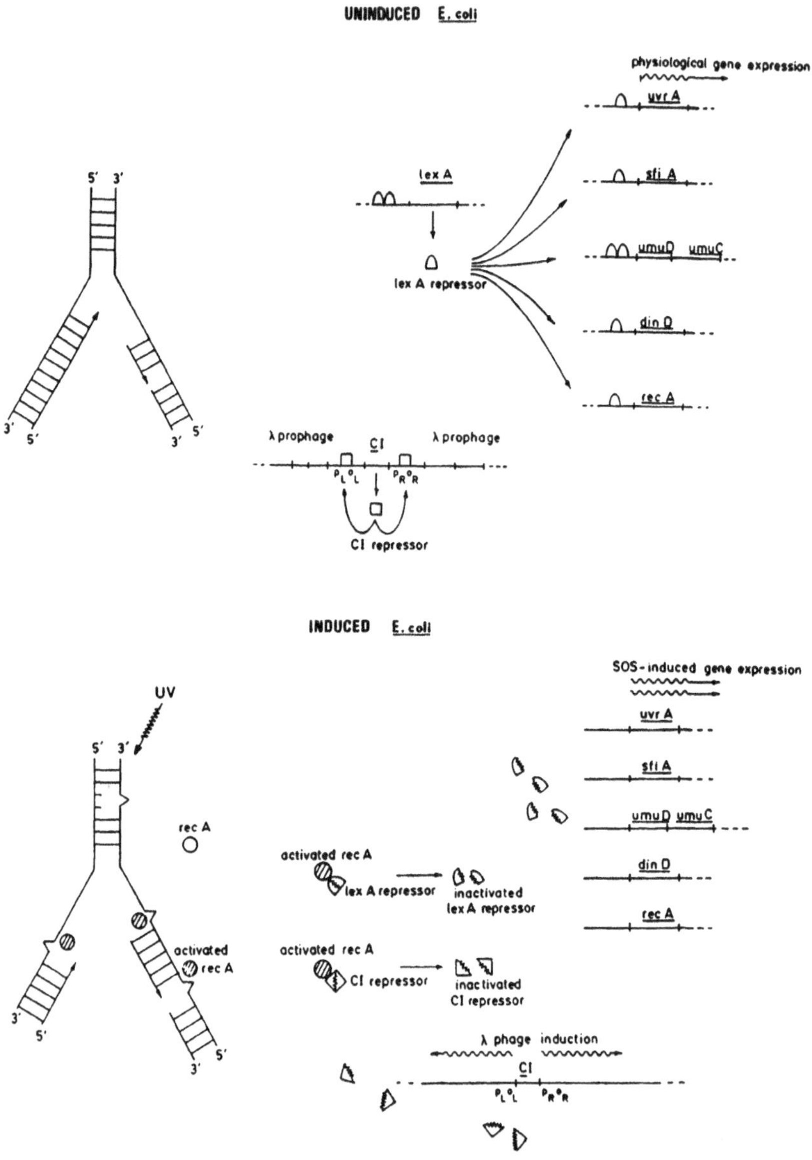

Fig. 1. SOS regulatory system in E. coli. In uninduced E. coli, the LexA protein represses the SOS genes by binding to their operator sequences. At the same time, the lysogenic lambda bacteriophage is repressed by CI repressor. In UV-induced E.coli, RecA is activated to behave as a protease which cleaves the LexA and CI repressors, thus allowing the expression of SOS genes and the induction of lambda prophage.

In E.coli, mutagens that produce non coding lesions of DNA interrupt the replication of DNA and induce a set of genes named SOS. These are 19 genes (table 1) whose expression is regulated by a common repressor, the Lex A protein (for review: Walker, 1984). Once the DNA is damaged, an unknown effector signal (probably single stranded DNA) is produced that activates the Rec A protein to behave as a protease (Rec A*) (Craig and Roberts, 1980, 1981; Roberts and Devoret, 1983; Cohen et al.,1983).
Rec A* catalyses the proteolytic cleavage of Lex A (Little et al., 1980; Horii et al., 1981), the SOS-induced UmuD and MucA proteins (Shinagawa et al., 1988; Burkhardt et al., 1988) and the repressors of certain temperate prophages (lambda, 434, P22) (Roberts et al, 1978; Sauer et al.,1982; Eguchi et al., 1988), thus allowing the full expression of SOS genes and the induction of prophages (Fig.1).

Table 1. SOS genes of E.coli

Gene	Function
uvr A	Excision of pyrimidine dimers and bulky lesions.
uvr B	As above.
uvr C	As above.
uvr D	DNA helicase II; roles in excision repair and methyl-directed mismatch repair.
rec A	Recombination; post-replication repair; as Rec A* role in repressors cleavage and mutagenesis.
rec N	DNA double-strand breaks repair.
ruv	Recombinational repair.
ssb	Single strand DNA binding protein; regulatory role in SOS induction.
umu D	Mutagenesis in bacteria and targeted mutagenesis in double strand phages (required for continued replication past the dimer-mismatch region?).
umu C	As above.
muc A	Plasmid-encoded analogue of umuD.
muc B	Plasmid-encoded analogue of umuC.
sfi A	Cell division inhibitor; filamentation.
lex A	SOS repressor.
him A	Site specific recombination.
din A	Unknown function.
din B	Unknown function.
din D	Unknown function.
din F	Unknown function.

The UV mutagenesis results from the SOS-dependent replication of UV-damaged DNA (for review: Witkin, 1976; Walker, 1984). UV photoproducts, such as pyrimidine dimers, are known to block the replication fork of DNA at, or just before the site of the lesions (Setlow, 1963; Benbow et al., 1974; Moore et al., 1981; Villani et al., 1978). According to the two-step model of UV mutagenesis (Bridges and Woodgate, 1984, 1985), the SOS functions Rec A* and Umu D,C are required to bypass the site of non coding lesions on the DNA template (Fig.2). In a first step, Rec A* interacts with the replication machinery (probably inhibiting the proofreading function of DNA polymerases) and allows the insertion of an incorrected base opposite to the site of the UV photoproduct. Successively, Umu D and Umu C force the progression of the replication machinery beyond the site of the lesion, restoring the

elongation of the DNA synthesis (Fig.2).
In conclusion, the interaction of Rec A* and Umu D,C with the damaged DNA template and the replication machinery of DNA results in the translesional synthesis of UV-damaged DNA templates (i.e. survival of the cell) together with the fixation of mutations opposite to the site of non coding lesions (UV targeted mutagenesis).
An increased number of mutations of unirradiated lambda bacteriophage in UV-irradiated E.coli hosts has also been reported (Ichikawa-Ryo and Kondo, 1975; Quillardet and Devoret, 1982; Wood and Hutchinson, 1984). This mutagenesis, defined "untargeted", is possibly due to the SOS-induced decrease in the fidelity of the replication machinery on undamaged portions of the genome.

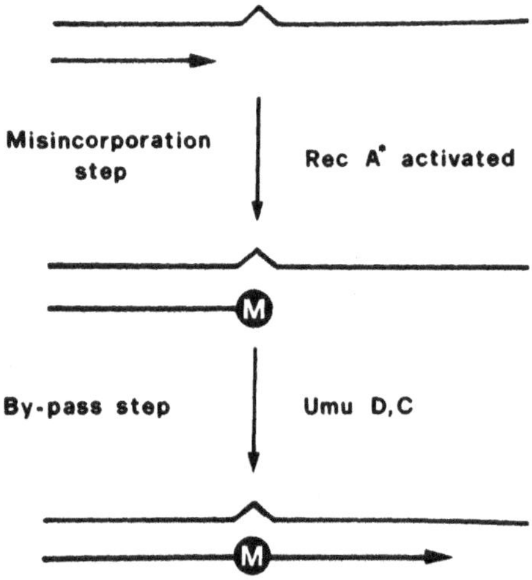

Fig. 2. Two-step model of UV mutagenesis. Bridges and Woodgate (1985) have proposed the two-step bypass synthesis of DNA-non coding lesions. In a first step, Rec A* influences the misincorporation of an incorrect base opposite the noninformational lesion. In a second step, UmuD and UmuC act by facilitating the use of the terminus of the strand containing the misincorporated base opposite the lesion as a primer for continued elongation.

UV MUTAGENESIS IN MAMMALIAN CELLS

Point mutations and chromosomal rearrangements are produced by UV irradiation in both eukaryotes and prokaryotes. However, while the fine genetics of E.coli led to the identification of a few DNA damage-inducible functions involved in the complex process of mutagenesis, in eukaryotes the mechanisms of UV mutagenesis are still poorly understood. In particular, there is no direct evidence that mammalian cells possess SOS-like functions that are involved in mutagenesis. Due to the differences in the genomic organization and

modality of DNA replication, the fixation of mutations in UV-damaged eukaryotic genome could require cell functions not necessarily inducible or, at least, acting differently from those discovered in E.coli.
There is evidence in support of UV-inducible functions in mammalian cells that could be related to the DNA repair and mutagenesis.
Enhanced virus reactivation and enhanced virus mutagenesis, similar to the W reactivation and W mutagenesis (Weigle, 1953), which are known to be part of the SOS response in E.coli, have been reported in mammalian cells. They consist in increased survival and mutagenesis of UV-irradiated viruses (papovavirus, herpes simplex, adenovirus and parvovirus) in UV-irradiated host cells as compared to unirradiated hosts (Das Gupta and Summers, 1978; Day and Ziolkowski, 1978; Lytle et al., 1980; Sarasin and Benoit, 1980; Mezzina et al., 1981; Cornelis et al., 1982; Gentil et al., 1982). Moreover, in analogy with the SOS-dependent prophage induction in E.coli (Witkin, 1976), induction of animal viruses from transformed mammalian cells after UV treatment has been demonstrated (Zamansky et al., 1980).
Genetic changes such as gene amplification (Lucke-Huhle and Herrlich, 1986) or increased gene expression (Angel et al., 1986; Fornace et al., 1988; Kartasova et al., 1988) represent the immediate UV response in mammalian cells. They are rapidly induced either directly by UV treatment of the cells (Angel et al., 1986; Lucke-Huhle and Herrlich, 1986; Fornace et al., 1988; Kartasova et al., 1988) or indirectly by the uptake of UV-damaged DNA (Herrlich et al.,1989). UV-induced RNA messengers (Angel et al., 1986; Fornace et al., 1988; Kartasova et al., 1988) and proteins (Mallick et al., 1982; Rahmsdorf et al., 1982; Schorpp et al., 1984; Buscher et al., 1988) have been reported. All the UV-induced functions identified to date (collagenase, DNA ligase II, metallothionein I and II, plasminogen activator) are not implicated directly in mutagenesis. However, there are nucleoproteins involved in transcription or replication (fos, NFkB).
Herrlich et al. (1989) have recently suggested that DNA damage could activate a cascade of events resulting in the enhanced expression and/or activation of transcription or replication nucleoproteins. These nuclear factors would be implicated in immediate UV responses like the amplification and/or the enhanced expression of determined genes. Long lasting or permanent consequences of these events at the genomic level would be gene rearrangements and mutations.

SPECTRUM OF UV-INDUCED MUTATIONS

The mutational spectrum of UV radiation (260 nm) was studied in E.coli (Coulondre and Miller, 1977 a,b), phages (Brandenburger et al., 1981; Le Clerc and Istock, 1982; Wood et al., 1984) and mammalian viruses (Bourre and Sarasin, 1983).
The most extensive analysis of UV-induced mutations in E.coli and mammalian cells was performed using the lacI system (Coulondre and Miller, 1977, a, b). In this genetic system, the chromosomal target of mutations is the lacI gene that codes for the repressor of the lac operon. By genetic and sequencing analysis of the mutated LacI gene, rescued from UV-irradiated E.coli (Coulondre and Miller, 1977b) or mammalian cells (Lebkowski et al., 1985), it was shown that the UV radiation induces targeted mutations, i.e. mutations occurring opposite the sites of specific photolesions.
The pyrimidine-pyrimidine sequence was identified as the principal site of occurrence of mutations, the pyrimidine on the 3' side being the preferential base for modifications (Coulondre et al., 1978; Miller, 1982). Moreover, the mutational hot spots occurred at the sites of UV-induced photoproducts represented by (6-4)pyC and pyrimidine dimers (Fig.3) (Brash and Haseltine, 1982). Some data indicated that the

(6-4)PyC lesion is the principal premutational lesion of UV irradiation (Brash and Haseltine, 1982; Wood, 1985; Glickman, 1986).
The types of UV-induced mutations were either base-pair substitutions or frameshifts (-1 base-pair), the GC ----> AT transition being largely favoured among all base-pair substitutions (for review: Miller, 1985).
The similarity of the spectrum of mutations produced in lacI after UV irradiation of E.coli (Coulondre and Miller, 1977b) and mammalian cells (Lebkowski et al., 1985) suggests that analogous mechanisms of UV mutagenesis are possibly involved in both prokaryotes and eukaryotes.

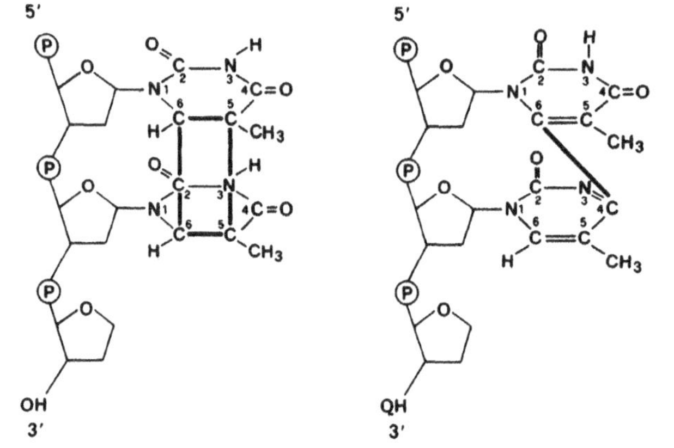

Fig. 3. UV photoproducts principally involved in mutagenesis. The cyclobutyl pyrimidine dimer is produced by saturation of the 5,6 double bonds of two adjacent pyrimidines, resulting in the formation of a 4-membered cyclobutyl ring linking the two pyrimidines. The (6-4)pyC lesion is produced by linkage between the C^6 position of the 5' pyrimidine and the C^4 position of the adjacent 3' pyrimidine.

GENOTOXIC POTENCY OF UV RADIATION

As consequence of DNA damage, a mutagen is able to produce point mutations and/or rearrangements of DNA in both prokaryotes and eukaryotes. Mutations can activate oncogenes and eventually initiate the neoplastic process. Thus, the evaluation of the genotoxic potency of a chemical is indicative of a risk for genetic events possibly involved in the development of cancer. This line of reasoning prompted several short-term tests in bacteria, yeast and mammalian cells to assess the risk of mutagenesis.
To compare the genotoxic potency of UV radiation to other physical and chemical carcinogens, we analysed three genetic and epigenetic effects induced in E.coli following UV irradiation, gamma irradiation or treatment with monofunctional and bifunctional alkylating agents (table 2; Quinto and Radman, 1987; Quinto et al., 1990).
A set of E.coli tester strains (Toman et al., 1985; Tenenbaum et al., 1988) were constructed that allowed the simultaneous detection of mutation (M), recombination (R) and SOS-induction (I) produced by DNA damage. For M, R and I, dose-effect curves were evaluated (Fig.4) where

the effective dose was represented by lethal hits overcoming the problem of physical dimensions of the dose (joule, rad, microgram, microliter) in the comparative analysis of the physical or chemical agents (Quinto and Radman, 1987). The M, R and I potency was measured as the integral under the yield-dose curve (Fig.4); it represents the total number of affected cells that had survived all the effective treatment (i.e. doses producing 0-100% lethality) with the chemical or physical agent (Quinto and Radman, 1987).

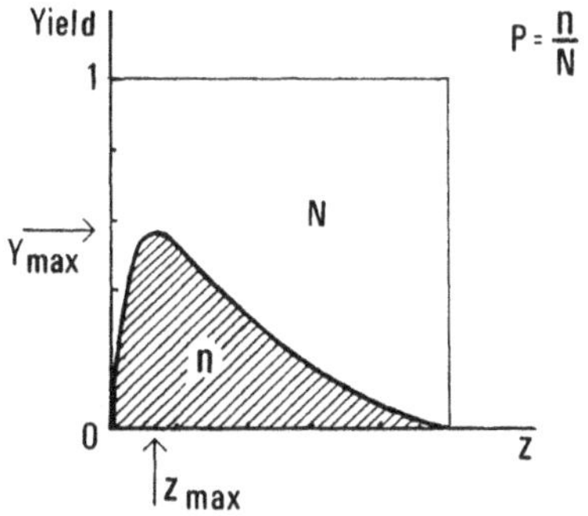

Fig. 4. Dose-response curve for the evaluation of the mutagenic, or recombinogenic, or SOS-inducer potency in the E.coli multitest. The potency (P) represents the cumulative yield of the mutated, or recombinated, or SOS-induced cell population (n/N in this figure) over the entire effective dose range of the treatment (Quinto and Radman, 1987). The yield (Y) is the fraction of treated cells giving rise to mutated, or recombinated, or SOS-induced colonies. The yield was calculated as :

$$\frac{n_x - n_o (S_x)}{N_o},$$

where n_x is the number of altered colonies (per ml of cell cultures) at dose x (the induced events), n_o is the number of altered colonies (per ml of cell cultures) at dose 0 (the spontaneous events), S_x is the survival fraction at dose x, and N_o is the number of colonies (per ml of cell cultures) at dose 0.

As shown in table 2, UV radiation is the strongest SOS-inducer, even though it is not the most mutagenic or recombinogenic agent among those examined. The absence of a correlation between the I and M potency was also evident for diethylsulfate (strong mutagen but weak SOS-inducer) and mitomycin C (weak mutagen but strong SOS-inducer) (see table 2). These data suggest that the lesions principally responsible for SOS-induction by causing the arrest of the DNA replication fork, are not necessarily the most premutational ones. In the case of UV radiation, evidence indicated that the (6-4)pyC lesions, and not the pyrimidine-pyrimidine dimers, which are known to block the DNA replication fork, are the most important premutational lesions (Brash and Haseltine, 1982; Wood, 1985; Glickman, 1986).

Table 2. The mutagenic, recombinogenic and SOS-inducer potency of UV radiation, gamma rays, monofunctional and bifunctional alkylating agents

Agent	P_M (x 10^{-5})	P_R (x 10^{-4})	P_I (x 10^{-4})
UV radiation (260 nm)	11.0	99.0	4500.0
gamma rays	3.8	18.0	48.0
mitomycin C	8.0	44.5	387.0
nitrogen mustard	<1	7.1	83.9
bis(2-chloroethyl)nitrosourea	<1	5.3	1.0
thiotepa	9.5	2.6	2.4
chlorambucil	<1	2.3	3.1
bis(2-chloroethyl)ether	<1	<1	<1
ethylnitrosourea	131.0	184.0	206.0
methylnitrosourethane	383.0	47.0	23.2
methylnitrosourea	473.0	35.0	13.8
methylnitrosoguanidina	348.0	37.7	27.1
diethylsulfate	91.0	43.5	<1
dimethylsulfate	11.8	52.0	32.0
ethylmethane sulfonate	28.3	18.0	<1
methylmethane sulfonate	36.0	68.5	39.0
epichlorohydrin	1.0	6.2	1.1

P_M = mutagenic potency; P_R = recombinogenic potency; P_I = SOS-inducing potency

ACKNOWLEDGMENTS

We thank J. Gilder, C. Maresca and N. Dello Stritto for help in preparing the manuscript. The work was supported by grants from A.I.R.C., C.N.R. and A.I.D.S. projects.

REFERENCES

Angel, P., Poting, A., Mallick, U., Rahmsdorf, H.J., Schorpp, M., and Herrlich, P., 1986, Induction of metallothionein and other mRNA species by carcinogens and tumor promoters in primary human skin fibroblasts, Molec.Cell.Biol., 6:1760.

Benbow, R.M., Zuccarelli, A.Y., and Sinsheimer, R.L., 1974, A role for single-strand breaks in bacteriophage ΦX174 genetic recombination, J.Mol.Biol., 88:629.

Bourre, F., and Sarasin, A., 1983, Targeted mutagenesis of SV40 DNA induced by UV light, Nature, 305:68.

Brandenburger, A., Godson, G.N., Radman, M., Glickman, B.W., van Sluis, C.A., and Doubleday, O.P., 1981, Radiation-induced base substitution mutagenesis in single-stranded DNA phage M13, Nature, 294:180.

Brash, D.E., and Haseltine, W.A., 1982, UV-induced mutation hotspots occur at DNA damage hotspots, Nature, 298:189.

Bridges, B.A., and Woodgate, R., 1984, Mutagenic repair in Escherichia coli X. The umuC gene product may be required for replication past pyrimidine dimers but not for the coding error in UV-mutagenesis, Mol.Gen.Genet., 196:364.

Bridges, B.A., and Woodgate, R., 1985, Mutagenic repair in Escherichia coli: products of the recA gene and of the umuD and umuC genes act at different steps in UV-induced mutagenesis, Proc.Natl.Acad.Sci.USA, 82:4193.

Burckhardt, S.E., Woodgate, R., Scheuermann, R.H., and Echols, H., 1988, UmuD mutagenesis protein of Escherichia coli: overproduction, purification, and cleavage by RecA, Proc.Natl.Acad.Sci.USA, 85:1811.

Buscher, M., Rahmsdorf, H.J., Liftin, M., Karin, M., and Herrlich, P., 1988, Activation of the c-fos gene by UV and phorbol ester: different signal transduction pathways converge to the same enhancer element, Oncogene, 3: 301.

Cohen, S.P., Resnick, J., and Sussman, R., 1983, Interaction of single-strand binding protein and RecA protein at the single-stranded DNA site, J.Mol.Biol., 167:901.

Cornelis, J.J., Su, Z.Z., and Rommelaere, J., 1982, Direct and indirect effects of ultraviolet light on the mutagenesis of parvovirus H-1 in human cells, The EMBO J., 1:693.

Coulondre, C., and Miller, J.H., 1977a, Genetic studies of the lac repressor III. Additional correlation of mutational sites with specific aminoacid residues, J.Mol.Biol., 117:525.

Coulondre, C., and Miller, J.H., 1977b, Genetic studies of the lac repressor IV. Mutagenic specificity in the lacI gene of E.coli , J.Mol.Biol., 117:577.

Coulondre, C., Miller, J.H., Farabaugh, P.J., and Gilbert, W., 1978, Molecular basis of base substitution hotspots in the lacI gene of E.coli, Nature, 274:775.

Craig, N.L., and Roberts, J.W., 1980, E.coli RecA protein-directed cleavage of phage lambda repressor requires polynucleotide, Nature, 283:26.

Craig, N.L., and Roberts, J.W., 1981, Function of nucleoside triphosphate and polinucleotide in Escherichia coli RecA protein-directed cleavage of phage lambda repressor, J.Biol.Chem., 256:8039.

Das Gupta, U.B., and Summers, W.C., 1978, Ultraviolet reactivation of herpes simplex virus is mutagenic and inducible in mammalian cells, Proc.Natl.Acad.Sci.USA, 75:2378.

Day, R.S. III, and Ziolkowski, C., 1978, Studies on UV-induced viral reversion, Cockayne's syndrome, and MNNG damage using adenovirus 5, in: "DNA repair mechanisms", P.C. Hanawalt, E.C. Friedberg and C.F.Fox ed., Academic, New York.

Defais, M., Fauquet, P., Radman, M., and Errera, M., 1971, Ultraviolet reactivation and ultraviolet mutagenesis of lambda in different genetic systems, Virology, 43:495.

Defais, M., Caillet-Fauquet, P., Fox, M.S., and Radman, M., 1976, Induction kinetics of mutagenic DNA repair activity in E.coli following ultraviolet irradiation, Mol.Gen.Genet., 148:125.

Eguchi, Y., Ogawa, T., and Ogawa, H., 1988, Cleavage of bacteriophage O80 CI repressor by RecA protein, J.Mol.Biol., 202:565.

Fornace, A.J.Jr., Alamo, I.Jr., and Hollaender, M.C., 1988, DNA damage-inducible transcripts in mammalian cells, Proc.Natl.Acad.Sci.USA, 85:8800.

Gentil, A., Margot, A., and Sarasin, A., 1982, Enhanced reactivation and mutagenesis after transfection of carcinogen-treated monkey kidney cells with UV-irradiated simian virus 40 DNA, Biochimie, 64:693.

Glickman, B.W., Schaaper, R.M., Haseltine, W.A., Dunn, R.L., and Brash, D.E., 1986, The C-C (6-4)UV photoproduct is mutagenic in Escherichia coli, Proc.Natl.Acad.Sci.USA, 83:6945.

Herrlich, P., van den Berg, S., Mai, S., Lavi, S., Kaina, B., Stein, B., Ponta, H., and Rahmsdorf, H.J., 1989, Mechanism of the UV response in mammalian cells, in: "Proceedings of the United Kingdom Environmental Mutagen Society/ DNA Repair Network", Sussex University, Brighton.
Horii, T. Ogawa, T., Nakatani, T., Hase, T., Matsubara, H., and Ogawa, H., 1981, Regulation of SOS functions: purification of E.coli LexA protein and determination of its specific site cleaved by the RecA protein, Cell, 27:515.
Ichikawa-Ryo, H., and Kondo, S., 1975, Indirect mutagenesis in phage lambda by ultraviolet preirradiation of host bacteria, J.Mol.Biol., 97:77.
Kartasova, T., Ponec, M., and van de Putte, P., 1988, Induction of proteins and mRNAs after UV irradiation of human epidermal keratinocytes , Exp.Cell.Res., 174:421.
Kondo, S., 1969, Mutagenicity versus radiosensitivity in Escherichia coli, in: "Proceedings of the XIIth International Congress on Genetics", Vol.II.
Lebkowski, J.S., Clancy, S., Miller, J.H., and Calos, M.P., 1985, The LacI shuttle: rapid analysis of the mutagenic specificity of ultraviolet light in human cells, Proc.Natl.Acad.Sci.USA., 82:8606.
Le Clerc, J.E., and Istock, N.L., 1982, Specificity of UV mutagenesis in the lac promoter of M13lac hybrid phage DNA, Nature, 297:596.
Little, J.W., Edmiston, S.H., Pacelli, L.Z., and Mount, D.W., 1980, Cleavage of the Escherichia coli LexA protein by the RecA protease , Proc.Natl.Acad.Sci.USA, 77:3225.
Lytle, C.D., Goddard, J.G., and Lin, C., 1980, Repair and mutagenesis of herpes simplex virus in UV-irradiated monkey cells, Mutation Res., 70:139.
Lucke-Huble, C., and Herrlich, P., 1986, Gene amplification in mammalian cells after exposure to ionizing radiation and UV, in: "Radiation carcinogenesis and DNA alterations", F.J.Burns, A.C.Upton and G.Silini ed., Plenum Press, Amsterdam.
Mallick, U., Rahmsdorf, H.J., Yamamoto, N., Ponta, H., Wegner, R.D., and Herrlich, P., 1982, 12-O-tetradecanoylphorbol 13-acetate-inducible proteins are synthesized at an increased rate in Bloom syndrome fibroblasts, Proc.Natl.Acad.Sci.USA, 79:7886.
Mezzina, M., Gentil, A., and Sarasin, A., 1981, Simian virus 40 as a probe for studying inducible repair functions in mammalian cells, J.Supramol. Struct. Cell.Biochem., 17:121.
Miller, J.H., 1982, Carcinogens induce targeted mutations, Cell, 31:5.
Miller, J.H.,1985, Mutagenic specificity of ultraviolet light, J.Mol.Biol., 182:45.
Miura, A., and Tomizawa, J., 1968, Studies on radiation sensitive mutants of E.coli III. Participation of the rec system in induction of mutation by ultraviolet irradiation, Mol. Gen.Genet., 103:1.
Moore, P.D., Bose, K.K., Rabkin, S.D., and Strauss, B.S., 1981, Sites of termination of in vitro DNA synthesis on ultraviolet-and N-acetylaminofluorene-treated 0X174 templates by procaryotic and eukaryotic DNA polymerases, Proc.Natl.Acad.Sci.USA, 78:110.
Mount, D.W., Low, K.B., and Edmiston, S., 1972, Dominant mutations (lex) in E.coli K12 which affect radiation sensitivity and frequency of ultraviolet light-induced mutations, J.Bacteriol.,112:886.
Quillardet, P., and Devoret, R., 1982, Damaged-site independent mutagenesis of phage lambda produced by inducible error-prone repair, Biochimie, 64:789.
Quinto, I., and Radman, M., 1987, Carcinogenic potency in rodents versus genotoxic potency in E.coli : a correlation analysis for bifunctional alkylating agents, Mutation Res., 181:235.

Quinto, I., Tenenbaum, L., and Radman, M., 1990, Genotoxicity profile of monofunctional alkylating agents in E.coli : quantitative correlations with carcinogenic potency in rodents, Mutation Res., 228:177.

Radman, M., 1974, Phenomenology of an inducible mutagenic DNA repair pathway in Escherichia coli: SOS repair hypothesis, in: "Molecular and environmental aspects of mutagenesis", L.Prakash, F.Sherman, M.Miller, C.Lawrence , and H.W.Tabor ed., Charles C. Thomas Publisher, Springfield.

Rahmsdorf, H., Mallick, U., Ponta, H., and Herrlich, P., 1982, A B-lymphocyte-specific high-turnover protein: constitutive expression in resting B cells and induction of synthesis in proliferating cells, Cell, 29:459.

Roberts, J.W., Roberts, C.W., and Craig, N.L., 1978, Escherichia coli recA gene product inactivates phage lambda repressor, Proc.Natl.Acad.Sci.USA, 75:4714.

Roberts, J.W., and Devoret, R., 1983, Lysogenic induction, in: "Lambda II", R.W.Hendrix, J.W. Roberts, F.W. Stahl, and R.A. Weisberg ed., Cold Spring Harbor Laboratory, Cold Spring Harbor, N.Y.

Sarasin, A., and Benoit, A., 1980, Induction of an error-prone mode of DNA repair in UV-irradiated monkey kidney cells, Mutation Res., 70:71.

Sauer, R.T., Ross, M.J., and Ptashne, M., 1982, Cleavage of the lambda and P22 repressors by RecA protein, J.Biol.Chem., 257:4458.

Schorpp, M., Mallick, U., Rahmsdorf, H.J., and Errlich, P., 1984, UV-induced extracellular factor from human fibroblasts communicates the UV response to nonirradiated cells, Cell, 37:861.

Setlow, R.B., Swenson, P.A., and Carrier, W.L., 1963, Thymine dimers and inhibition of DNA synthesis by ultraviolet irradiation of cells, Science, 142:1464.

Shinagawa, H., Iwasaki, H., Kato,T., and Nakata, A., 1988, RecA protein dependent cleavage of UmuD protein and SOS mutagenesis, Proc.Natl.Acad.Sci.USA, 85:1806.

Tenenbaum, L., Quinto I., and Faelen, M., 1988, The E.coli multitest: a set of strains to characterize diverse genotoxic effects, Mutation Res., 203:415.

Toman, Z., Dambly-Chaudière , C., Tenenbaum, L., and Radman, M., 1985, A system for detection of genetic and epigenetic alterations in E.coli induced by DNA damaging agents, J.Mol.Biol., 186:97.

Villani, G., Boiteux, S., and Radman, M., 1978, Mechanisms of ultraviolet-induced mutagenesis: extent and fidelity of in vitro DNA synthesis on irradiated templates, Proc.Natl.Acad.Sci.USA, 75:3037.

Walker, G.C., 1984, Mutagenesis and inducible responses to deoxyribonucleic acid damage in Escherichia coli, Microbiol.Rev., 48:60.

Weigle, J.J., 1953, Induction of mutation in a bacterial virus, Proc.Natl.Acad.Sci.USA, 39:628.

Witkin, E.M., 1967, Mutation proof and mutation prone modes of survival in derivatives of E.coli B differing in sensitivity to ultraviolet light, Brookhaven Symp.Biol., 20:17.

Witkin, E.M., 1969, Ultraviolet induced mutation and DNA repair, Annu.Rev.Microbiol., 23:487.

Witkin, E.M., 1976, Ultraviolet mutagenesis and inducible DNA repair in Escherichia coli , Bacteriol.Rev., 40:869.

Wood, R.D., and Hutchinson, F., 1984, Non-targeted mutagenesis of unirradiated lambda phage in Escherichia coli host cells irradiated with ultraviolet light, J.Mol.Biol., 173:293.

Wood, R.D., Skopek, T.R., and Hutchinson, F., 1984, Changes in DNA base sequence induced by targeted mutagenesis of lambda phage by ultraviolet light, J.Mol.Biol., 173:273.

Wood, R., 1985, Pyrimidine dimers are not the principal pre-mutagenic lesions induced in lambda phage DNA by ultraviolet light, J.Mol.Biol.,184:577.
Zamansky, G.B., Kleinman, L.F., Black, P.H., and Kaplan, J.C., 1980, Reactivation of herpes simplex virus in a cell line inducible for simian virus 40 synthesis, Mutation Res., 70:1.

LASER OUTPUT PARAMETERS

NECESSARY FOR HAZARD ASSESSMENT*

James Franks

Laser-Microwave Division
US Army Environmental Hygiene Agency
Aberdeen Proving Ground, MD

INTRODUCTION

Normally, measurement of commercial laser output parameters is not necessary since the manufacturer has provided a hazard classification, and class-dependant control procedures are well defined in laser safety standards. However, if you are a laser manufacturer and must classify lasers or if you are a laser safety officer and potential for overexposure to unprotected personnel exists, then measurement may be required.

RADIANT POWER (Φ) OR RADIANT ENERGY (Q)

Radiant power measured in watts, or radiant energy measured in joules, are the most important parameters to measure for laser classification. Power is used to describe the output of continuous lasers while energy is usually used to describe the output of single-pulse lasers. Repetitively pulsed lasers may be described with either quantitiy. Standards differ, however, on what portion of the total power or energy emitted, is necessary to determine hazard classification. In the US Army's TB MED 524 (US Army, 1979) the total output of the laser is measured while the United States Food and Drug Administration (USFDA, 1985), which regulates the manufacture of lasers in the United States, would measure the total energy or power through a 7 mm aperture, unless use of collecting optics is anticipated. The 7 mm aperture is used irregardless of wavelength, even for far infrared wavelengths. The American National Standards Institute (ANSI) Standard for the Safe Use of Lasers requires a 1 mm, 7mm or 11 mm aperture (ANSI 1986) depending on the laser wavelength. A measurement of power or energy through a known aperture is a measure of irradiance or radiant exposure as long as the laser beam diameter is at least as large as the aperture. These different aperture diameters result in some interesting differences between classification for the different standards. For example, a carbon dioxide laser rangefinder with a pulse repetition frequencey (PRF) of 15 Hz, a beam diameter of 10 cm, radiant energy of 65 mJ, and a beam divergence of 0.5 mrad, would be classified as Class 1

*The opinions or assertions contained herein are the private views of the author and are not to be construed as reflecting the official views of the U.S. Department of the Army or the U.S. Department of Defense.

Light, Lasers, and Synchrotron Radiation
Edited by M. Grandolfo *et al.*, Plenum Press, New York, 1990

by the USFDA, Class 3b using ANSI Z136.1-1986 and Class 4 using the US Army standard. Keep in mind, however, that the USFDA standard regulates manufacturers while the other standards are for users. So, we have the manufacturer telling us that the laser has no risk associated with it because it is Class 1, ANSI says that the laser is an acute hazard, and the US Army says this is a high risk system. The true risk assessment probably lies closest to Class 1 or what may be called Class 3a, that is, the output radiant exposure is below the limit but if we could somehow focus the beam we may then cause an injury. For this reason it might be a good idea to eliminate the requirement that the output should lie between 1 and 5 times the Class 1 limit for Class 3a, particularly for far infrared wavelengths, where glass optics do not transmit, or refract incident laser radiation.

RADIANT EXPOSURE (H) OR IRRADIANCE (E)

Radiant exposure is energy measured through a limiting aperture and is usually given in units of J/cm^2. Irradiance is power that is measured through a limiting aperture and is given in units of W/cm^2. The limiting aperture is the maximum size over which the measurement should be averaged. The size of the limiting aperture is 7 mm for visible and near infrared wavelengths, 1 mm for ultraviolet and infrared from 1.4 to 100 µm and 11 mm for 100 µm to 1 mm (ANSI 1986). These aperture sizes may not be the same size as the collecting aperture for hazard classification. Radiant exposure and irradiance are criteria used to determine if a particular laser is a diffuse reflection hazard and therefore Class 4. These are also the quantities that should be measured to determine if an individual received an exposure exceeding the exposure limits in a laser incident investigation. Far-field beam divergence can also be calculated from radiant exposure or irradiance measurements made at several distances. Minimum attenuation requirements for laser eye protection may also be determined from radiant exposure or irradiance measurements.

BEAM DIVERGENCE (ϕ)

Beam divergence describes how the laser beam diameter changes with distance. It is usually not important in predicting potentially hazardous areas for collimated beams in a laboratory environment since intrabeam viewing hazard distances for highly collimated beams normally would extend far beyond the boundaries of the laboratory. It does become important however, when determination of the so-called Nominal Ocular Hazard Distance (NOHD) is desired. The NOHD is the distance within which the radiant exposure or irradiance falls below the exposure limit. NOHD's can vary from 10's of meters to kilometers. The relationship between beam divergence and NOHD is inverse square so that decreasing the beam divergence by a factor of 2 for example, would increase the NOHD by a factor of 4. There are several methods of measuring beam divergence described in Mr. Marshall's paper, Laser Beam Diagnostics, that appears elsewear in this volume.

BEAM DIAMETER (a)

Changes in beam diameter over a given distance determines the divergence. Measuring the change in beam diameter over a given distance is a measurement of divergence. It is necessary to determine initial beam diameter (a) to calculate NOHD. Beam diameter can be defined in many different ways that are also described in Mr. Marshall's paper.

PULSE WIDTH (Δt)

The primary use of pulse width in hazard evaluation is in determining exposure duration for a single laser pulse. Exposure duration is one important factor in the determination of exposure limit. The important pulse width ranges as can be seen from the ANSI standard (ANSI 1986) are < 1 ns, where the standard does not apply, 1ns to 18 μs for 400 nm to 1050 nm and 1 ns to 50 μs for 1051 nm to 1400 nm. So that often times it may not be actually necessary to measure the pulse width but only to know that it is within a certain range. Application of the standard would require pulse width measurement for lasers whose pulse width are near the ends of these ranges. Of course, for lasers with pulse widths longer than 50 μs to 100 μs, depending on the wavelength, the pulse width is necessary to calculate an exposure limit. Knowledge of the pulse width for pulses less than 1 ns is also necessary since the guidance in the ANSI standard (ANSI 1986) for these ultrashort pulses is to limit peak power to that limit for a 1 ns pulse. When viewing laser sources that are considered extended sources (subtend an angle > than alpha minimum) it is also necessary to know the pulse width to determine what the alpha minimum angle is.

WAVELENGTH (λ)

Wavelength is a laser parameter that is usually known apriori. As a consequence, wavelength is not usually measured. However, when it is required, a monochromator is the most convenient way of measuring an unknown wavelength.

PULSE REPETITION FREQUENCY (PRF)

PRF must be determined for repetitively pulsed lasers since the PRF correction factor C_p, depends upon the total number of pulses in the exposure. For example, a laser that operates at a PRF of 20 Hz will emit 200 pulses in a 10 s exposure. C_p for an exposure to 200 pulses 1/200 which equals 0.27. Therefore, PRF is a critical bit of information in application of a laser safety standard.

SOLID ANGLE (Ω)

Laser sources that are not well collimated, such as diode arrays or diffuse reflections, may use a radiance criteria rather than an irradiance criteria for the exposure limit. The radiance could be measured directly with the appropriate instrument or it may be calculated from irradiance measurements and a measurement of the solid angle subtended by the source along the direction of observation. The solid angle could be calculated from a physical measurement of the diameter of a diffuse laser spot, or source diameter could be determined photographically from a negative and microdensitimeter scan.

OTHER CONCERNS

Military rangefinders are operating at wavelengths outside the ocular hazard region of 400 nm to 1400 nm. These lasers are sometimes referred to as "eye- safe" because the exposure limits are much higher outside this spectral region. For example, the exposure limit for exposure to a single, 1 ns pulse at 1540 nm is 1 J/cm^2 as compared to 5 $\mu J/cm^2$ at 1064 nm. A common technique for generation of the 1540 nm wavelength is Raman shifting of the 1064 nm wavelength. This, in effect, shifts the operating wavelength from 1064 nm to 1540 nm and transforms a laser which is normally a serious retinal hazard to one that is more or less "eye-safe". A problem arises, however, when the 1064 nm primary wavelength is not sufficiently filtered from the 1540 nm radiation. Assuming equal amounts of the two wavelengths, the 1064 nm contribution must be attenuated by a factor of about 200,000 to maintain an eye-safe status. A measurement problém arises when attempting to measure small amounts of 1064 nm radiation in the presence of the much larger amount of 1540 nm radiation. A reasonable technique is to use a silicon detector with a filter that passes 1064 nm and attenuates 1540 nm radiation. A Schott filter glass-like BG-36 is ideal for this type of measurement. A 3 mm thickness of this material has a transmission of about 80 per cent at 1064 nm and 1 per cent at 1540 nm. So that a 9 mm thickness would have approximately 50 per cent transmission at 1064 nm and .0001 per cent transmission at 1540 nm. This filter in combination with a silicon detector should provide at least 7 orders of magnitude of rejection at 1540 nm and provide reasonable assurance that the energy being measured is at the 1064 nm wavelength. The only test that is left is to make sure that there is no visible wavelength contribution from the optical pump used to excite the laser. A filter glass which absorbs 1064 nm radiation but passes visible wavelengths can be used for this test. Schott KG-3 glass 3mm thick transmits 85 per cent in the visible and about .0015 per cent at 1064 nm.

Determination of the extent of the Nominal Hazard Zone (NHZ) also requires measurement of the laser beam divergence, beam polarization, and characteristics of the reflector if it is determined that the laser is a specular or diffuse reflection hazard.

Laser beams propagated in the outdoors also require knowledge of an appropriate atmospheric attenuation coefficient.

CONCLUSION

Laser Safety Officers must have knowledege of many laser output parameters to effectively evaluate potential hazards from lasers. However, applying control procedures for the predetermined hazard classification may eliminate the need for most measurement.

REFERENCES

American National Standards Institute (ANSI) Standard for the Safe Use of Lasers, Z136.1, 1986.

Department of the Army, Control of hazards to health from laser radiation, Washington, DC, Department of the Army, TB MED 524, 20 June 1985.

United States Food and Drug Administration, Performance standard for laser products, Washington, DC, US Government Printing Office, 21 Code of Federal Regulations 1040.10(e)(3); 1985.

CONTROL OF OUTDOOR LASER HAZARDS*

James Franks

U.S. Army Environmental Hygiene Agency
Aberdeen Proving Ground, MD, 21010-5422

INTRODUCTION

Tank-mounted laser rangefinders (LRF) were fielded in the US Army in the early 70's. Since that time laser target designators (LTD) and direct fire simulators (DFS) have been developed. Fire-control lasers (LRFs and LTDs) are either Class 3 or Class 4 systems according to the laser hazard classification scheme used by the U.S. Army. These fire-control lasers can cause serious retinal injury from exposure to the direct beam and may only be operated with restrictive procedures in the training environment. DFS are training devices that are used to simulate the firing of direct fire weapons in the tactical training environment. The DFS are considered safe for field use and the use of DFS constitute the biggest occupational exposure to lasers in the US Army.

OCCUPATIONAL EXPOSURE LIMITS

Exposure limits (EL) have changed since 1969 when the US Army first published limits for a single Q-switched laser exposure (US Army 1969). The EL was $0.01 \mu J/cm^2$ for laboratory or indoor use and $0.05 \mu J/cm^2$ for outdoor applications for visible and near infrared wavelengths. The EL for exposure to visible and near infrared wavelengths now (Army 1985) varies with wavelength but for a single exposure to a Q-switched pulse at a wavelength of 1064 nm is $5.0 \mu J/cm^2$ and at 694.3 nm is $0.5 \mu J/cm^2$. This reduction in EL for outdoor laser applications by a factor of 100 would theoretically reduce the Nominal Ocular Hazard Distance (NOHD) by a factor of 10. A typical 1064 nm LRF of today with an NOHD of 7 km would have a 1969 NOHD of 70 km neglecting atmospheric attenuation. This, however is only significant if lines-of-sight (LOS) longer than 7 km are present. Most ground locations in a European environment do not have LOS longer than 7 km unless you are located at a high elevation. Airborne lasers can of course have LOS much greater than 7 km.

----

*The opinions or assertions contained herein are the private views of the author and are not to be construed as reflecting the views of the U. S. Department of the Army or the U. S. Department of Defense.

EL for exposure to repetitively pulsed Q-switched lasers like LTD's did not exist in 1969 and were first published in the US by the American National Standards Institute (ANSI) in 1973 (ANSI 1973). These EL have also changed since 1973 but are based on the single pulse exposure limit of 5.0 $\mu J/cm^2$. A correction factor (C_p) has been developed to account for a decrease in threshold of injury for exposure to multiple pulses (ANSI 1973 and US Army 1974). This early correction factor depends only upon PRF while a new PRF correction factor (US Army 1985 and ANSI 1986) is based upon the total number of pulses in the exposure. Little change has occurred in the PRF correction factor for low PRF's.

Although the multiple pulse EL has not changed dramatically since 1973, the way in which the EL is applied has. For example, the worst-case exposure to LTD's is a continuous exposure at the maximum system PRF. (A worst-case exposure now would be considered a 10 s exposure while a worst-case exposure in 1973 was considered an eight-hour expsoure) An NOHD would be calculated on the basis of a continuous exposure and this NOHD would be appropriate for personnel in the target itself such as aircraft crewmembers. However application of this continuous exposure criteria to an accidental single pulse exposure is overly conservative. The approach adopted by the US Army is to protect authorized personnel who may be in the target or target area out to the multiple pulse NOHD but unauthorized personnel who may be exposed to a single pulse accidentally, because of the relative motion of the laser beam, need only be excluded from the beam path out to the single pulse NOHD.

CORNEAL EXPOSURE TO INBAND WAVELENGTHS

Intrabeam viewing of the direct or specularly reflected laser beam through direct view magnifying optics increases the corneal exposure for inband wavelengths. This increase is proportional to the square of the magnifying power of the optic and the transmission of the optic at the laser wavelength (US Army 1985). This implies that the NOHD for unaided viewing would be increased when viewing through a direct view optic with a magnifying power greater than 1. As a guideline, the typical magnifying power used prior to 1985 was 11.4 corresponding to an optic with an 80 mm collecting aperture and a 7 mm exit pupil. This resulted in extended NOHD's out to 70 km for an airborne LTD (US Army 1985). Extended NOHD's listed in tables now (Franks 1986) are based on a 50 mm collecting optic with a 7 mm exit pupil which corresponds to a 7.1 magnifying power.

The most important change in approach to use of magnifying optics came, however, when the Food and Drug Administration (FDA) of the US Department of Health and Human Services amended the Federal Performance Standard for Laser Products (FDA 1985) in 1985. This revision states that the worst-case assumption of optically aided viewing should not be applied unless such viewing conditions are likely. What this means in practice is that when LOS greater than the unaided NOHD extend into an uncontrolled area you need not assume that magnifying optics will be used to view the direct beam. Prior to this policy statement by the FDA the US Army approach was to assume that when these long LOS were present into an uncontrolled area that positive steps had to be taken to assure that the direct beam or specular reflections of the direct beam could not be directed into these uncontrolled areas. Now, the policy is to always use a backstop, if possible, but if not, control the area out to the unaided NOHD and only consider magnifying optics if you know they are being used.

ATMOSPHERIC PROPAGATION

As a laser beam passes through the atmosphere the beam is absorbed or scattered depending on the relative size of the atmospheric constituent with respect to the laser wavelength. The attenuation will vary with wavelength and local atmospheric conditions. For visible and near infrared wavelengths the attenuation and scattering can be correlated to the meteorological visibility. If the visibility V is known then the atmospheric attenuation coefficient can be computed by

$$\mu = 3.9 \times V^{-1}$$

For a given value of visibility, the attenuation coefficient is also a function of wavelength and it changes by a factor of 2.5 in the 400nm-1400nm spectral range at sea level with lower values at longer wavelengths (Middleton 1952). The value used by the Army when assigning NOHD's for a particular system is 5×10^{-7} cm^{-1} at 1064 nm. This corresponds to an Exceptionally Clear day with a 40 km meteorological visibility. Under almost all ground to ground lasing conditions this is a conservative value. Atmospheric attenuation does not play a significant role in determination of NOHD for NOHD's less than 4 km.. Atmospheric attenuation becomes important when long LOS exist and optically aided viewing is expected.

SCINTILLATION

Scintillation is a statistical phenomena that may cause hot spots in the beam. Changes in temperature across the beam path cause changes in the local index of refraction which can focus and defocus the beam. The irradiance profile may show localized variations of several orders of magnitude but the variation above the average value is approximately a factor of 20 (Sliney and Wolbarsht, 1980). Most laser safety standards say you must consider scintillation but don't give details as how to handle it. STANAG 3606 says there is a probability that the EL will be exceeded beyond the NOHD so use a backstop (STANAG 1983). The ANSI standard says to use a worst-case analysis but gives no guidance as to what this is (ANSI 1986). The British have used fixed scintillation gains of 12* for air to ground operations. However, the Army does not consider scintillation significant for <u>accidental</u> exposure for several reasons:

1. The probability of accidental exposure is about 1 in 10^5 (Sliney and Wolbarsht, 1980). This probability assumes that the beam can be pointed at random to plus or minus 10 mrad of the target location and that the target is looking at the laser transmitter when it lases. A British study * finds that the probability that scintillation will cause a gain of 10 over the EL at the NOHD is about 6 in 10^3. The total probability of being exposed accidentally to a level that is 10x above the EL at the NOHD is therefore 1 in 6×10^8. A level that is 10x above the EL at 1064 nm (20 J) would still be a factor of 7.5 below a level which would cause a minimal injury (150 J).

2. The US Army requires use of backstops for ground to ground lasing unless groundspace is controlled out to the NOHD (Army 1985).

3. It is still not clear how the turbulon will be imaged on the retina. It may well be that a minimal retinal image will not be formed for eyes focused at infinity (Sliney and Wolbarsht, 1980).

* Presentation by William Smerdon (Head of Mathmatics Group, Ferranti Defence Systems Ltd., Edinborgh, Scotland, and Consultant to the U.K. Military Laser Safety Committee) to U. K./U. S. International Exchange Group on laser hazard area trace (LHAT) modeling, August 1986, Dover, NJ.

4. Scintillation saturates at approximately 1000 meters downrange so that the magnitude of fluctuations does not increase beyond this distance but actually decreases (Sliney and Wolbarsht, 1980). Personnel within 1000 m would be wearing laser eye protectors with more than enough attenuation to account for scintillation.

5. The beam divergence actually is increased by scintillation so that the average value of irradiance or radiant exposure is decreasing over what it would be in a nonturbulent medium (Sliney and Wolbarsht, 1980).

In conclusion, since backstops are normally used for ground to ground lasing and the probability of being accidentally exposed to a level that would cause an injury is so small it is felt that scintillation should only be considered when intentional exposure to unprotected personnel occurs at the NOHD.

NOMINAL HAZARD ZONE

The capability to produce a diffuse reflection hazard is a criteria for a Class 4 hazard classification (US Army 1985). Currently only the ruby laser rangefinder exceeds this criteria and no object could be placed within the beam path out to a distance of 10 meters unless those who could view the object wore laser eye protectors. ANSI 1986 now describes an area called a Nominal Hazard Zone or NHZ (ANSI 1986). Assuming that a laser is Class 4 because it may produce a diffuse reflection hazard the extent of the NHZ is defined by a distance r where

$$r_{NHZ} = \sqrt{\frac{Q}{EL \cdot \pi}} \qquad \text{Q is radiant energy}$$

$$= \sqrt{\frac{0.1}{\pi \cdot 5\times10^{-7}}} \qquad \text{for a ruby LRF}$$

$$= 2 \text{ m}$$

$$= \sqrt{\frac{0.1}{\pi \cdot 5\times10^{-6}}} \qquad \text{for a YAG LRF}$$

$$= 0.8 \text{ m}$$

So that not only does an diffuse object need to be placed within a minimum distance of the laser transmitter (10 m for the ruby LRF), you must also be within a rather short distance from the object before you may be exposed to a potential hazard. Eye protectors are not required at distances outside the NHZ.

CLASS 1 OR EYE-SAFE LASERS

An "eye-safe" laser to the laser safety specialist, is a Class 1 laser, i.e. it can not exceed the EL under any exposure condition. Other laser workers, who say a laser is "eye-safe" are really mean that the laser has an operating wavelength longer than 1400 nm. The EL for exposure to a single Q-switched pulse is 10 mJ/cm^2 at wavelengths of 1400 nm to 10^6 nm, except at a wavelength of 1540 nm where the EL for a single Q-switched pulse is 1 J/cm^2. Compared to visible and near infrared lasers the EL differ by at least 5 orders of magnitude. The defining aperture in this eye-safe spectral region is 1 mm so that for a single pulse exposure the Accessible Emission Limit for Class 1 is a radiant energy of 7.9 mJ at 1540 nm and 79μ J at 10,600 nm(CO_2). A Class 1 LRF has been developed that operates at 1540 nm (AEHA 1985). There is also interest in having a training mode in currently fielded fire-control lasers that would be less hazardous than the existing system.

REFERENCES

American National Standards Institute, 1973, American National Standard for the safe use of lasers, ANSI Z136.1-1973, ANSI, 1430 Broadway, New York, N.Y.10018.

American National Standards Institute, 1976, American National Standard for the safe use of lasers, ANSI Z136.1-1976, ANSI, 1430 Broadway, New York, N.Y.10018.

American National Standards Institute, 1980, American National Standard for the safe use of lasers, ANSI Z136.1-1980, ANSI, 1430 Broadway, New York, N.Y.10018.

American National Standards Institute, 1986, American National Standard for the Safe Use of Lasers, ANSI Z136.1-1986, ANSI, 1430 Broadway, New York, N.Y.10018.

Departments of the Army and Navy, 1969, Control of hazards to health from laser radiation. Washington, DC: Departments of the Army and Navy, TB MED 279, NAVMED P-5052-35; February 1969.

Department of the Army. 1985, Control of hazards to health from laser radiation. Washington DC: Department of the Army; TB MED 524; 20 June 1985.

Franks, J.K., 1986, Nonionizing Radiation Protection Study No. 25-42-0356-86, Camp Shelby, MS. US Army Environmental Hygiene Agency.

Marshall, W. J., 1980, Hazard analysis on Gaussian shaped laser beams, AIHA Journal 41:547-551.

Middleton, W.E.K., 1952, Vision through the atmosphere. Canada: University of Toronto Press.

Sliney, D. H., Wolbarsht, M.W. 1980, "Safety with Lasers and Other Optical Sources," Plenum Press, New York, NY.

North Atlantic Treaty Organization, 1983, Evaluation and control of laser hazards. Brussels, Belgium: Military Agency for Standardization (MAS) Army Board, Stanag 3606 LAS(Edition 4).

US Food and Drug Administration, 1985, Performance standard for laser products. Washington, DC: U. S. Government Printing Office; 21 Code of Federal Regulations 1040.10 (e)(3).

PROTECTIVE SCREENS AND FILTERS FOR IR PROTECTION

Paul Eriksen

National Institute of Occupational Health
Lersø Parkallé 105
DK-2100 Copenhagen
Denmark

INTRODUCTION

The best known adverse effects of infrared (IR) radiation are dry eyes and burn or scalding of the skin as acute effects, and cataract as a long term effect. This is discussed in greater detail in other lectures. Severe acute effects of IR radiation are only seldom seen in the industrial environment but one often meets situations with dry, irritated eyes and situations of obvious discomfortable viewing conditions. Most of these situations can be alleviated quite easily. Of greatest concern is the possibility of accelerating cataract formation in the eye's lens because this may call for an early surgical operation.

The purpose of this lecture is to present information, data and examples on protective filters and screens that will attenuate hazardous IR radiation to a safe level. In practice a safe level is a level below the proposed threshold limit values (TLVs) recommended by the American Conference of Governmental Industrial Hygienists (ACGIH, 1988). Or by the World Health Organization (WHO) in cooperation with the International Nonionizing Radiation Committee of the International Radiation Protection Association (WHO, 1982). For broadband infrared radiation (λ > 700 nm) the TLV is an irradiance of 10 mW/cm^2 (= 100 W/m^2). However, for IR laser radiation the TLV depends on wavelength, exposure duration and exposure pattern.

INDUSTRIAL IR SOURCES

In this lecture I shall primarily discuss filters for broadband (non-laser) radiation although several of the filters that I shall mention will provide excellent protection against some IR lasers. My reason for focusing on non-laser sources is that most of the practical protection problems that you meet in industry concerns broadband sources.

Broadband IR radiation is found not only in the traditional industries, i.e., in steel mills, glass works and foundries, but also in factories and shops where drying processes, metal-spraying and metal-cutting processes are utilized. Much to our luck several of these sources come close to being blackbody sources (fig.1). This fact can be used in many situations to calculate the actual exposure quite accurately, and thereby any necessary

Light, Lasers, and Synchrotron Radiation
Edited by M. Grandolfo *et al.*, Plenum Press, New York, 1990

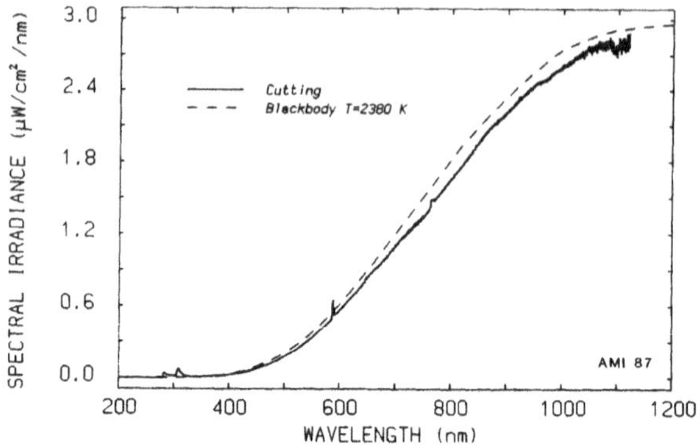

Fig.1. This graph shows the measured spectral irradiance at a distance of 0.6m from a metal-gas-cutting process. Also shown is the relative spectral output of a blackbody with a temperature of 2380 K. Apart from a few spectral peaks the "fit" is very good. A wavelength dependent emissivity is evident however (see also the appendix).

protective measure. For this reason a good knowledge of the characteristics of blackbody emitters is very helpful, and I've put a few of these in an appendix. It is my experience that many industrial exposure situations requiring IR protection can be easily and elegantly solved by considering the various IR filters that are available.

GLASSES AND PLASTICS

Most common transparent glass and plastic materials transmit optical radiation with wavelengths between approximately 0.3 and 3 μm. Only special and expensive glasses, such as CaF_2, ZnSe, Ge or Sapphire, transmit appreciably above 3 μm (fig.2). This means that the most common transparent or tinted glasses and plastics exhibit some degree of protection against far-IR, or IR-C radiation, at least for non-laser sources. To protect against high power or high energy pulsed lasers special consideration must be given to possible filter damage caused by the high incident irradiance or radiant exposure. Above a certain damage threshold that is easily exceeded with industrial type IR-lasers, plastic materials will burn through whereas glass materials will crack and thereby still retain at least some degree of protection.

FILTERS FOR IR PROTECTION

There are basically two different types of filters that can be used to protect against IR radiation: heat absorbing filters and heat reflecting filters. However, some of the available plain colored glass filters can be used for protection against near IR radiation over a limited range, for example against laser diode radiation. Apart from protecting against

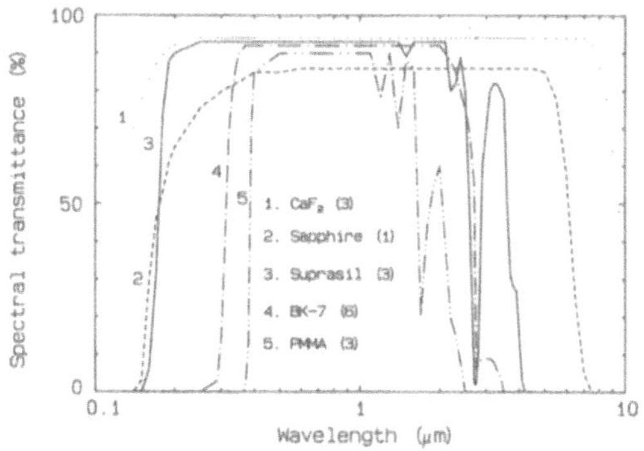

Fig.2. Spectral transmittance of four glass types and a transparent plastic material (acrylic, PMMA or Poly-MethylMetAcrylate). The thickness in mm is shown in (). Notice that of the "filters" shown, only Calcium Flouride and Sapphire has appreciable transmittance above 3 µm. Suprasil is a quartz type used in the UV range whereas BK-7 is an "ordinary" glass type.

hazardous radiation it is desirable that such filters should have as high a luminous transmittance as possible without causing discomfort. For assessment of comfort or discomfort the parameter of interest is the source luminance. The border between comfortable and discomfortable viewing conditions is not a sharp one. It depends upon the source size, viewing angle, individual vision and the luminance of the surrounding scene. As a rule of thumb a luminance of more than 10^4 cd/m^2 is considered to be discomfortable. In summary, filters should attenuate IR levels to below 10 mW/cm^2, and attenuate light such as to keep the source luminance (as seen behind the filter) below about 10^4 cd/m^2.

Welding Filters

By tradition many IR exposure situations in industry, particularly in steel mills and foundries, are handled by simply demanding that workers wear dark welding filters. One of the reasons is that there is an international standard that advices to choose certain welding filters depending upon the temperature of the molten material (steel, iron etc.). The result of such a demand is that the workers can hardly see the surrounding scene because the luminous transmittance of such filters is very low (table I). This may well lead to other dangerous situations. Welding filters are charcterized by a shade number N. For most regular welding operations shade numbers 9-12 are common, and shade number 10 is by far the most used. The shade number N and the luminous transmittance τ_v are related by the formula: $N = 1 - \log_e \tau_v$. There are requirements to the UV transmittance and IR transmittance of welding filters. Used as IR protecting filters shade numbers 5-8 are normally demanded (prEN 169, 1986; ISO 4850, 1979).

Table 2 shows clearly that welding filters are not a very good solution if you must have a high luminous transmittance. If we want the IR to be attenuated 10 times the sacrifice is a luminous transmittance of only about

TABLE 2. Requirements to welding filters as IR protective filters.

shade number	percent luminous transmittance	percent maximum avg. IR-A transmittance
1.7	43 - 58	40
2.0	29 - 43	28
2.5	18 - 29	15
3	9 - 18	12
4	3 - 9	6.4
5	1 - 3	3.2

10 percent. It gets worse if we demand more IR attenuation. Another shortcoming of welding filters is that their spectral transmittance is such that it is difficult, if not impossible, to distinguish between colors when wearing them, 'everything is green' when seeing through them. This is a serious drawback in many situations. Thus welding filters are neither a flexible nor a desirable solution to IR protection. Unfortunately they are widely used by tradition, because they are cheap and because they are readily available.

Heat Absorbing Filters

The purpose of heat absorbing filters, or heat absorbing glass, is to absorb heat, i.e. IR radiation, and to pass visible radiation relatively unattenuated. This means that they are almost transparent and that they will withstand strong IR radiation. Most glass manufacturers sell heat absorbing filters, e.g. Schott, Corning, HOYA, etc.. A well known heat absorbing

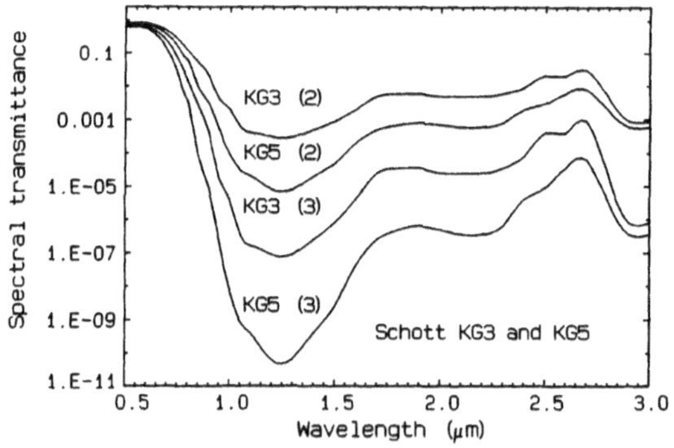

Fig.3. Spectral transmittance of Schott KG3 and KG5 heat absorbing glasses. Thickness in mm is shown in (). They are almost transparent to light but offers substantial attenuation of IR radiation. Note the logaritmic ordinate, from 1 (100%) to 10^{-11}.

filter from Schott is the KG3 filter. Fig.3 shows the spectral transmittance of two Schott filters between 0.5 and 3.0 μm. Their luminous transmittance is shown in Table 2.

Table 2. Percent luminous transmittance of heat absorbing glasses KG3 and KG5 from Schott for different black body sources.

	black body temperature[a]		
	2856 K	3200 K	6500 K
KG3, 2mm	58	80	76
KG3, 3mm	43	75	69
KG5, 2mm	59	81	76
KG5, 3mm	44	76	70

[a]2856K is the color temperature of the CIE standard A-illuminant (tungsten lamp). 3200K is the color temperature of most tungsten-halogen lamps. 6500K is the daylight color temperature (Xenon lamp).

The fact that the heat absorbing glasses are almost transparent and at the same time offer significant attenuation of IR radiation make them an ideal choice for eyeprotectors. When seeing through these glasses you can easily discriminate colors. This is a substantial improvement compared to welding filters. The only problem that may remain is if the source luminance is so high that it causes discomfort or even pain. Then the visible radiation must be attenuated to a comfortable level. However, this is a nice problem because it is easy to solve. Either by combining the heat absorbing glass with a neutral density filter (grey-filter) of appropriate density, or by coating the front surface of the heat absorbing filter with a thin layer of (reflecting) aluminum for example. The first method is the easiest and quickest method of obtaining a suitable, practical and affordable protective eyewear. Generally heat absorbing filters can offer flexible and simple solutions to problems in IR protection.

Heat Reflecting Filters

One drawback of heat absorbing filters is that as they heat up under the impact of strong IR radiation they re-emit radiation. However, the spectral distribution of the reemitted radiation is shifted strongly to longer wavelengths. Nevertheless it may be a nuisance that may lead to discomfort. If this is the case a heat reflecting filter is the solution. Heat reflecting filters consist of a thin metallic layer coated on one side of a glass surface. The metal may be aluminum, brass, gold, rhodium or other suitable metals or dielectrics. Fig.4 shows the reflectance (and transmittance) of thin layers of aluminum and gold on a glass substrate. A special kind of heat reflecting filter, where the coating is applied to a thin sheet of plastic, is available for your home or car to keep the inside cooler.

Notice from fig.4 that gold has a very high IR reflectance. This is the reason why delicate optics (mirrors) in infrared instruments are gold-coated. An inconvenience with coated glass is that it easily scratches. This is particularly a problem in an industrial environment. One can make the coating more durable by applying a second transparent protective coating. Still it may be damaged easily in harsh environments and one may have to use a transparent piece of plastic as a front "glass". Just like in the case of heat absorbing filters heat reflecting filters may be combined with neutral density filters to obtain suitable eye protectors.

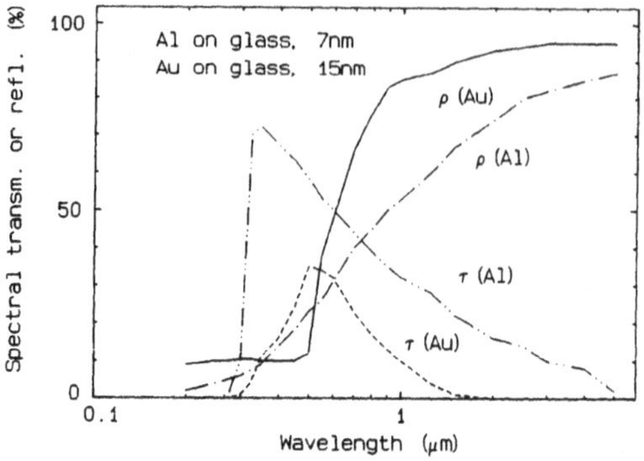

Fig.4. Spectral reflectance (ρ) and transmittance (τ) of thin layers (7nm and 15nm) of aluminum (Al) and gold (Au) on a glass substrate.

If you compare fig.4 and fig.3 you will see that the heat absorbing filters are superior to heat reflecting filters when it comes to discriminating color through the filters. However, special heat reflecting filters that have high and spectrally "flat" transmittance in the visible can be constructed: they're called "hot mirrors". Fig.5 shows the spectral transmittance of a hot mirror.

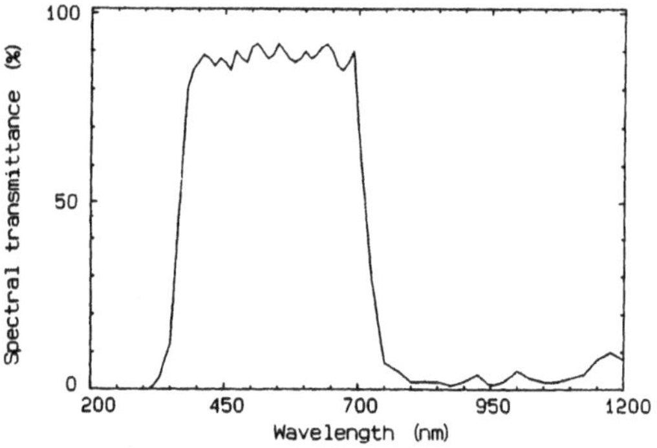

Fig.5. Spectral transmittance (τ) of a hot mirror. The infrared radiation is reflected from the front surface so the mirror appears to be "hot".

Hot mirrors are constructed from multiple layers of thin coatings on a substrate (Fowles, 1975; Dobrowolski, 1978). As you can see from fig.5 it will be easy to discriminate color through such a filter. Hot mirrors are used extensively in lamp systems to get rid of the heat produced by the lamp(s). They are not readily available in larger dimensions but must be made on special order. Neither are they cheap. Cold mirrors are opposite to hot mirrors but are similarly constructed, they reflect UV and visible radiation but transmit IR.

EXAMPLE: SPECIFYING SUITABLE EYE PROTECTION

In an industrial process molten metal was poured from a huge furnace into a big circular "pan" 3 m in diameter but only some 20 cm deep. The temperature of the molten metal was about 1200 C. The pan was 0.5 m above floor level and was in a horizontal plane. A few minutes after the molten metal filled the pan it developed a crust on its surface and the temperature dropped quickly. Two workers was supervising the process and from time to time they had to poke into the pool of molten material with 3-4 m long poles. They had been adviced to wear shade number 8 welding filters which have a luminous transmittance of about 0.1% and a maximum average IR-A transmittance of 0.4%. They were complaining that they were unable to see the surrounding scene which was neither sage nor satisfactory.

From the table in the appendix we see that the total radiance of a 1500K melt is 9 $W/(cm^2sr)$. At a distance of 4 m away from the edge of the pan and 1.7 m above the floor (man's height) the projected area of the pan is 1.2 m^2 (area times cosine to the angle between vertical and the workers eye). The pan is therefore viewed under a solid angle of $1.2/(5.5)^2 = 0.04$ sr. Then the total irradiance at the workers eye is the product of radiance and solid angle, 9 x 0.04, which is 360 mW/cm^2. Most of this total irradiance is in the infrared (appendix, table). Allowing for an emissivity of the melt of say 0.7 we should end up with a total irradiance of about 250 mW/cm^2. We measured a total irradiance of 130 mW/cm^2. This demonstrates how close calculations come to actual measurements when black body radiation is involved. A 2 mm thick Schott KG3 filter (fig.3) is sufficient to protect against hazardous IR radiation, it will attenuate the IR by more than 100 times which will be safely below the TLV of 10 mW/cm^2. According to the table in the appendix the luminance of the melt should be 8000 cd/m^2. We measured it to be 10000 cd/m^2. We therefore suggested a filter combination of a 2 mm thick KG3 and a neutral density filter of 10% transmittance (optical density = 1). This should yield comfortable viewing conditions and provide sufficient IR protection. Notice from the table in the appendix that the ultraviolet radiation is not of concern at this (low) black body temperature. From column 9 in the table the effective radiance is 8.3×10^{-10} $W/(cm^2sr)$. When multiplied by the solid angle of 0.04 sr the effective UV irradiance is 0.33×10^{-10} W/cm^2. Even for an exposure time of 10^4s (2.8 h) the effective UV dose is far below the TLV of 3 mJ/cm^2.

REFERENCES

ACGIH, 1988. "Threshold Limit Values and Biological Exposure Indices for 1988-1989," American Conference of Governmental Industrial Hygienists, Cincinnati, OH.

WHO, 1982. "Environmental Health Criteria 23. Lasers and Optical Radiation", World Health Organization, Geneva.

ISO 4850, 1979. "Personal eye-protectors for welding and related techniques - Filters - Utilisation and transmittance requirements," International Organization for Standardization, Geneva.

prEN 169, 1986. "Personal eye-protectors for welding and related techniques. Transmittance requirements," Commission Europeen de Normalisation, Paris.

Dobrowolski, J.A., 1978. Coatings and filters. In: "Handbook of Optics." Driscoll, W.G., Vaughan, W. (eds), McGraw-Hill, New York.

Fowles, G.R., 1975. "Introduction to Modern Optics." 2ed., Holt, Rinehart and Winston, New York.

The spectral radiance of a blackbody emitter of temperature T is given by the well known formula

$$L_\lambda = (2hc^2 / \lambda^5) / (\exp(hc/\lambda kT) - 1)$$

where h is Planck's constant, c is the velocity of light and k is Boltzmann's constant and L_λ is in SI units. To allow for a wavelength dependent emissivity $\varepsilon(\lambda)$ should be multiplied on the right hand side, the blackbody is then called a greybody. The list below shows several useable data on blackbody emitters.

Black body radiance calculations.

- column 1: Blackbody temperature (Kelvin)
- column 2: Specral radiance/exitance peaks at this wavelength in micrometers
- column 3: Total radiance of blackbody. Units in $W/(cm^2 sr)$
- column 4: Radiance within wavelength-interval 0.2-1.4 micrometers. Units in $W/(cm^2 sr)$
- column 5: Radiance within wavelength-interval 0.2-0.4 micrometers. Units in $W/(cm^2 sr)$
- column 6: Radiance within wavelength-interval 0.4-0.75 micrometers. Units in $W/(cm^2 sr)$
- column 7: Radiance within wavelength-interval 0.75-1.4 micrometers. Units in $W/(cm^2 sr)$
- column 8: Radiance within wavelength-interval 1.4-3.0 micrometers. Units in $W/(cm^2 sr)$
- column 9: Spectral radiance weighted with ACGIH (1989) S-spectrum (producing "effective" UV radiance)
- column 10: Spectral radiance weighted with Erythema-spectrum (McKinlay&Diffey) (producing erythema-effective radiance)
- column 11: Spectral radiance weighted with ACGIH (1989) A-spectrum (Aphake eye) (producing aphake-effective radiance)
- column 12: Spectral radiance weighted with ACGIH (1989) B-spectrum (Blue light) (producing blue-light-effective radiance)
- column 13: Spectral radiance weighted with ACGIH (1989) R-spectrum (Burn hazard) (producing burn-effective radiance)
- column 11: Spectral radiance weighted with CIE photopic eye response (color vision) (producing source luminance)

TEMP	Lpeak	Ltotal	.2-1.4	.2-.4	.4-.75	.75-1.4	1.4-3	LweighS	LweighE	LweighA	LweighB	LweighR	LweighV
K	mu	W/cm^2sr	W/cm^2sr	W/cm^2sr	W/cm^2sr	W/cm^2sr	W/cm^2sr	W/cm^2sr	W/cm^2sr	W/cm^2sr	W/cm^2sr	W/cm^2sr	cd/m^2
1000	2.898	1.8E+00	1.4E-02	3.4E-12	1.1E-05	1.4E-02	4.8E-01	2.7E-16	8.3E-16	5.3E-09	5.3E-09	3.0E-03	2.7E+00
1250	2.318	4.4E+00	1.5E-01	5.8E-09	6.5E-04	1.5E-01	1.8E+00	1.6E-12	2.8E-12	9.0E-07	8.7E-07	3.3E-02	3.1E+02
1500	1.932	9.1E+00	7.7E-01	8.5E-07	1.1E-02	7.5E-01	4.4E+00	8.3E-10	1.2E-09	4.2E-05	3.8E-05	1.8E-01	7.7E+03
1750	1.656	1.7E+01	2.6E+00	3.1E-05	7.9E-02	2.5E+00	8.7E+00	8.4E-08	1.1E-07	7.7E-04	6.5E-04	6.9E-01	7.9E+04
2000	1.449	2.9E+01	6.6E+00	4.8E-04	3.7E-01	6.2E+00	1.5E+01	2.8E-06	3.5E-06	7.6E-03	5.7E-03	2.0E+00	4.6E+05
2500	1.159	7.1E+01	2.7E+01	2.3E-02	3.4E+00	2.4E+01	3.2E+01	3.9E-04	5.0E-04	2.2E-01	1.3E-01	1.1E+01	5.6E+06
2856	1.015	1.2E+02	5.8E+01	1.6E-01	1.1E+01	4.7E+01	4.8E+01	4.8E-03	6.2E-03	1.2E+00	6.1E-01	2.9E+01	2.0E+07
3000	0.966	1.5E+02	7.6E+01	3.1E-01	1.6E+01	5.9E+01	5.5E+01	1.1E-02	1.5E-02	2.3E+00	1.0E+00	4.2E+01	3.0E+07
3200	0.906	1.9E+02	1.1E+02	7.2E-01	2.6E+01	8.0E+01	6.5E+01	3.2E-02	4.2E-02	4.8E+00	2.0E+00	6.6E+01	5.1E+07
3500	0.828	2.7E+02	1.7E+02	2.1E+00	4.9E+01	1.2E+02	8.2E+01	1.2E-01	1.7E-01	1.3E+01	4.7E+00	1.2E+02	1.0E+08
4000	0.725	4.6E+02	3.2E+02	9.1E+00	1.2E+02	2.0E+02	1.1E+02	7.6E-01	1.1E+00	5.0E+01	1.5E+01	3.0E+02	2.5E+08
4500	0.644	7.4E+02	5.6E+02	2.9E+01	2.4E+02	3.0E+02	1.5E+02	3.1E+00	4.7E+00	1.4E+02	3.5E+01	6.2E+02	5.2E+08
5000	0.580	1.1E+03	9.1E+02	7.5E+01	4.1E+02	4.2E+02	1.8E+02	9.8E+00	1.5E+01	3.4E+02	7.2E+01	1.1E+03	9.2E+08
5500	0.527	1.7E+03	1.4E+03	1.7E+02	6.6E+02	5.6E+02	2.2E+02	2.5E+01	4.1E+01	7.0E+02	1.3E+02	1.9E+03	1.5E+09
6000	0.483	2.3E+03	2.0E+03	3.2E+02	9.9E+02	7.2E+02	2.6E+02	5.5E+01	9.5E+01	1.3E+03	2.1E+02	3.0E+03	2.2E+09
6500	0.446	3.2E+03	2.9E+03	5.8E+02	1.4E+03	8.9E+02	3.0E+02	1.1E+02	1.9E+02	2.2E+03	3.2E+02	4.3E+03	3.1E+09

WELDING SAFETY MEASURES AND WELDING FILTERS

Per G. Söderberg

Dept. of Ophthalmology (Head: Björn Tengroth) and
Dept. of Medical Biophysics (Head: Rudolf Rigler)
Karolinska Institutet, Box 60 400, S-104 01 Stockholm
Sweden

INTRODUCTION

The welder can quickly handle numerous welding tasks under continuous quality control. For this reason manual welding will allways be needed as a complement to automatic welding. During welding the eyes are exposed to such an extreme environment that the biological defense is insufficient. The potential hazards for the eyes during welding activity decompose into risk for damage due to the welding process and risk for damage due to activity directly coupled to the welding. Both of these risks should be considered for protection. Contact lens wear during welding requires special safety precautions.

SOURCES OF HAZARD DURING WELDING

Welding can be defined as fusion of metal or plastic by high temperature. The energy needed to produce the high temperature is obtained by gas combustion, "Gas welding", by passing a strong electric current, "Arc welding", or from high power optical radiation "Laser welding".

Risks for damage due to the welding process

The deposition of large amounts of energy at the welding site causes three potential hazards. Encapsulated gas at the welding site explodes when the temperature rises and the explosion eject small glowing or melted metal fragments, welding sparks. The energy deposed is partly reemitted as optical radiation. Finally, the temperature rise creates nitrous gases and ozon, and if the pieces of metal to be welded are surface coated numerous species of vapours may be released.

Risks for damage due to activity directly coupled to the welding

Very often the welder want to adjust the piece of metal welded. This is most commonly done by grinding or by pounding. During the grinding small fragments of glowing chip are ejected. If hardened steel such as a hammer is pounded against steel small steel splinters may be ejected from the hardened steel.

EYE DAMAGE FROM WELDING

Damage from welding sparks

A welding spark that reaches the surface of the eye etches a cavity into the superficial epithelium and sticks. In the conjuctiva such damage causes irritation and foreign body sensation and the vessels of the conjunctiva dilate around the lesion. A foreign body in the corneal epithelium induces irritation and pain, sensitivity for light, and occasionally decreased visual acuity. The foreign body usually requires extraction under local anesthesia. After extraction of the welding spark the edge of the cavity has usually become pigmented. This destroyed tissue should be abraded. The remaining lesion normally heal without any scarring within two days. However, if the groove penetrates into the lamina limitans anterior of the cornea a white scar will occur. Such scars may cause glare and, if within the optical zone, decreased visual acuity. In rare cases a secondary infection occurs. If so the risk for white fibrotic scars increases.

Radiation damage

Biological damage from optical radiation depends on the wave band of the optical radiation, the irradiance at the tissue, and the time of exposure. Optical radiation may damage tissue by inducing photochemical damage, thermal damage or photomechanical disruption (Hillenkamp 1986).

Photochemical damage is evoked primarily by high energy photons, blue light and ultraviolet radiation (UVR), when target molecules have absorbtion lines corrresponding to the incident radiation. The absorbtion of the radiation energy induces changes in the structure of biomolecules in the target tissue thus interfering with the biological processes. The absorbtion of low energy photons, red photons and infrared radiation (IRR) induces temperature increase which at high temperature destroys the tertiary and the quartenary conformation of large biomolecules, denaturation, thus wrecking biological structure and function. If the temperature rises above 100 ^{o}C water transforms into vapour that expands and may cause mechanical disruption. At very high power densities the electromagnetic field of the optical radiation causes plasma formation creating a shockwave that disrupts tissue. Ultraviolet photons at very high power densities cause surface ablation (Kreuger et al. 1985).

The spectral radiance at gas welding depends on the temperature (Fig. 1). The spectral radiance during arc welding is much more complex (Fig. 2). There is a great variation as a function of time because the energy deposited alters. Furthermore, the composition of the material welded and inert gases, if adopted, modify the spectrum (Marshall et al. 1977, Hörnell et al. 1972). Both the site of gas welding and the site of arc welding are from a physical viewpoint isotropic radiators, indicating that the intensity of the radiation decreases with the square of the distance.

Energy derived from high power lasers may be used for welding, cutting, and surface shapening. Currently employed devices for welding and cutting are IRR lasers. For surface shapening ultraviolet radiating excimer lasers are used. The intensity of the radiation from high power lasers is at short distances essentially independent of the distance from the laser because of the paralellity of the beam. The wavelength depends on the choice of laser. The protection against laser beams is extensively treated elsewhere (Sliney and Wolbarsht 1985).

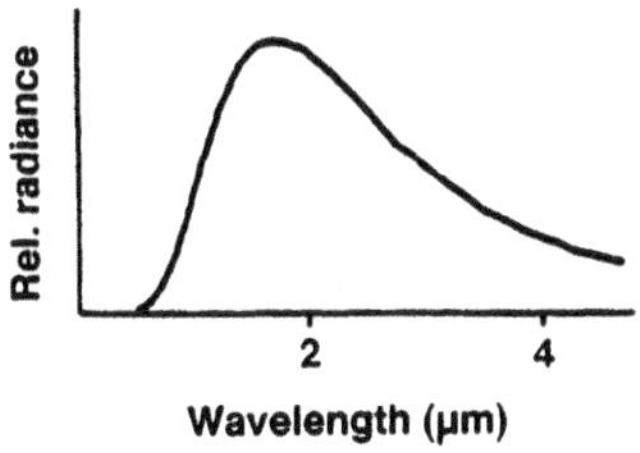

Fig. 1. Spectral radiance during gas welding at 1700 K as predicted by Planck's radiation law.

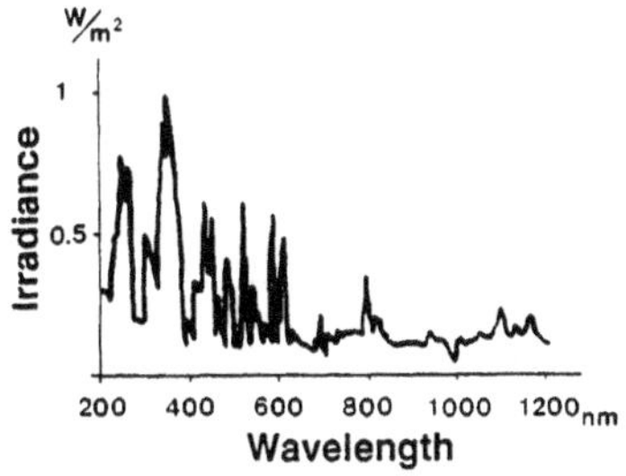

Fig. 2. An example of spectral radiance during arc welding.

<u>Radiation damage to the eye during welding</u>

<u>Infrared radiation.</u> IRR emitted at gas welding and arc welding is of such low intensity that no acute damage can be caused in the eyelids or in the anterior segment of the eye. The large proportion of radiation emitted in the visual spectrum triggers reflex closing of the eyelids, thereby preventing exposures of the retina and the choroid to toxical doses. However, at forced viewing of a welding site, such as welding without any protection there is a potential risk for thermal damage in the retina and in the choroid. This holds especially for arc welding with inert gases which may add spectral components in the infrared region (Brittain 1988). A small thermal damage centrally in the macula may be critical for the visual acuity while a wide spread peripheral lesion has no impact at all. Occupational longerm low dose exposure of the eye to IRR in the steel and glass industry has been found to correlate with the development of cataract (Lydahl 1984). No such data is available for welders. Even if there is a potential risk for the lens the life time

exposure to IRR for a welder is probably insufficient to cause damage considering the extensive use of eye protectors with suitable filters.

Visual radiation. Intense light from gas or arc welding has no impact on the eye lids or the anterior segment of the eye. Viewing of intense light sources may cause a reversible feeling of glare and after images. Absorbtion of high doses of visible radiation at the surface of the retina induces temperature rise wich may evoke thermal damage. However, discomfort of seeing occurs when the illumination level is approximately 1000 times lower than that causing retinal burns (Tengroth and Vulcan 1977). For this reason protection of the eye is not needed if the welding filter provide a comfortable luminance. However, blue light protection should be considered by protection standards since high doses of blue light causes photochemical lesions in the retina (Ham and Mueller 1976)

Ultraviolet radiation. During arc welding UVR constitutes a large proportion of the emitted radiation (Fig. 2). UVR evokes a symptom complex that has been described to occur in several environments rich of UVR; snow blindness, ophthalmia nivalis, industrial photophthalmi, photoelectric keratoconjuctivitis, arc eye syndrom, atom bomb keratitis, and welders flash. In the eye lids the UVR evokes a sun burn. Hours after an exposure an aching sensation and reddening of the skin occurs. A sufficiently high exposure of the eye lids to UVR induces blisters (Hausser 1969). The eye lids become markedly swollen due to accumulation of water in the loose connective tissue. The reddening of the skin is after some days substituted by increased pigmentation. Rupture of blisters should be avoided. Exposure of the conjuctiva and the cornea to UVR evokes a sensation of beeing swollen in the eye hours after the exposure and later irritation. The tear production increases and the eye white turns intensely red because of dilation of the conjuctival vessels. Later, 6-12 hrs after the exposure, the victim has a sensation of sand in the eyes and finally an intense pain develops associated with extreme sensitivity for light. Slit lamp microscopic examination of the cornea reveals increased amount of tear fluid on the surface of the eye, swollen superficial cell layers in the conjuctiva and the cornea, and sometimes sloughing off of superficial epithelial sheats in spots. Light microscopic examination demonstrates decreased mitotic rate, degeneration of cell nuclei in the epithelial cells and loss of superficial epithelial cells (Lerman 1980). The very intense pain easily disappears with a local anesthetic applied as an ointment. The symptoms usually disappear without sequele 24 to 36 hrs after the exposure.

Experimental exposure of the eye to UVR induces cataract (Widmark 1891, Verhoeff and Bell 1915/16, Söderberg 1988) and several epidemiological investigations demonstrate a correlation between exposure to UVR and cataract (Hiller et al. 1983, Talor et al. 1988). Since most full time welders have had numerous episodes of welders flash their eyes must have been exposed to high doses of UVR. Despite this, data is lacking on a possible correlation between welding and cataract.

The ultraviolet radiation entering the eye is to a great extent filtered out by the lens. Thus, the internal structures of the eye receive only limited doses of UVR primarily in the long wavelength region. The knowledge about effects of UVR to the vitreous is very limited (Ueno and Chakrabarti 1989). The doses of UVR reaching the retina (Boettner and Wolter 1962, Maher 1978) during welding are probably to low to have any significance.

Due to the very high power densities used for laser welding there is a potential risk for heat effects such as coagulation or even carbonization. Furthermore, mechanical damage within the eye from shockwaves is theoretically possible but very unlikely. If photoablative lasers are used for industrial surface modification there is a potential for accidental photoablative effects in the eye surface.

Damage to the eye from toxic vapours released during welding

Nitrous gases and ozon are produced during arc welding. These gases may cause irritation in the eyes during welding but no longterm effects on the eyes have been reported. Furthermore, when coated surfaces are welded it is essential to realize the potential toxic effects of the coatings as these are vapourized.

DAMAGE TO THE EYE IN ACTIVITY COUPLED TO WELDING

Apart from direct injury to the eye from the welding process a large number of eye trauma are evoked during work directly related to the welding. Often, a slight adjustment is needed through grinding. Thereby, released glowing grinding chip etch into the surface of the eye. The clinical syndrom and the treatment of these traumata is identical to that for welding sparks. Furthermore, if the piece of metal that has been welded is treated by pounding small, very sharp, steel splinters may be ejected from the metal. Steel splinters, reaching the eye, usually perforates the eye bulb. The steel splinter then has to be extracted, usually with a magnet, and a severe loss of visual acuity commonly follows.

EYE PROTECTION DURING WELDING

The safest eye protection during welding is the automatic welding isolated from the human being. However, still a great deal of welding activity has to be manual. For manual welding the classical eye protection is a head fixated or a handhold passive filter. The latest development is a head fixated or a handhold active filter.

Passive welding filters

The passive welding filter should protect against mechanical damage and optical radiation and provide visibility without glare. Mechanical protection is usually obtained by fitting the optical filter in a closed mask. Unfortunately, the welding mask is usually awkward to wear and therefore taken off immediately after the welding. Thus, the eye is exposed to grinding chips and pounding splinters.

Standard for passive welding filters. In Europe the nomenclature of welding filters has been standardized by Comitte' Europeen de Normalisation, CEN (CEN 1986). The standard is designed to make it easy for the welder to choose the appropriate filter. Filters are coded on an arbitrary scale from 1.2-16. A low number corresponds to a low optical density. The adequate filter can be read from a table provided in the standard. The choice of filter is based on the selection of welding procedure and the distance to the welding site. The coding system consider the transmittance of optical radiation in the ultraviolet, in the blue, and in the infrared part of the spectrum to protect the eye from damage and the luminous transmittance in the visual part of the spectrum to provide comfortable luminance.

The spectral transmittance, $\tau_e(\lambda)$, is defined as the power of radiant energy leaving an optical medium, $\Phi_e(\lambda)$, devided by the power of the incident energy, $\Phi_e(\lambda)o$, as a function of wavelength, λ (Eq. 1).

$$\tau_e(\lambda) = \frac{\Phi_e(\lambda)}{\Phi_e(\lambda)o} \qquad \text{Eq. 1}$$

The luminous transmittance, τ_v, is defined as the luminous flux, Φ_v, transmitted by the filter, devided by the incident luminous flux, $\Phi_v o$, within the wavelength region 380-780 nm. The spectral luminous flux, $\Phi_v(\lambda)$, is defined by

$$\Phi_v(\lambda) = \Phi_e(\lambda)\, V(\lambda) \qquad \text{Eq. 2}$$

Here, $V(\lambda)$ is the spectral luminous efficiency (CIE 17). Then,

$$\tau_v = \frac{\int_{380}^{780} \Phi_e(\lambda)\, \tau(\lambda)\, V(\lambda)\, d\lambda}{\int_{380}^{780} \Phi_e(\lambda)\, V(\lambda)\, d\lambda} \qquad \text{Eq. 3}$$

The transmittance of the filters in the visible wavelength region shall be determined using a black body radiator at 2856 K to allow adoption of the luminous efficiency function (CIE 17).

For protection against UVR damage the coding system for welding filters define a maximum spectral transmittance of radiant power at 313 nm, and at 365 nm. It is stated that in the wavelength region 210-313 nm the maximum spectral transmittance should be less than that defined for 313 nm, and in the wavelength region 313-365 nm the spectral transmittance should be below that defined for 365 nm. To provide comfortable luminance a maximum and a minum luminous transmittance is defined for the wavelength interval 380-780 nm. In the interval 365-400 nm the maximum spectral transmittance should be below the maximum luminous transmittance. To protect for blue light damage in the retina the spectral transmittance in the wavelength interval 400-480 nm should never exceed the spectral transmittance at 480 nm. For protection against IRR a maximum mean spectral transmittance is defined for the wavelength interval 780-1400 nm.

To express the dampening of luminous flux in a filter the shade number is given. The shade number, N, is defined as

$$N = 1 + \frac{7}{3} \lg \frac{1}{\tau_v} \qquad \text{Eq. 4}$$

The choice of code number depends on the radiance from the welding site. For gas welding the radiance depends on the gas flow rate, the flux and the material welded. For arc welding the radiance is determined by the arc current, the flux, the material and if applied inert or active gases.

The fixed combinations of optical density for UVR, light and IRR, corresponding to certain code numbers, are inconvenient considering the quick development of new welding procedures with unique spectral radiances. Thus, it may be required to select a code number correlated with un unnecessary dark filter to protect sufficiently against UVR damage.

The main problem with the passive filters is that in order to provide comfortable luminance during welding the filters are so dense that they allow almost no visibility before the welding has been initiated. Consequently, all welders observe the welding site without any protection thus accumulating a high enough UVR dose during a work shift to acquire a welders flash.

Active welding filters

In order to help the welder to initiate the welding process at good luminance without being exposed to toxic radiation photosensitive filters have been developed. These filters have the same requirements for dampening optical radiation but the filter in the visual part of the spectrum is an almost transparent liquid crystal that closes at the initiation of the welding process. The current problem with this very promising concept is that the recovery time is slightly extended, thus impeding the welder to restart after interrupting the welding process.

CONTACT LENS WEAR AND WELDING

The publication of one case of a corneal ulcer in a contact lens wearer after exposure to an electric arc (Kersley 1970) triggered a debate about contact lens wearing during welding (Mereau and Ulysse 1985). Several reviews of the circumstances around the incidence strongly has questioned whether the exposure to the electric arc was of any significance (CLAO Board 1983, Mereau and Ulysse 1985). Experimental studies indicate that absorbtion of IRR in the water on the surface of the contact lens may induce a temperature rise (Lövsund et al. 1979). However, to cause drying out of the lens very high doses have to be applied without any blinking.

In a recent review (Chiappino et al. 1983) it is concluded that current knowledge does not provide a firm bases to prohibit contact lens wearing during welding. However, it is pointed out that the welding environment is unsuitable for contact lens wearing. Particles and irritating vapours, released during the welding, may stick to the lens and cause irritation. In welding under pressure, e.g. work under water, the contact lens is a potential hazard. During welding with pressure helmets strong air flows may extract the lens. If there is a gas bubble under the contact lens during pressure increase the contact lens may cause suction trauma to the corneal surface. Conversely, a contact lens on the cornea during decrease of pressure may impede dissolution of gases dissolved under the high pressure. Thereby, formed gas bubbles may disrupt the corneal tissue.

FINAL REMARKS

A wide spectrum of eye problems may occur during welding or in activity directly coupled to welding. Current welding filters provide sufficient protection when used. However, the low visibility through the filter before the welding process is inititated forces the welder to commence without radiation protection. For this reason welders flash is still a common occupational problem among professional welders. Likewise, eye damage from grinding chips and steel splinters during activity directly coupled to the welding is still a considerable inconvenience. Some of these problems may be possible to overcome with education especially of the increasing group of hobby welders. Furthermore, a continuous development of the design of the protection devices is necessary. The active welding filter is a promising example of crucial improvement.

ACKNOWLEDGEMENTS

I am indebted to Björn Tengroth for reviewing the manuscript and to Lotta Vig for help with the preparation of the manuscipt.

REFERENCES

Brittain, G. P. H., 1988, Retinal burns caused by exposure to MIG-welding arcs: report of two cases. Brit. J. Ophthalmol., 72:570.

Boettner, E. A., Wolter, J. R., 1962, Transmission of the ocular media, Invest. Ophthalmol., 1:776.

CEN, 1986, Personal eye-protection. Filters for welding and related techniques; Transmittance requirements and recommended use. Comite' European de Normalisation, Ref. No. EN 169:1986 E.

Chiappino, G., Miglior, M., and Terrana, T., 1983, Lenti a contatto e particolari attivita lavorative, Med. Lav., 74:257.

CIE 17, Comission International d'Eclairage.

CLAO Board, 1983, Policy statement on arc welding and contact lens wear, Clao J., 9:343.

Ham, W. T., and Mueller, H. A., 1976, Retinal sensitivity to damge from short wavelength light, Nature, 260:153.

Ham, W. T., Ruffolo, J. J., Mueller, H. A., and Guerry Du Pont III, 1980, II Quantitative dimensions of intense light damage as obtained from animal studies. The nature of retinal radiation damage: dependence on wavelength, power level and exposure time, Vis. Res., 20:1105.

Hausser, K. W., and Vahle, W., 1969, The biological effects of ultraviolet radiation. Sunburn and suntanning. Pergamon Press. Oxford.

Hillenkamp, F., 1986, Biophysical mechanisms of damage induced by light, in: Hazards of light, myths and realties, eye and skin, p. 27, Pergamon Press, Oxford.

Hiller, R., Sperduto, R. D., and Ederer, F., Epidemiologic associations with cataract in the 1971-1972 national and nutrition examination survey, Am. J. Epidemiol., 118:239.

Hörnell, Å., and Vulcan, J., 1972, Spektralfördelningen hos svetsljusbågar i våglängdsområdet 200-1200 nm. Gradual Thesis, Institution of electrophysics II, Chalmers Institute of Technology, Gothenburg.

Kersely, H. J., 1977, Arc flash and the contact lens wearer, Br. Med. J., 2:639.

Kreuger, R. R., Trokel, S. L., and Schubert, H. D., 1985, Interaction of ultraviolet laser light with the cornea, Invest. Ophthalmol. Vis. Sci., 26:1455.

Lerman, S., 1980, Biologic and chemical effects of ultraviolet radiation, in: Radiant energy and the eye, p. 115, Macmillian publishing Co. Inc., New York.

Lydahl, E., 1984, Infrared radiation and cataract, Thesis, Acta Ophthalmol. (Copenh.), Suppl. 166.

Lövsund, P., Nilsson, S. E., Lindh, H., and Öberg, P. Å., 1979, Temperature changes in contact lenses in connection with radiation from welding arcs, Scand. J. Work. Environ. Health., 5:271.

Maher, E. F., 1978, Transmission and absorbtion coefficients for ocular media of the rhesus monkey, Report SAM-TR-78-32, USAF School of Aerospace Medicine, Aerospace Medical Division (AFSC).

Marshall, W. J., Sliney, D. H., Lyon, T. L., Krial, and N. P., Del Valle, P. F., 1977, Nonionizing radiation protection special study no. 42-0312-77 Evaluation of the potential retinal hazards from

optical radiation generated by electric welding and cutting arcs, USA Environmental Hygiene Agency, Aberdeen Proving Ground, MD 21010.
Mereau, P, and Ulysse, J. F., 1985, Port de lentilles de contact en milieu de travail. Nouvelle mise au point, Documents pour le medecin du travail, Fiche medico-technique, TC 13: 5.
Sliney, D. H., and Wolbarsht, M., 1985, Safety with lasers and other optical sources, New York.
Söderberg, P. G., Acute cataract in the rat after exposure to radiation in the 300 nm wavelength region. A study of the macro-, micro-, and ultrastructure, Acta Ophthalmol. (Copenh.), 66:141.
Taylor, H. R., West, S. K., Rosenthal, F. S., Munoz, B., Newland, H. S., Abbey, H., and Emmett, E. A., 1988, Effect of ultraviolet radiation on cataract formation, New Engl. J. Med., 319:1429.
Tengroth, B., and Vulcan, J., 1977, Welding light, Strahlentherapie, 153:267.
Ueno, N., and Chakrabarti, B, 1989, UV-light induced liquefaction and generation of damaging photosensitizers in the vitreous of simulated aphacic eyes, Invest. Ophthalmol. Vis. Sci., Suppl., 30:460.
Verhoeff, and Bell, 1915/16, The pathological effects of radiant energy upon the eye, Proc. Am. Acad. Art Sci., 51:629.
Widmark, J, 1891, Ueber die Durchlässigkeit der Augenmedien fur ultraviolette Strahlen, Beitr. z. Ophthalmol., 460.

LASER EYE PROTECTORS*

Penelope K. Galoff

Laser Microwave Division
United States Army Environmental Hygiene Agency
Aberdeen Proving Ground, Maryland USA 21010-5422

BACKGROUND

Almost since the first laser was developed it has been known that laser eye protection has been necessary. Our eyes, because of their unique ability to focus light, are the most susceptible part of the human body to laser radiation. Protecting our eyes from laser light would be very easy if we did not need to see through the protection. A piece of metal or wood would perform just fine. But since we are required to perform tasks that require the use of our eyes, developing useful laser eye protectors can be a very complicated exercise.

Early laser eye protectors were dark, heavy, and gave the wearer tunnel vision. These were developed by taking common welders goggles and replacing the protective shield with a material that was known to attenuate the ruby and neodymium: YAG laser wavelengths, BG-18. This type of laser eye protection is still used in some areas where good visual acuity is not required. It is easy to understand why there was and still is a great demand for better laser eye protectors.

Today there are many new features that laser eye protectors provide and many characteristics that need to be considered when selecting laser eye protection. Before any of these traits can be assessed, one must first know some basics about the laser system they are protecting their eyes against.

EYE PROTECTION PARAMETERS: WAVELENGTH

First, it is necessary to know the wavelength of the laser. This may seem like a simple enough task since most lasers only lase at one specific wavelength. But this is not true for all laser systems. A very common example is the argon laser. This laser will provide two primary wavelengths depending upon the optics that are used in the laser. The common blue-green color argon laser is actually lasing at these two primary

* The opinions or assertions contained herein are the private views of the author and are not to be construed as reflecting the views of the Department of the Army or the Department of Defense.

wavelengths, the 488 nm line and the 514 nm line. These two lines can be run in combination with each other or exclusive of one another.

There are laser systems that can provide multiple wavelengths. These are called dye lasers. These laser systems, with the proper pumping source and correct dye can produce almost any wavelength desired. Determining your laser wavelength can be simple, such as reading the manufacturers manual (most commonly used method) or more complicated, one which is out of the realm of this discussion. You can be assured that the wavelength provided by the manufacturer is accurate for the purposes of determining proper laser eye protection.

Once wavelength is determined it is also important to know the first disadvantage to current laser eye protectors. That is that laser eye protectors are wavelength specific. In other words, they only work at a specified laser wavelength or, at best, a wavelength region which may or may not include more than one useful wavelength. We will use the ruby laser wavelength of 694.3 nm as our example laser. Figures 1 and 2 are attenuation curves for two different types of laser eye protectors. From Figure 1 it can easily be understood that protection is only provided for approximately 10 - 20 nms around the 694.3 nm wavelength. Figure 2 shows that the 694.3 nm wavelength is protected along with those within the wavelength band between 650 nm and 1100 nm. If, as in the previous example using the argon laser, the laser was running in the combination mode and your laser eye protector used the technology shown in Figure 2, it is conceivable that you could have protection for only one of the argon laser wavelengths.

EYE PROTECTION PARAMETERS: OPTICAL DENSITY

The next question to ask is how much protection is needed. This is called the optical density (O.D.) of the filter material. To determine how much optical density is required for protection you need to know a little more information about your laser. The information you need will come from determining your anticipated exposure conditions. Use this information to determine the Maximum Permissible Exposure (MPE) for your particular laser system and use situation. This process can be found in the American National Standards Institute (ANSI) Z136.1 - 1986, Safe Use of Lasers. Then the equation used in the determination of the optical density required for safe laser use is as follows:

$$\text{O.D.} = \log 10 \frac{H}{\text{MPE}} = -\log 10\ t \qquad [1]$$

where H is the potential eye exposure expressed in the same units as the appropriate MPE and optical density is wavelength specified.

To determine the optical density of a protective filter the equation used is given as:

$$\text{O.D.} = \log 10\ (Pi\ /Pe) \qquad [2]$$

where Pi is the optical power incident upon the filter and Pe is the optical power exiting the filter. Although this measurement looks relatively simple in principle, there are many factors that must be taken into account when actually performing optical density measurements (Eriksen and Galoff, 1989). These factors may provide for inaccurate measurements resulting in a lower optical density for the material measured. Some laser eye protective materials may have some angle of incidence dependence thus anticipated viewing angles should be evaluated as well.

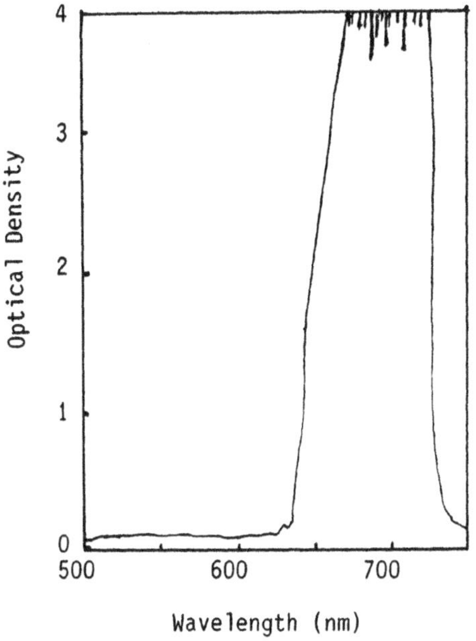

Figure 1. Spectral Attenuation of a Reflective Filter.

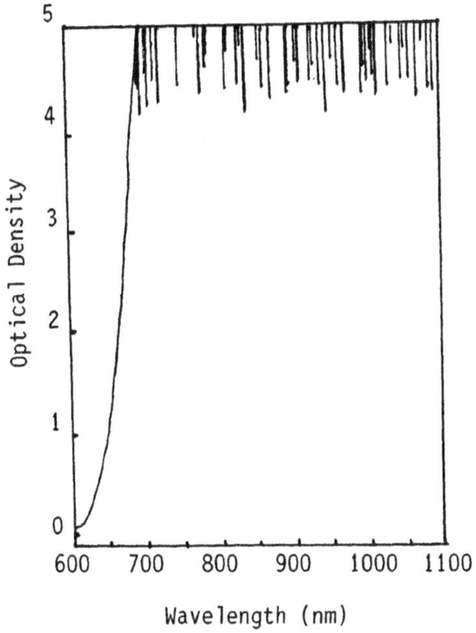

Figure 2. Spectral Attenuation of an Absorption Filter.

METHODS OF CONSTRUCTION

There are two groupings of laser eye protection, these are fixed and dynamic filters. Currently dynamic filters have not yielded adequate response times, can not discriminate between lasers and other light sources, and do not provide required rejection efficiencies, thus they are not available. Of the four types of fixed filter technology; absorption, reflection, diffraction, and refraction, only two are presently on the market. Diffraction filters are not available due to problems with the high angle of incidence dependency, and visual distortion due to problems with interpupillary distance and eye relief. Refraction filters are not as well studied as other fixed filter technologies and are also not available due to similar problems with angle of incidence dependency. This leaves us with the two basic methods of construction of laser eye protection filters, absorption and reflection.

Absorption

The most common and easily manufactured filter is the absorptive filter. Absorption filters utilize organic and inorganic dyes to absorb the laser light. These dyes usually absorb over a broad spectral region (Figure 2). This is where the problems begin. If the wavelength of the laser is in the visible or near the visible portion of the spectrum, this large area of absorption could very likely lower the overall transmittance of the filter. This making color and contrast sensitivity lower and overall task performance more difficult. When more than one wavelength is protected against, it is easy to see how the use of these broadband absorbers could only compound the problem. In extreme cases, this can lead to nonuse of the protective filter.

The dyes themselves have their own problems. Some of the organic dyes when used with certain types of plastics can darken and their area of protection can shift. This is the result of a chemical reaction which is stimulated by ultraviolet exposure, heat and high humidity. It is important to note that although the band of protection does shift, it has never been observed to shift completely out of region where the protection is desired. This is primarily due to the darkening of the filter. The filter becomes very dark and is not used or is discarded. Stabilizers are currently being researched to solve this problem and progress has been made that will slow down this darkening effect.

Organic dyes have been shown to bleach or saturate under certain high peak energy conditions. This has primarily been observed when the laser source is a Q-switched device. The amount of energy that is contained in such a short pulse of light can actually saturate the absorption dyes and cause it to fail, thus allowing the laser light to pass directly through the filter totally unobstructed.

Reflection

Reflective filters can use dielectric coatings on a glass or a plastic substrate to reflect narrow wavelength bands. Multiple thin film layers are laid down so that they are a quarter of a wavelength apart for the desired wavelength protection. This than interferes with the incoming laser light, reflecting it back before it can reach the eyes. This may sound like the best method for developing laser eye protection because it solves a lot of the problems that are present with absorption filters. The problem with dielectric coatings is its angle of incidence sensitivity.

The wavelength of operation shifts as the angle of incidence shifts. Manufacturing this type of device is extremely expensive and can only be used under specialized conditions.

SELECTING EYEWEAR

Now that you know the wavelength of your laser and the optical density that you need to protect your eyes from overexposure, you must decide which laser eye protection is right for you given your individual exposure and use conditions.

Spectacles versus Goggles

Spectacles come in a variety of lens sizes to accommodate for various interpupillary distances. This would be important if a prescription is to be provided along with the laser eye protection. Prescriptions can only be ground into glass filters. Commonly, glass filters are used in spectacles and plastic filters in goggles, although the reverse is available for those who wish. A spectacle would be recommended for situations where personnel are long term employees. Spectacles are recommended when laser eye protection will be an investment and which will be custom fit for each employee. For cases where individuals will be more transient, a goggle may be a better choice. The goggle provides for a more universal fit and individuals with prescriptions can usually wear their glasses along with the goggle. Prices are comparable between plano spectacles and goggles. Prescriptions in the glass filters will be an added charge.

Glass versus Plastic

The advantages of glass filters when compared to plastic materials are: scratch resistance, less fogging, better stability against solarization and bleaching, readily ground to optical prescriptions, generally better optical quality and generally better visual transmittance for the same optical density afforded at the wavelength of interest. The advantages of plastic are lighter weight, and a more universal fit.

Visual Transmittance

This may very well be the most important characteristic when picking laser eye protection. Remember when the laser is a visible wavelength, the filter is going to block your ability to see that wavelength and possible the entire color. This may have a large impact on your ability to perform a task. How well do personnel need to see to do their task? For a pilot required to fly an aircraft, high visual transmittance is a must and cost per eyewear will be higher. For the person working in a lab where all procedures are routine and personnel are well trained, an extremely high visible transmission would not be required and comfort would be of greater concern. Comfort being the difference between personnel wearing the eyewear or not.

Lateral Protection

Lateral protection is usually required in all cases with only a few exceptions. The case of fitting pilots with laser eye protection is a good example. Lateral protection is not required in some cases, where the aircraft is not the target vehicle for the laser system, when the side shields interfere the pilots' helmet. All laboratory conditions encountered thus far require side shields.

Laser Filter Damage Threshold

This may be important in an industrial environment. Lasers usually found in this type of setting are high powered laser systems. Burn through time for the filters as well as the frames may be of concern. What is important to remember here is that you are not only dealing with an eye hazard, but of more concern would be the skin and fire hazards from these high powered laser systems.

User Acceptability

Unfortunately, probably the users highest concern. Finding a goggle or spectacle that the user will accept as comfortable and good looking may be your hardest decision. Education may be your best alternative. Letting your personnel know the risks involved with not wearing their eyewear may make it a little easier to get the personnel to wear odd looking eyewear.

CONCLUSIONS

Laser eye protectors are essential to provide adequate protection against accidental or intentional exposure to the direct or specularly reflected beam from the growing number of laser systems. Although far from ideal, presently available laser eye protection can cover much of the optical spectrum. Understanding why the "perfect laser eye protection" is not available and what the risks of nonuse present, make it easier to select laser eye protection for your specific needs.

REFERENCES

American National Standards Institute, "American National Standard for the Safe Use of Lasers, ANSI Z136.1-1986," ANSI, 1430 Broadway, New York, NY 10018.

Chisum, G.T., Concepts in laser eye protection. In: Proceeding of the International Symposium on the Biological Effects of Lasers and Protection Standards, F-92260 Fontenay-auxRoses: Commisariat a L'Energie Atomique; 1988.

Galoff, P.K., Sliney, D.H., Evaluation of laser eye protectors in the ultraviolet and infrared. In: Proceeding of the International Symposium on the Biological Effects of Lasers and Protection Standards, F-92260 Fontenay-auxRoses: Commisariat a L'Energie Atomique; 1988.

Eriksen, P., Galoff, P.K., Measurements of Laser Eye Protective Filters. In: Proceeding of the First Symposium on the Biologival Effects, Hazards and Protection from Nonionizing Radiation in Outdoor Applications, Health Physics Journal Vol 56, No. 5, May 1989.

Sliney, D.H., Wolbarsht, M., Safety wiht Lasers and Other Optical Sources, Plenum Press, New York, 1980.

VISIBLE AND INFRA-RED RADIATION (NON-LASER) PROTECTION STANDARDS

A. F. McKinlay

National Radiological Protection Board
Chilton, Didcot, Oxfordshire, UK

INTRODUCTION

The subject of this Chapter is a summary of recommended exposure limits applicable to broad-band (non-laser) sources of light and infra-red radiation. The biological effects on which these limits are based and aspects of practical hazard assessment are dealt with in other parts of this volume.

Whereas recommended exposure limits for laser radiation enjoy a large measure of international 'official' agreement (IEC 1984) and apart from certain minor differences have been incorporated in national standards by many countries there are as yet no internationally agreed limits for light and infra-red radiation from broad-band optical sources. However, the American Conference of Governmental Industrial Hygienists (ACGIH) have made proposals for 'Threshold Limit Values' TLVs) for light and infra-red radiation which they have promulgated by means of their annual published booklets on TLVs for chemical substances and physical agents in the workroom (ACGIH 1987). The status of these recommended limits is of 'intended changes' whereby they are recommended for use until such time when scientific evidence may necessitate a change and they are intended for hazard assessment in the workplace. In the absence of such evidence they are then considered for official adoption. An example of such a change (addition) is provided by the publication of ACGIH of 'intended changes' for 1988-89 of additional recommended limits for aphakes which effectively extend retinal 'blue-light' hazard analysis down to 305 nm in the ultra-violet region (ACGIH 1988).

The recommended limits referred to in this Chapter are those of the 1988 ACGIH publication.

RECOMMENDED EXPOSURE LIMITS

Exposure limits for non-laser sources would be expected to differ from those applicable to lasers for several reasons. For example: whereas lasers generally (but not always) emit radiation at a single or over a narrow range of wavelengths, non-laser sources generally emit a continuous spectrum; in the case of the latter separate consideration must be given to the hazards associated with each spectral region: apart from the specific conditions of extended source viewing, such as from diffuse target reflections, exposures to lasers are generally limited by considering small

Light, Lasers, and Synchrotron Radiation
Edited by M. Grandolfo *et al.*, Plenum Press, New York, 1990

retinal images (intra-beam viewing); non-laser sources are generally extended sources producing resolved (large) retinal images: the radiance from a non-laser source (which is directly related to the retinal image irradiance) is physically limited by the design of the source and is unlikely to suddenly and sustainingly exceed its design value: and; the conditions for exposure to light from non-laser sources are likely to result in a minimal 3 mm diameter pupil whereas for laser safety a fully dilated 7 mm diameter pupil is assumed and no variation (relaxation) in limits is permitted for small pupil diameters.

Photochemical and thermal retinal hazards are considered separately though the use of 'blue light' B_λ and 'burn' R_λ hazards functions, table 1. In addition the 'blue light' hazard to aphakes is considered by the use of the 'aphake photic' hazard function A_λ. The B_λ, R_λ and A_λ hazard functions are illustrated in figure 1.

The exposure limits apply to exposure in an 8 hour working day and require data on the spectral radiance (L_λ) of the source or the corneal spectral irradiance (E_λ) resulting from exposure to the source.

In this context radiance is an appropriate measure of hazard for radiations focused onto the retina and forming a resolved image. ie, from 400 to 1400 nm in the intact eye. In principle this corresponds to the extended source viewing conditions used in laser hazard evaluations. For viewing conditions resulting in a minimal image on the retina, corneal irradiance is the appropriate measurement parameter. This corresponds in principle to the 'intra-beam viewing conditions' used in laser hazard evaluations.

BLUE-LIGHT RETINAL HAZARD EVALUATION

The exposure limits to protect against photochemical retinal injury are expressed in terms of radiance ($W\ m^{-2}\ sr^{-1}$) or temporally integrated radiance ($J\ m^{-2}\ sr^{-1}$) as a weighted spectral summation over the range of wavelengths 400 to 700 nm and are denoted mathematically as

$$\sum_{400}^{700} L_\lambda \cdot B_\lambda \cdot \Delta\lambda \leqslant 10^2 \quad (t > 10^4\ \text{s})$$

$$\sum_{400}^{700} L_\lambda \cdot t \cdot B_\lambda \cdot \Delta\lambda \leqslant 10^6 \quad (t \leqslant 10^4\ \text{s})$$

where L_λ is the spectral radiance, B_λ the blue light hazard function and t the exposure time (s). ACGIH designate the weighted mathematical product of $L_\lambda B_\lambda$ as L(BLUE). This enables the calculation of the value of a maximum permissible exposure time (t_{max}) in seconds for a given value of L(BLUE) from

$$t_{max} = \frac{10^6\ J\ m^{-2}\ sr^{-1}}{L(BLUE)}$$

For monochromatic radiation these limits are less restrictive than the exposure limits for laser radiation. This is because a light-adapted pupil diameter (2-3 mm) is assumed, and the tabulated values of the function B_λ gradually increase from 0.1 at 400 and 500 nm to a maximum value of 1.0 at wavelengths of 435 and 440 nm. In the laser standards the 'blue sensitivity'

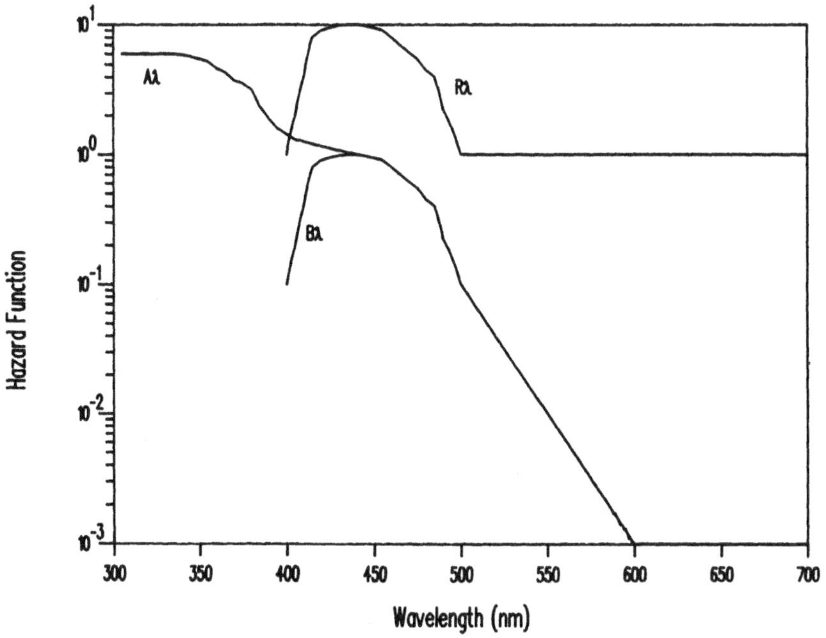

Figure 1. Blue-light B_λ, Aphake photic A_λ and Burn R_λ hazard functions

is assumed constant from 400 to 550 nm and the pupil diameter is for the dark-adapted eye (7 mm).

It is important to note that whereas in their previously published recommendations ACGIH extended the values of Bλ and thereby the upper wavelength limit of the summations above to 1400 nm, no values are now provided above 700 nm. As photochemical injury is unproven for infra-red wavelength radiation these new wavelength limits appear appropriate. For an optical source that subtends an angle α (angular subtense) of less than 11 mrd the above limits are relaxed and are expressed in terms of corneal irradiance (W m^{-2}) or corneal radiant exposure (J m^{-2}) and are denoted mathematically as

$$\sum_{400}^{700} E_\lambda \cdot B_\lambda \cdot \Delta\lambda \leqslant 10^{-2} \quad (t > 10^4 \text{ s})$$

$$\sum_{400}^{700} E_\lambda \cdot t \cdot B_\lambda \cdot \Delta\lambda \leqslant 10 \quad (t \leqslant 10^4 \text{ s})$$

The maximum permissible exposure time t_{max} in seconds for a given value of the weighted mathematical product $E_\lambda B_\lambda$ [E(BLUE)] is calculated from the expression

$$t_{max} = \frac{10^2 \text{ J m}^{-2}}{E(\text{BLUE})}$$

For aphakes the potential photochemical retinal hazard extends into the ultra-violet region and unless the person exposed has been fitted with an intra-ocular lens with proven appropriate UVR-absorption the calculations described above should be carried out using values of the aphake hazard function A_λ instead of the blue-light hazard function B_λ and with a lower wavelength of 305 nm. With regard to the increased transmission in the ultra-violet A and part ultra-violet B, (305 - 400 nm) of the anterior ocular media of infants and young children it may well be appropriate to use the A_λ values for the evaluation of potential photochemical retinal hazards to children.

THERMAL RETINAL (BURN) HAZARD EVALUATION

The burn hazard function R_λ is limited to exposure durations of < 10 s where the spectral radiance weighted against R_λ and summed over the wavelength range 400 to 1400 nm should not exceed

$$\sum_{400}^{1400} L_\lambda \cdot R_\lambda \cdot \Delta\lambda \leqslant \frac{10^4}{\alpha t^{1/4}}$$

Where α is the angular subtense of the largest dimension of the source at the eye and t is the exposure duration. The power law of $t^{-1/4}$ is intermediate between the empirical power laws of the laser standards for small and large retinal images (intra-beam and extended source viewing). With its limitation to viewing durations of < 10 s the use of the burn hazard function is directed towards accidental viewing conditions and not those involving fixated staring. For such conditions and where there is a strong visual stimulus present the natural protective aversion responses will

Table 1. Blue-light B_λ, Aphake photic A_λ and Burn R_λ hazard functions

Wavelength λ (nm)	B_λ	A_λ	R_λ
305	-	6.00	-
310	-	6.00	-
315	-	6.00	-
320	-	6.00	-
325	-	6.00	-
330	-	6.00	-
335	-	6.00	-
340	-	5.88	-
345	-	5.71	-
350	-	5.46	-
355	-	5.22	-
360	-	4.62	-
365	-	4.29	-
370	-	3.75	-
375	-	3.56	-
380	-	3.19	-
385	-	2.31	-
390	-	1.88	-
395	-	1.58	-
400	0.1	1.43	1.0
405	0.2	1.30	2.0
410	0.4	1.25	4.0
415	0.8	1.20	8.0
420	0.9	1.15	9.0
425	0.95	1.11	9.5
430	0.98	1.07	9.8
435	1.00	1.03	10.0
440	1.00	1.00	10.0
445	0.97	0.97	9.7
450	0.94	0.94	9.4
455	0.90	0.90	9.0
460	0.80	0.80	8.0
465	0.70	0.70	7.0
470	0.62	0.62	6.2
475	0.55	0.55	5.5
480	0.45	0.45	4.5
485	0.40	0.40	4.0
490	0.22	0.22	2.2
495	0.16	0.16	1.6
500	0.10	0.10	1.0
500–600	$10^{\left[\frac{450-\lambda}{50}\right]}$	$10^{\left[\frac{450-\lambda}{50}\right]}$	1.0
600–700	10^{-3}	10^{-3}	1.0
700–1049	-	-	$10^{\left[\frac{700-\lambda}{500}\right]}$
1050–1400	-	-	0.2

prevent thermal retinal injury. The empirical nature of the derivation of the time varying term $t^{-1/4}$ results in recognised significant uncertainties in the appropriateness of the recommended values at short exposures such as might be experienced from pulsed xenon sources. Current limits may be too restrictive at short exposures and ACGIH currently have these aspects of their exposure recommendations under review (Sliney 1989).

For conditions where there is no strong visual stimulus present, as for example in the direct viewing of a 'heat lamp' ACGIH recommend that the radiance of the source L_λ ($W\ m^{-2}\ sr^{-1}$) as viewed by the eye should be limited to

$$\sum_{770}^{1400} L_\lambda \cdot \Delta\lambda \leqslant \frac{0.6 \times 10^4}{\alpha}$$

Under these viewing conditions a 7 mm fully dilated pupil is assumed.

To avoid chronic effects to the lens of the eye (thermal cataract) ACGIH recommend that the infra-red radiation (λ > 770 nm) should not exceed $10^2\ W\ m^{-2}$.

REFERENCES

ACGIH, 1987, "Threshold Limit Values for Chemical Substances and Physical Agents in the Workroom Environment", The American Conference of Governmental Industrial Hygienists, ACGIH, Cincinnati, USA.

ACGIH, 1988, "Threshold Limit Values for Chemical Substances and Physical Agents in the Workroom Environment", The American Conference of Governmental Industrial Hygienists, ACGIH, Cincinnati USA.

Sliney, D. H., 1989, Personal Communication.

ULTRAVIOLET STANDARDS

Branko F.M. Bosnjakovic

Radiation Protection Directorate
Ministry of Housing, Planning and Environment
Leidschendam, The Netherlands

INTRODUCTION

According to a dictionary definition, standardisation is something that is established by authority, custom, or general consent as a model or example to be followed (Webster, 1966). In a world of rapidly changing life-styles, markets, opportunities and risks, there is a increasing call for standardisation.

Standards can be market-oriented (to promote and enable technical and economical progress) and protection-oriented (to limit and avoid health and environmental risks). In the present context, the interest is focussed at protection-oriented standards, keeping in mind that some standards, e.g. those concerning devices, may serve at the same time technical-economic purposes. Even so, "standard" is a general term, incorporating both regulations and guidelines (Repacholi, 1985), and can be defined to be a set of specifications or rules to promote the health and environmental safety. A regulation is promulgated under a legal statute and is referred to as a mandatory standard. A guideline generally has no legal force and is issued for guidance only - a voluntary standard in principle.

In practice, differences between regulations and guidelines may become blurred. Compliance with a regulation depends for instance on the firmness of the legal authority to enforce it, and on the litigation practice concerning compensation. A guideline may similary acquire far-reaching power if social or market forces demand it.

When summarising standards, it seems reasonable to give some thoughts to the causal chain potentially leading to adverse effects. Such a chain consists of a source, an agent, an object to be protected, and adverse effects of exposure. We are concerned here with ultraviolet radiation as agent. Traditionally, human health has been the primary object of protection, but increasingly the protection of ecosystems is called for. The definition of a source need not be limited to devices emitting UV radiation, but may be extended to any activities leading to increased exposure to UV.

Standards can apply both to sources and to exposures. Exposure standards are often called primary standards, and source-oriented standards are considered as secondary standards, but such a distinction is not always justified and necessary.

The process of setting standards is always based on two inputs:

- a technical-scientific basis e.g. describing the source characteristics and the relationship between exposure and adverse effects;
- value judgements (e.g. which effect is adverse and which not, what is the acceptable probability of a chance effect etc.) which by their nature are ethical and/or political.

Both inputs for standards contain many uncertainties which must be coped with.

APPROACHES TO SETTING STANDARDS

Protection-oriented standardisation has often evolved from practical experience of occupational hygienists and medical practitioners. Many existing standards reflect this origin and are not explicit with respect to the strategic aim and concepts involved in protection.

Practice-oriented ad hoc standards were based on a combination of technical-scientific knowledge and value judgements of the experts involved, and were not questioned for long time by the public or the political authorities. The rationale for exposure limits was generally formed by threshold values for evidently adverse health injuries.

Such an approach needs to be repeated for each class of sources, each agent and different exposure circumstances. It may become questionable in case that stochastic effects - without a clear-cut threshold - may occur, as is the case with radiation induced cancer.

A great step forward in the protection philosophy was achieved by the International Commission on Radiological Protection (ICRP, 1977) in developing the basic principle for the limitation of exposure to ionizing radiation of workers and members of the public. ICRP's approach is based on three basic principles: justification, optimisation and individual exposure limits. Justification requires that a practice be adopted only if a positive net benefit can be demonstrated. Optimisation requires that all exposures be kept as reasonably low as achievable, social and economic factors being taken into account (ALARA principle). Dose limits require that individual exposure (integrated over all controllable non-medical and non-natural sources) should not exceed a certain value. They are different for workers and the members of the public. In the ICRP philosophy, the application of the first two principles - justification and ALARA is seen as primary tool in achieving low enough exposure, whereas the dose limits are seen as boundary condition - or safety net - in order to avoid excessive individual exposures.

The recommendations of the ICRP have been widely adopted by national authorities throughout the world. Still, more recently both within the ICRP and outside, need has been arising for a more unified risk approach in protection. There is a growing consensus that not only planned exposures but also the probability of serious accidents must be limited. In parallel, there is a growing need for common risk quantities applicable in different risk limitations schemes for ionizing and non-ionizing radiations, chemicals etc. (Brinkman et al., 1988).

Whereas optimisation at the work-place may be straight-forward, in many environmental circumstances the causal relationship between source and exposure is too remote, or too complicated, or too irreversible (such as is the case with ozone layer depletion) to be able to rely on traditional protection approaches.

The international trend in protection is towards more uniform risk limits, applicable to both anticipated situations and accident prevention, with an emphasis on source-related measures, and with view of protecting not only human health, but also all relevant ecosystems and even economic values.

Of course there are limits to the uniformity of protection approaches. From numerous studies, but also from daily experience, it is well-known that the perception, and thus the acceptance, of certain risks depends on the situation. In general, one may state that for consciously and voluntarily self-imposed risks, with clear economic or psychological reward for the person in question, the threshold of non-acceptance is much more easily taken than for risks which are imposed without knowledge or consent by, and no clear reward for, the person in question.

BASIC SITUATIONS OF EXPOSURE

Exposure to UV occurs under very different circumstances which may require very different protection approaches (Bosnjakovic, 1988b). If the broad acceptability of risk is chosen as basis for the classification of exposure situations, the main dividing line is between intentional and unintentional exposure.

When considering anticipated exposures (thus excluding accidents), the exposure of patients to UV must be always considered as intentional. Medical uses of UV are numerous and have been extensively treated by Czerski (1988). Similarly, exposures in the work-place must always be considered as unintentional. Exposures of the public take up an intermediate position: some of them are intentional (sunbathing and exposure in tanning booths for cosmetic purposes), the rest is unintentional.

A characteristic feature of UV radiation, in common with ionizing radiation, is the fact that the most ubiquitous and intense source of it is natural in origin - the sun. Obviously there is no way of controlling this source, but this does not exclude the possibility to influence or control the activities or behaviour which lead to enhanced exposure to solar UV.

This is certainly the goal in case of the use and emission of ozone-layer depleting substances. Theoretically, control could be envisaged of the occupational exposure to solar radiation, as is the case with outdoor professions (farmers, fishermen). If it does happen at all, it is due to custom and not because of regulations. This is significant since it can be estimated that about 80% to 90% of exposure to UV is due to outdoor activities.

Thus, one has to keep in mind that the standards developed or proposed cover in practice only that fraction of the average exposure which is due to artificial sources, either used intentionally, for cosmetic purposes, or with a potential for unintentional exposure.

UV STANDARDS

In general the standards may serve the following purposes:

- o for unintentional exposure (work-place, public) from artificial sources
  - to avoid deterministic effects i.e. generally acute effects with a threshold
  - to limit the occurrence of stochastic effects i.e. effects generally without a clearcut threshold where only the probability of occurence depends on the amount of (chronic) exposure
  - to ensure, by technical requirements on devices, that the probability of accidental exposure be limited;
- o for intentional exposures (patients, public) from artificial sources
  - to achieve the intended therapeutic or cosmetic effect with a minimum amount of adverse effects
  - to ensure that the probability of accidental exposure be limited.

International agreements or treaties forbidding or phasing out activities leading to unacceptable global increases of the UV levels, are discussed elsewhere (Bosnjakovic, 1989). They are generally not called standards.

There are five well-established types of adverse effects to the eye and the skin from UV exposure which have been considered in the development of UV standards (Sliney, 1987). For the skin: erythema (sunburn), accelerated ageing of the skin and photocarcinogenesis (skin cancer induction). For the eye: photokeratoconjunctivitis (acute inflammation of the cornea and conjunctiva as in the "snowblindness" or "welders' flash") and cataractogenesis (lens cataract). There is growing evidence for two additional adverse effects: immunological changes in the skin and retinal degradation which may possibly occur not only from short wavelength light but also from the trace amount of UV-A that reaches the retina as well. Sunburn, snowblindness and lens cataract, as deterministic effects, are generally considered to have a threshold. Skin cancer induction and probably skin ageing are stochastic effects, probably without a clear-cut threshold. The question of potential hazards to the eye from chronic exposure is still surrounded with controversy.

The susceptibility to UV injury of the skin varies widely with racial factors, thickness of the stratum corneum and degree of pigmentation. The threshold for erythema from UV-B radiation may vary by a factor of 10 or more. Thresholds for eye injury have narrower distributions.

As far as the stochastic effects are concerned, a dose-effect relationship is established in a quantitative way only for non-melanoma skin cancer (Van der Leun, 1988; Roza et al., 1989). It is much less well known for the induction of melanoma, and skin-ageing.

LIMITATION OF INTENTIONAL EXPOSURE

In some countries regulations have been promulgated on the construction and the use of UV sources for irradiation. The status of the regulations differs from country to country, ranging from legal requirements, via official standards to recommendations.

Some of the regulations are still in the draft stage. The list in Table 1 is probably not complete, but is thought to give a representative overview of the type of regulations to be encountered in (Western) industrialised countries (Bosnjakovic, 1988a).

Table 1. Overview of regulation on the UV appliances for skin tanning

Country	Year	Spectral region (nm)	Type of regulation	Limitation of irra-diance	Limitation of radiant exposure	Timer obligatory	Upper time limit	Emergency switch obligatory	Eye protection required
USA	1980 1985	180-320 200-400	)legal product)requirement	yes	no	yes	yes	no	yes
Canada	1980	180-320	legal product requirement	yes	no	yes	yes	no	yes
Sweden	1982	200-400	licensing and legal product requirement	yes	no	yes	yes	no	yes
UK	1982	100-400	recommendation	no	no	yes	no	yes	yes
FRG	1982 1987	200-400	draft industrial standard; recommendation	yes	yes	yes	yes	no	no
Australia	1983	100-400	standard	yes	yes	yes	yes	yes	yes
The Netherlands	1986	200-400	recommendation	yes	yes	yes	no	yes	yes
France	1987	280-410	project for regulation	yes	yes	yes	no	no	yes

References to table 1

USA (1980): 21 CFR, part 1040.20, Sunlamp products and ultraviolet lamps intended for use in sunlamp products (May 7, 1980)

USA (1985): 21 CFR part 1040.20, Sunlamp products: performance standard: final rule, 50 Federal Register 36548-36552 (September 6, 1985).

Canada (1980): P.C. 1980-1652, part XI. Sunlamps (June 19, 1980)

Sweden (1982): SSI FS 1982: 1. The regulations of the Swedish National Institute of Radiation Protection concerning sunlamps (February 1982)

UK (1982): Guidance note GS 18 from the Health and Safety Executive, Commercial ultra-violet tanning equipment (December 1982).

FRG (1982): Entwurf DIN 5050. Nichttherapeutische UV-Bestrahlungsgeräte für den menschlichen Körper; Messung, Kennzeichnung, Typeeinteilung (November 1982).

FRG (1987): Empfehlungen zur Begrenzung gesundheitlicher Strahlenrisiken bei der Anwendung von Solarien und Heimsonnen. Bundesgesundheidsblatt 30 (1), 19-30.

Australia (1985): Australian Standard 2635-1983. The installation, maintenance and operation of solaria for cosmetic purposes (1983).

The Netherlands (1986):Gezondheidsraad (Health Council of the Netherlands). UV Radiation, Human exposure to ultraviolet radiation. Report 1986/9E. Gezondheidsraad, P.O. Box 92517, 2509 LM The Hague.

France (1987): Muel, B, Césarini, J. -P. and Bastie, J. The French project for sunlamp regulations. Excerpta Medica, Int. Congr. Ser. No. 744, 497-501.

The Swedish legislation is very strict. In order to own and operate an UV source a license is required. None of the other countries has a licensing system. The general aim is to prevent acute, undesirable effects (and in some cases to limit the risk of chronic effects) without undue curtailment of the desired cosmetic effect. Most standards put restrictions on the irradiance (total or spectral) and some on the annual radiant exposure (sometimes expressed in the maximum radiant exposure per session plus the number of sessions per year). Most standards limit the irradiance time and require a timer. Some standards require an emergency switch. Eye protection must be available and used in most standards. Labels on the apparatus with instructions and warnings are required in most cases.

These labels may include

- information on general risks of exposure to UV
- information on sensitivity due to medication or cosmetics
- request to protect eyes
- recommendation to consult a physician in case of special skin sensitivity, disease or treatment
- a special warning for persons who have undergone eye lens surgery to wear glasses.

A safety performance standard for sunlamp products was first brought into effect in the USA. This American standard became effective in 1980 and has been amended in 1985. Standards have also been issued by several other countries.

The standards issued have gradually become more and more elaborated and new ideas and concepts have been incorporated in the course of time. While the first US standard contained only a few technical points and little guidance concerning appropriate exposure time and schedules, the amended version is accompanied by a policy document with detailed criteria concerning these matters.

Considering the uncoordinated effects in various countries in order to establish regulations and national standards for UV emitting diveces, the IEC (International Electrotechnical Commission) took in 1982 the decision to amend the existing publication on particular requirements for UV and IR skin treatment appliances for household use, by including radiation protection concepts.

The IEC draft proposal (Kivisäkk, 1987) provides primarily a classification scheme containing four types of applicances. The raditionale for this seems to be to allow countries wanting to issue additonal installation and application requirements for the different types to do so, or even exclude some of the types from the market.

The classification shall be made on the basis of the irradiance, measured at exposure distance in W/m^2 and weighted according to a skin action spectrum which covers the whole UV wavelength range.
The representatives of IEC and CIE have agreed informally that the skin action spectrum, in the numerical form proposed by McKinlay and Diffey (1987), should be recommended (fig. 1). In the meantime, CIE (Comité International de l'Eclairage) has recently approved and published this new reference action spectrum for skin erythema (CIE, 1987).

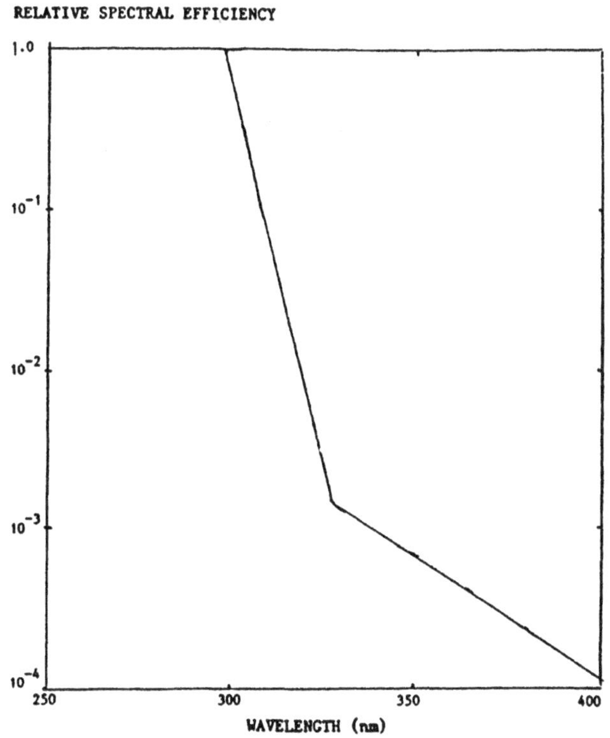

Fig. 1. Skin action spectrum, as proposed by McKinlay and Diffey (1987)

The new reference erythema action spectrum is considered to have the following advantages in comparison with the old CIE erythema action spectrum. It has been formulated from statistical considerations of recent scientific data. It additionally covers the UV-A which is particularry important for sources which emit mainly UV-A.
It can be represented by relatively simple functions over three clearly defined and radiologically significantly different spectral regions.
It should be kept in mind that the CIE action spectrum is meant to apply to any UV emitting light sources, and not only those used for intentional exposure. This action spectrum is, however, not intended to be used when assessing the carcinogenic potential of an irradiation. CIE does not make any recommendations concerning the acceptable levels of exposure.

The limits for the different IEC types of sources, according to the most recent draft of the IEC Technical Committee No. 61,. are as follows:

	< 320 nm	320 - 400 nm
Type I	< 0,0005	⩾ 0,15
Type II	0,0005 - 0,15	⩾ 0,15
Type III	< 0,15	< 0,15
Type IV	⩾ 0,15	< 0,15
(The sun	0,0095	0,02)

All values are in W/m², weighted by the skin action spectrum of McKinlay and Diffey (1987).

The IEC working group, responsible for the proposal, did consider the carcinogenic properties as the most serious negative effect of UV radiation. Thus, the CIE skin action spectrum is considered to represent in a sufficiently correct way the sensitivity of the skin to both erythema and skin cancer. Manufacturers of sunlamps therefore are to be obliged to provide an exposure schedule based on a maximum yearly radiant exposure (probably 20 kJ/m² per skin area), weighted according to the CIE skin action spectrum, in the instructions for the user.
This is less than the average difference in the annual dose from solar radiation between an outdoor and or an indoor worker and is considered by the IEC working group as a reasonable contribution for an individuel who does not receive excessive exposures from other sources.

It may be doubted, however, whether 100 MED of UV is really necessary to achieve and sustain the desired pigmentation. No agreement has been reached so far on the definition of a standard pigmentation action spectrum.

LIMITATION OF UNINTENTIONAL EXPOSURE

Unintentional exposure from artificial UV sources can occur both in the work-place and for the members of the public. If exposure to UV radiation cannot be avoided, but is considered, at least in principle, to be justified, the exposure should be limited as far as reasonably achievable (ALARA principle). To prevent unacceptable health risks to the individual, individual exposure limits are of importance. As skin cancer induction by UV radiation is assumed to have no threshold dose, these exposure limits cannot be interpreted as boundaries below which no effect will occur.

Whereas in case of intentional, cosmetic exposure to UV it may be assumed that the eyes are well protected, this may not be the case if the exposure is unintentional. Therefore, a general standard has to take into account the protection of both eye and skin. The American Conference of Governmental Industrial Hygienists (ACGIH) was the first body to formulate the corresponding threshold limit values since 1972. This standard has been adopted by the US National Institute for Occupational Safety and Health (NIOSH, 1972), and the UK National Radiological Protection Board (Diffey, 1980). The most recent, modified version of the ACGIH standard was published in 1988 (ACGIH, 1988).

The underlying principles and exposure limits have been further developed by the IRPA International Non-ionizing Radiation Comittee INIRC (IRPA, 1985; IRPA, 1989) and by the Health Council of the Netherlands (1986). In the latter document further reduction is explicitly considered not to be necessary below a level which corresponds to the exposure needed for vitamin D production (5 to 10 kJ/m^2 per year). Both these documents require the ALARA principle to be applied. The purpose of the IRPA/INIRC guidelines is to provide limits of exposure to UV and represent conditions under which it is expected that nearly all individuals may be repeatedly exposed without adverse effects.

The Dutch Health Council states more explicitly that the exposure limit recommendations are based on two principles:

- harmful effect for which there is a threshold must be avoided;
- the risk of chronic effects for which there is no threshold dose, is to be restricted to a reasonable value.

The exposure limits apply to exposure from artificial sources and the sun during a normal (working) day, and are adequate to protect lightly pigmented individuals, both in the working and general population. They do not apply to:

- lasers;
- very short exposure duration;
- some rare, photosensitive individuals;
- individuals concomitantly exposed to photosensitising agents;
- and to aphakic individuals (persons with a lens removed).

In fig. 2, data are presented on the induction of erythema and of photokeratitis as a function of the wavelength of the UV radiation. The exposure limit proposed by the Dutch Health Council is also shown in the figure for comparison.

The exposure limit has a minimum value of 30 J/m^2 at 270 nm. In the wavelength region up to 315 nm, it is based upon combined effects on the skin and the eye, and is practically equal in all three documents.

For UV-A wavelengths between 315 nm and 400 nm, two limits must be considered: a wavelength dependent exposure limit, mainly derived from the erythema action spectrum, and a limit for the overall irradiance and/or radiant exposure integrated over all UV-A wavelengths, mainly derived to limit against ocular injury.

The wavelength dependent exposure limits in the UV-A region are very similar in ACGIH (1988), IRPA (1989) and Dutch Health Council (1986) recommendations. Wavelength dependent exposure limits recommended by IRPA for the whole UV-B and UV-A range are given in Table 2.

Table 2

Ultraviolet radiation exposure limits and spectral weighting function (IRPA/INIRC 1989)

Wavelength* (nm)	EL (Jm^{-2})	Wavelength* (nm)	EL (Jm^{-2})
180	2,500	310	2,000
190	1,600	313	5,000
200	1,000	315	1.0×10^4
205	590	316	1.3×10^4
210	400	317	1.5×10^4
215	320	318	1.9×10^4
220	250	319	2.5×10^4
225	200	320	2.9×10^4
230	160	322	4.5×10^4
235	130	323	5.6×10^4
240	100	325	6.0×10^4
245	83	328	6.8×10^4
250	70	330	7.3×10^4
254	60	333	8.1×10^4
255	58	335	8.8×10^4
260	46	340	1.1×10^5
265	37	345	1.3×10^5
270	30	350	1.5×10^5
275	31	355	1.9×10^5
280	34	360	2.3×10^5
285	39	365	2.7×10^5
290	47	370	3.2×10^5
295	56	375	3.9×10^5
297	65	380	4.7×10^5
300	100	385	5.7×10^5
303	250	390	6.8×10^5
305	500	395	8.3×10^5
308	1,200	400	1.0×10^6

* Wavelengths chosen are representative; other values should be interpolated at intermediate wavelengths.

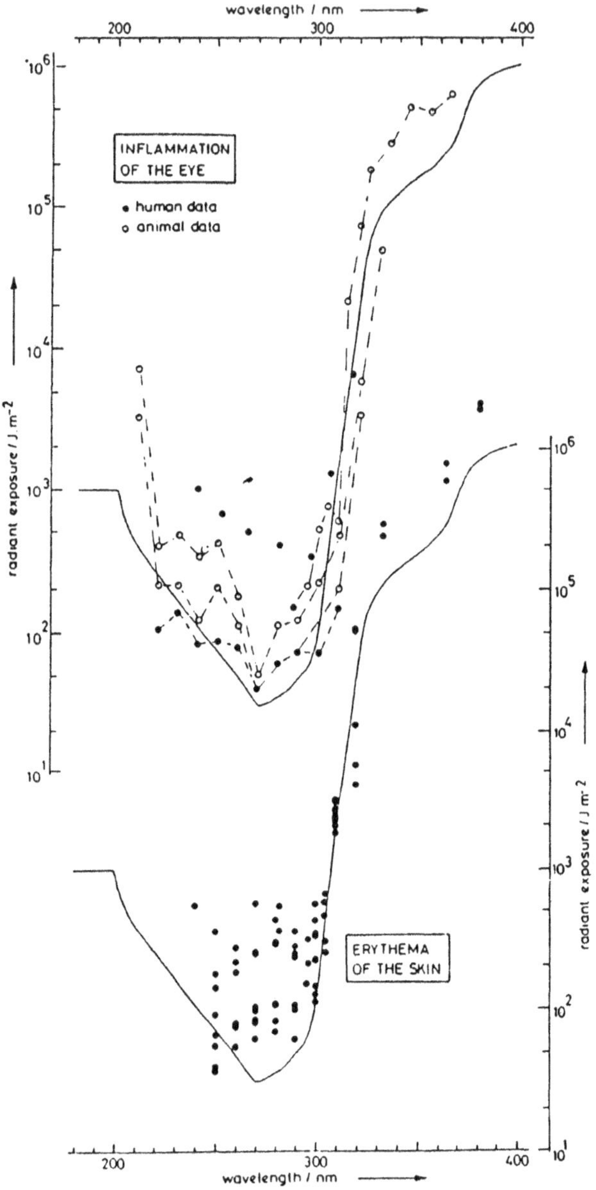

Fig. 2. Threshold doses for inflammation of the eye and erythema of the skin as a function of wavelength. Full curve: exposure limit for unintentional UV irradiation proposed by the Health Council of the Netherlands (Passchier and Van der Leun, 1987).

The UV-A integrated ocular limit is not identical in the three standards. IRPA recommends that the daily ocular radiant exposure be below 10 kJ/m^2. ACGIH recommends that the radiant exposure be limited to 10 kJ/m^2 for exposure durations below 1000 seconds, and that the irradiance be limited to 10 W/m^2. Health Council of the Netherlands recommends that chronic exposure of the eye should not exceed 1 W/m^2 on the average in order to avoid chronic cataract.

The latter value is a reference value which can be compared with the exposure to radiation from the sun, which is also chronic in character. For a clear mid-summer day in the Netherlands, the irradiance from the sun does not exceed 50 W/m^2. About half of that value is due to direct radiation, and the other half to scattered radiation. Assuming that only scattered radiation is incident on the eye, only a small percentage of the available irradiance reaches the lens of the eye, i.e. 1 to 2 W/m^2. The irradiance averaged over many years will be lower by more than an order of magnitude.

Passchier and Van der Leun (1987) have estimated the skin cancer risk associated with the exposure limits for unintentional exposure to UV. A radiant exposure equal to the proposed eye-skin exposure limit will, depending on the spectral distribution of the radiation, correspond to a carcinogenic effective radiant exposure not exceeding 60 J/m^2. Exceptionally a daily value of 150 J/m^2 might be reached. The average exposure will be smaller than the exposure limit, a value of 20 J/m^2 per day being a reasonable estimate. This is a factor 10 smaller than the difference between an outdoor and indoor worker in the Netherlands.

In case an indoor worker is exposed daily during working life to an extra erythema effective radiant exposure of 20 J/m^2, his or her skin cancer risk will be increased by about 25%. The skin cancer risk of an outdoor worker is a factor 5 larger than that of an indoor worker. The risk of skin cancer induction in case of chronic exposure to UV radiation approximating the eye-skin exposure limit appears to be much less than the extra risk caused by working outdoors.

It may be disputed whether this reasoning is convincing. To start with, the mentioned standards for unintentional exposure to UV do not exclude the sun as cause of exposure. Strictly speaking, exposure of outdoor workers should also comply with these standards and it would be erroneous to use the present excessive exposure of outdoor workers as reference.

Secondly, the choice of what is the acceptable cancer mortality risk is an unsettled question. Some countries are developing uniform risk limits for all environmental agents. For the exposure of the members of the public, the Netherlands (1985) adopted the policy that the individual mortality risk due to a single source should not exceed a limit of 10^{-6} per annum. For each group of comparable agents, e.g. chemicals, ionizing radiation, the cumulative individual annual risk should not exceed 10^{-5}. The average individual skin cancer mortality risk in North-Western Europe amounts to nearly or above 10^{-5} per annum (Bosnjakovic, 1988b). Therefore it seems doubtful whether a 25% increase would be accomodated under Dutch policy.

CONCLUSIONS

Ultraviolet standards developed so far have still mainly an ad hoc character, with a steady tendency to be based on more general principles. The main problems in developing convincing uniform standards are as follows:

- the dominating source of UV radiation is the uncontrollable solar radiation
- this natural source shows a wide geographical variation
- there is a wide variation of human susceptibility to UV
- there is a lack of knowledge concerning the susceptibility of ecosystems to UV exposure
- there is still a lack of universally accepted principles of risk limitation with respect to cancer mortality.

Nevertheless, international agreement is emerging concerning the human exposure to industrial and consumer UV emitting devices. In this context, an important role is being played by CIE, IEC and IRPA. In view of the possible consequences of the ozone layer depletion, a wider interest for the risks of solar UV radiation is expected to exist in the future.

References

ACGIH, 1988.
TLV's Threshold Limit Values and Biological Exposure Indices for 1988-1989. American Conference of Governmental Industrial Hygienists, Cincinnati, Ohio.

Bosnjakovic B.F.M., 1988a.
Ultraviolet radiation - risk limitation and protection of the public. In: Non-ionizing radiations. Proceedings of the international non-ionizing radiation workshop (M.H. Repacholi, Editor).
Melbourne, 5-9 April, 1988.
Yallambie, Victoria, Australia.

Bosnjakovic B.F.M., 1988b.
Ultraviolet and ionizing radiation: a comparison from the protection point of view.
Tagungsband "Nichtionisierende Strahlung" (Red. Norbert Krause).
Fachverband für Strahlenschutz.
Sekretariat: H. Brunner, PSI
CH-5232 Villegen PSI.

Bosnjakovic B.F.M, 1989.
Towards a solution of the ozone layer depletion problem, in "Optical sources, laser and synchrotron radiation: biological effects and hazard potential" (M. Grandolfo, A. Rindi, D.H. Sliney, eds.).
Plenum Press, New York and London.

Brinkman R, Bosnjakovic B.F.M. and De Jongh P, 1989.
The Dutch risk management policy and its impact for radiation protection. Proceedings of the "Seminar on applications, perspectives and limitations of comparative risk assessment and risk management".
Nice, France, 26-30 September 1988 (M. Olast and J. Sinnaeve, Eds.), EUR 11465 EN, p. 322-329.
Commission of the European Communities, Luxembourg.

CIE, 1987.
A reference action spectrum for ultraviolet induced erythema in human skin. CIE Research Note prepared by TC 6-02.
CIE-Journal, Vol.6., No.1, p. 17-22.
Czerski P., 1989.

Optical radiation medical devices: regulatory approaches and safety of non-coherent sources, in: "Optical sources, laser and synchrotron radiation: biological effects and hazard potential" (M. Grandolfo, A. Rindi and D.H. Sliney, eds.)
Plenum Press, New York and London.

Diffey, B.L., 1980.
Ultraviolet radiation physics and the skin.
Phys. Med. Biol. 25, 405.

Health Council of the Netherlands, 1986.
UV radiation: human exposure to UV radiation.
Report 1986/9E.
Gezondheidsraad, P.O. Box 90519, 2509 LM The Hague.

ICRP, 1977.
International Commission on Radiological Protection
ICRP Publication 26.
Oxford, Pergamon Press.

IRPA/International Non-Ionizing Radiation Committee, 1985.
Guidelines on limits of exposure to UV radiation of wavelengths between 180 nm and 400 nm (Coherent optical radiation).
Health Physics 49, 1985, p. 331-340.

IRPA/International Non-Ionizing Radiation Committee, 1989.
Proposed change to the IRPA 1985 guidelines on limits of exposure to ultraviolet radiation.
Health Physics 56, 1989, (to be published).

Kivisäkk E., 1987.
Intentional exposure to UV radiation: risk reduction and present regulations. In: Human Exposure to UV radiation: risks and regulations (W.F. Passchier and B.F.M. Bosnjakovic, Eds.), p. 443-454.
Excerpta Medica, International Congress Series 744.
Amsterdam - New York - Oxford.

Van der Leun J.C., 1989.
Effects of increased UV-B on human health. In: Atmospheric ozone research and its policy implications (T. Schneider et al., eds.).
Elsevier Science Publishers B.V. Amsterdam.

McKinlay A.F. and B.L. Diffey, 1987.
A reference action spectrum for UV induced erythema in human skin.
In: Human Exposure to UV radiation: risks and regulations
(W.F. Passchier and B.F.M. Bosnjakovic, eds.), p. 83-87.
Excerpta Medica, International Congress Series 744.
Amsterdam - New York - Oxford.

The Netherlands, 1985.
Environmental Program of the Netherlands 1986-1990,
Ministry of Housing , Physical Planning and Environment,
Department of Information and International Relations, The Hague.

NIOSH, 1972.
Criteria for a recommended standard: occupational exposure to UV-radiation
US Department of Health, Education and Welfare
NIOSH, HSM 73-11009
Washington, D.C.

Passchier W.F. and J.C. van der Leun, 1987.
Human Exposure to ultraviolet radiation: limits for unintentional UV exposure, in: "Human exposure to UV radiation: Risks and regulations", W.F. Passchier and B.F.M. Bosnjakovic, eds.
Elsevier Science Publishers B.V. (Biomedical Division).

Repacholi M.H., 1985.
Standards on static and ELF electric and magnetic fields and their scientific basis, in: "Biological effects and dosimetry of static and ELF electromagnetic fields" (eds. M. Grandolfo, S.M. Michaelson, A. Rindi).
Plenum Press, New York and London.

Roza L., R.A. Baan, J.C. van der Leun, L. Kligman, A.R. Young, 1989.
UV-A hazards in skin associated with the use of tanning equipment, Photochem. Photobiol., B: Biology, 3 (1989) 281-287.

Sliney D.H., 1987.
Unintentional exposure to ultraviolet radiation: risk reduction and exposure limits, in: "Human exposure to ultraviolet radiation: Risks and regulations" (W.F. Passchier and B.F.M. Bosnjakovic, eds.), p. 425-437. Excerpta Medica, Amsterdam, New York, Oxford.

Webster, 1966.
Webster's Third New International Dictionary.
G&C Merriam Co., New York

IRPA/INIRC GUIDELINES ON ULTRAVIOLET

Jürgen H. Bernhardt

Institute for Radiation Hygiene
of the Federal Health Office
D 8042 Neuherberg

INTRODUCTION

In 1974, the International Radiation Protection Association (IRPA) formed a working group on non-ionizing radiation (NIR), which examined the question of protection against adverse health effects of exposure to ultraviolet radiation. At the 1977 IRPA Congress in Paris, this working group became the International Non-Ionizing Radiation Committee (IRPA/INIRC).

The IRPA/INIRC, in cooperation with the Environmental Health Division of the World Health Organization (WHO), has undertaken responsibility for the development of health criteria documents on non-ionizing radiation. These form part of the WHO Environmental Health Criteria Programme, which is funded by the United Nations Environmental Programme (UNEP). The documents include an overview of the physical characteristics, measurement and instrumentation, sources and applications of the NIR, a thorough review of the available standards and their rationale, and evaluations of the health risks of human exposure to NIR. These criteria then become the scientific data base for the development of exposure limits and codes of practice.

A document entitled Environmental Health Criteria 14, Ultraviolet Radiation (UNEP 1979) was published in 1979. The document contains a review of the biological effects reported from exposure to ultraviolet radiation and serves as the scientific rationale for the development of these guidelines. The purpose of these guidelines is to deal with the basic principles of protection against non-coherent ultraviolet radiation, so that they may serve as guidance to the various international and national bodies or individual experts who are responsible for the development of regulations, recommendations or codes of practice to protect the workers and the general public from the potentially adverse effects of ultraviolet radiation. The limits in these guidelines were based on the scientific data and no consideration was given to economic impact or other nonscientific priorities. However, from presently available knowledge, the limits should provide a healthy working or living environment from exposure to ultraviolet radiation (UVR) under all normal conditions.

The IRPA Associate Societies as well as a number of competent institutions and individual experts were consulted in the preparation of the guidelines.

The guidelines were published in Health Physics in 1985 (IRPA, 1985). Later, between 1987 and 1988, INIRC proposed a change to the guidelines. Recent studies of skin and ocular injury action spectra in the UV-A-range provided sufficient data to update the exposure limits (McKinley and Diffey, 1987; Parrish et.al., 1982; Urbach and Gange (eds), 1986). For UV-A wavelengths between 315 nm and 400 nm the exposure limits (EL) were updated (IRPA, 1989). No change was proposed relative to the IRPA EL for actinic UVR (< 315 nm).

The wording in this chapter is mainly received from the original text (IRPA, 1985). In order to avoid overlappings with other chapters, the rationale for the limits are shortened to the most relevant facts. For more details consider the original text (IRPA, 1985) and, furthermore, the relevant chapters in this book.

PURPOSE AND SCOPE

The purpose of these guidelines is to provide guidance on limits of exposure to UVR in the spectral region between 180 nm and 400 nm and represent conditions under which it is expected that nearly all individuals may be repeatedly exposed without adverse effects. This guideline has been limited to wavelengths greater than 180 nm where UVR is transmitted through air.

The EL values for exposure of the eye or the skin may be used to evaluate potentially hazardous exposure from UVR; e.g. from arcs, gas and vapor discharges, fluorescent lamps, incandescent sources, and solar radiation. The limits do not apply to UV lasers. Most incoherent UV sources are broadband, although single emission lines can be produced from low-pressure gas discharges. These values should be used as guides in the control of exposure to both pulsed and continuous sources where the exposure duration is not less than 0.1 μs. These ELs are below levels which would be used for UV exposures of patients required as a part of medical treatment or for cosmetic purposes.

Exposure during sun bathing and tanning under artificial sources may well exceed these limits but exposed individuals should be advised that some health risk is involved with such activity. Eye protection is always required during therapeutic exposures. Nevertheless, occasional exposures to conditioned skin may not result in adverse effects. Furthermore, the ELs are exceeded by noonday sunlight overhead at 0° - 40° latitude within 5 - 10 minutes in the summertime. The ELs should be considered absolute limits for the eye, and "advisory" for the skin because of the wide range of susceptibility to skin injury depending on skin type. The ELs should be adequate to protect lightly pigmented individuals (i.e. white Caucasian) with greatest sensitivity and genetic predisposition.

The spectral band designations of the CIE is used, i.e., UV-A is from 315 to 400 nm, UV-B is from 280 to 315 nm, and UV-C is from 100 to 280 nm. The UVR exposure should be quantified in terms of an irradiance E (W/m^2 or W/cm^2) for continous exposure or in terms of a radiant exposure H (J/m^2 or J/cm^2) for time-limited (or pulsed) exposures of the eye and skin. The geometry of exposure to UVR is very important. For example, the eyes (and to a lesser extent the skin) are somewhat anatomically protected against UVR exposure from overhead sources (e.g. the sun overhead). The limits should be applied to exposure directed perpendicular to those surfaces of the body facing the radiation source, measured with an instrument having cosine angular response.

EXPOSURE LIMITS

The EL for exposure to UVR incident upon the skin or eye where irradiance values are known and the exposure duration is controlled are as follows:

For the near-ultraviolet UV-A spectral region (315 to 400 nm) the total radiant exposure incident upon the unprotected eye should not exceed 1.0 J/cm^2 (10 kJ/m^2) within an 8-hour period and the total 8-hour radiant exposure incident upon the unprotected skin should not exceed the values given in Table 1. Values for the relative spectral effectiveness, S_λ, are given up to 400 nm to expand the action spectrum into the UV-A for determining the EL for skin exposure.
For the actinic UV spectral region (UV-C and UV-B from 180 to 315 nm), the radiant exposure incident upon the unprotected skin or eye within an 8-hour period should not exceed the values given in Table 1.

To determine the effective irradiance of a broadband source weighted against the peak of the spectral effectiveness curve (270 nm), the following weighting formula should be used:

$$E_{eff} = \Sigma E_\lambda \cdot S_\lambda \cdot \Delta_\lambda \qquad (1)$$

where:

E_{eff} = effective irradiance in E/m^2 $[J/(s \cdot m^2)]$ normalized to a monochromatic source at 270 nm
E_λ = spectral irradiance from measurements in $W/(m^2 \cdot nm)$
S_λ = relative spectral effectiveness (unitless)
Δ_λ = bandwidth in nanometers of the calculation or measurement intervals.

Permissible exposure time t_{max} in seconds for exposure to actinic UVR incident upon the unprotected skin or eye may be computed by dividing 30 J/m^2 by the value of E_{eff} in W/m^2:

$$t_{max} = \frac{30}{E_{eff}} \qquad (2)$$

SPECIAL CONSIDERATIONS

These EL values are intended to apply to UVR exposure of the working population, but with some precaution also apply to the general population. However, it should be recognized that some rare, highly photosensitive individuals exist who may react adversely to exposure at these levels. These individuals are normally aware of their heightened sensitivity. Likewise, if individuals are concomitantly exposed to photosensitizing agents, a photosensitizing reaction can take place. It should be emphasized that many individuals who are exposed to photosensitizing agents (ingested or externally applied. chemicals, e.g. in cosmetics, foods, drugs, industrial chemicals, etc.) probably will not be aware of their heightened sensitivity. Lightly pigmented individuals conditioned by previous UVR exposure (leading to tanning and hyperplasia) and heavily pigmented individuals can tolerate skin exposure in excess of the EL without erythemal effects.

Table 1. IRPA exposure limits for ultraviolet radiation and spectral weighting function

Wavelength[a] (nm)	EL ($J\ m^{-2}$)	EL ($mJ\ cm^{-2}$)	Relative Spectral Effectiveness S_λ	Wavelength[a] (nm)	EL ($J\ m^{-2}$)	EL ($mJ\ cm^{-2}$)	Relative Spectral Effectiveness S_λ
180	2,500	250	0.012	310	2,000	200	0.015
190	1,600	160	0.019	313[b]	5,000	500	0.006
200	1,000	100	0.030	315	1.0×10^4	1.0×10^3	0.003
205	590	59	0.051	316	1.3×10^4	1.3×10^3	0.0024
210	400	40	0.075	317	1.5×10^4	1.5×10^3	0.0020
215	320	32	0.095	318	1.9×10^4	1.9×10^3	0.0016
220	250	25	0.120	319	2.5×10^4	2.5×10^3	0.0012
225	200	20	0.150	320	2.9×10^4	2.9×10^3	0.0010
230	160	16	0.190	322	4.5×10^4	4.5×10^3	0.00067
235	130	13	0.240	323	5.6×10^4	5.6×10^3	0.00054
240	100	20	0.300	325	6.0×10^4	6.0×10^3	0.00050
245	83	8.3	0.360	328	6.8×10^4	6.8×10^3	0.00044
250	70	7.0	0.430	330	7.3×10^4	7.3×10^3	0.00041
254[b]	60	6.0	0.500	333	8.1×10^4	8.1×10^3	0.00037
255	58	5.8	0.520	335	8.8×10^4	8.8×10^3	0.00034
260	46	4.6	0.650	340	1.1×10^5	1.1×10^4	0.00028
265	37	3.7	0.810	345	1.3×10^5	1.3×10^4	0.00024
270	30	3.0	1.000	350	1.5×10^5	1.5×10^4	0.00020
275	31	3.1	0.960	355	1.9×10^5	1.9×10^4	0.00016
280[b]	34	3.4	0.880	360	2.3×10^5	2.3×10^4	0.00013
285	39	3.9	0.770	365[b]	2.7×10^5	2.7×10^4	0.00011
290	47	4.7	0.640	370	3.2×10^5	3.2×10^4	0.000093
295	56	5.6	0.540	375	3.9×10^5	3.9×10^4	0.000077
297[b]	65	6.5	0.460	380	4.7×10^5	4.7×10^4	0.000064
300	100	10	0.300	385	5.7×10^5	5.7×10^4	0.000053
303[b]	250	25	0.190	390	6.8×10^5	6.8×10^4	0.000044
305	500	50	0.060	395	8.3×10^5	8.3×10^4	0.000036
308	1,200	120	0.026	400	1.0×10^6	1.0×10^5	0.000030

[a] Wavelengths chosen are representative; other values should be interpolated at intermediate wavelengths.
[b] Emission lines of a mercury discharge spectrum.

PROTECTIVE MEASURES

Protective measures will differ depending upon whether the UVR exposure occurs indoors or outdoors. The use of hats, eye protectors, facial shields, clothing, and sun-shading structures are practical protective measures. As with any indoor, industrial hazard, engineering control measures are preferable to protective clothing, goggles, and procedural safety measures. Glass envelopes for arc lamps will filter out most UV-B and UV-C. Where lengthly exposure to high power glass-envelope lamps, and quartz halogen lamps will occur at close proximity, additional glass filtration may be necessary. Light-tight cabinets and enclosures and UVR absorbing glass and plastic shielding are the key engineering control measures used to prevent human exposure to hazardous UVR produced in many industrial applications. Shields, curtains, baffles, and a suitable separation distance are used to protect individuals against the UVR emitted by open-arc process such as arc welding and arc-cutting. There is a need for operational rules to protect potentially exposed individuals. Operators should be trained to follow these general rules properly. Ventilation may be required exhausting ozone and other airborne contaminants produced by UV-C radiation.

MEASUREMENT

Although UVR radiometers exist, attempts to produce relatively inexpensive field safety survey meters which respond directly to UV-B and UV-C radiation (following the S_λ function) have not been fully successful. However, relatively expensive instruments exist which respond to UV-B and UV-C radiations according to the relative spectral effectiveness, S_λ. Spectroradiometric measurements of the source which can then be used with the S_λ weighting function to calculate E_{eff} are often necessary for measurements more accurate than with simple, direct-reading safety meters. Whichever measurement technique is applied, the geometry of measurement is important. All the preceding ELs for UVR apply to sources which are measured with an instrument having a cosine-response detector oriented perpendicular to the most directly exposed surfaces of the body when assessing skin exposure and along (or parallel to) the line(s) of sight of each exposed individual when assessing ocular exposure. The use of UV film badges makes it possible to integrate UV exposure on specific body sites which move respect to the UVR source; however, the spectral response of such film badges still does not accurately follow S_λ.

RATIONALE FOR THE LIMITS OF EXPOSURE TO UVR

A comprehensive review of UVR effects has been published by UNEP, 1979 and the interested reader is referred to that document and, furthermore, to the relevant chapters in this book. The following text is a summary of the physical and biological factors used to derive the UVR guidelines as it has been published in the original text (IRPA, 1985).

Erythema

Erythema is the most commonly observed effect on skin after exposure to UVR. This effect was first quantitatively documented as a wavelength dependent effect by Hausser and Vahle (1928). These and other quantitative studies since that time have confirmed that the erythemal threshold varies with anatomical site, wavelength, and time between exposure and assessment. In addition, the variation in published threshold values is due to differences in the clinical definition and estimate of minimal erythema and radiometric measurement techniques. Erythema is a photochemical response of the skin normally resulting from overexposure to wavelengths in the UV-C and UV-B regions (180-315 nm). Exposure to UV-A alone can produce erythema, but only at very high radiant exposures (> 100 kJ/m^2). The UV-A added prior to UV-B exposure may slightly intensify the erythemal response (photoaugmentation, Willis et al., 1972). The opposite effect where one previous exposure desensitizes the skin also occurs, and may be more pronounced for simultaneous exposure of UV-A and UV-B (Van der Leun and Stoop, 1969). Erythema induced by the longer UV-B wavelengths (295-315 nm) is more severe and persists for a longer period than that for shorter wavelengths. The increased severity and time course of the erythema may result from the deeper penetration of these wavelengths into the epidermis. In general it is accepted that UVR releases a number of diffusing mediators, which in turn carries the inflammatory effect into deeper skin layers. The assessment of a threshold for the maximum sensitivity of the skin to erythema varies from 250 nm to 297 nm depending upon the criteria of assessment and the period following the exposure. Action spectra for different grades of erythema are quite different. For the most severe grade of erythema this maximum sensitivity occurs between 290 and 300 nm. The minimal erythemal doses (MED) reported in more recent studies for untanned, lightly pigmented skin range from 60 to 300 J/m^2 (Everett et al., 1965; Freeman et al., 1966; Parrish et al., 1982). These MED data suggest that for this type of skin, the EL values are approximately 1.3 to 6.5 times less than the MED values. Skin pigmentation and "conditioning" (thickening of the stratum coneum and tanning) may result in an increase of the MED by at least one order of magnitude.

Delayed effects of the skin

Chronic exposure to sunlight, especially the UV-B component, accelerates the skin aging process and increases the risk of developing skin cancer (Van der Leun, 1984). The solar spectrum is greatly attenuated by the earth's ozone layer, limiting terrestrial UV to wavelengths greater than approximately 290 nm. The UV-B irradiance at ground level is a strong function of the sun's elevation angle in the sky. This is due to the change of UV attenuation with atmospheric path length (time of day and season). Several epidemiologic studies have shown that the incidence of skin cancer is strongly correlated with latitude, altitude, and cloud cover (Cutchis, 1974; Urbach et al., 1974). Exact quantitative dose-response relationships have not yet been established although fair-skinned individuals, especially of Celtic origin, are much more prone to develop skin cancer. Skin cancer is typically a disease of outdoor workers such as farmers and seamen. Only a few quantitative studies have examined work populations chronically exposed to artificial sources of UV-B to determine whether there is an increased skin cancer risk in the occupational environment. Squamous cell carcinoma is the most common type. This is localized at exposed sites (e.g. hands and back of the neck).

Photokeratoconjunctivitis

Actinic UVR (UV-B and UV-C) is strongly absorbed by the cornea and conjunctiva. Overexposure of these tissues causes photokeratoconjunctivitis, commonly referred to as welder's flash, arc-eye, etc.. Pitts has characterized the course of ordinary clinical photokeratitis (Pitts and Tredici, 1971; Pitts, 1974). The latent period varies inversely with the severity of exposure ranging from 1/2 to 24 hours but usually occurs within 6-12 hours. Conjunctivitis tends to develop more slowly and may be accompanied by erythema of the facial skin surrounding the eyelids. The individual has the sensation of a foreign body or sand in the eyes and may experience photohobia and lacrimation. The acute symptoms last from 6 to 24 hours and discomfort usually disappears within 48 hours.

Although exposure rarely results in permanent ocular injury, the individual is visually incapacitated during this 48-hours period. Threshold data for photokeratitis in humans have been established by Pitts and Tredici (1971) for 10 nm wavebands from 220 to 310 nm. The guideline ELs between 200 nm and 305 nm are about 1.3 to 4.6 times less than the threshold for minimal change. The maximum sensitivity of the human eye was found to occur at 270 nm. The wavelength response (action spectrum) between 220 and 310 nm does not vary as greatly as in the case of erythema with the thresholds varying from 40 to 140 J/m^2. Corneal injury from UV-A wavelengths requires levels exceeding 100 kJ/m^2 (Zuclich, 1980).

Cataract

Wavelengths above 295 nm can be transmitted through the cornea and are absorbed by the lens. Pitts et al.(1977) have shown that both transient and permanent opacities of the lens (cataracts) can be produced in rabbits and monkeys by exposure to UVR having wavelengths in the 295-320 nm band. Thresholds for transient opacities ranged dramatically with wavelength, from 1.5 to 126 kJ/m^2. Thresholds for permanent opacities were typically twice those for transient opacities. Whether chronic exposure at lower levels will produce lenticular opacities has not been determined.

Retinal effects

The cornea and crystalline lens normally sufficiently shield the retina from acute effects from UVR exposure. Normally, less than 1 percent of UV-A reaches the retina, shorter UV-B wavelengths being totally attenuated. Upon removal of the crystalline lens, Ham and colleagues (1982) demonstrated acute retinal injury at levels of the order of 50 kJ/m^2 at the retina.
In recent years there has been a rapidly growing population of aphakic individuals who have had one or both crystalline lenses removed following cataract. Many of these aphakics have received intraocular lenses of glass or plastic. Aside from a few with implants designed to absorb UV-A, these persons would not be adequately protected against retinal injury from UV-A exposure at the EL. Such persons should be fitted with UV-A protective lenses if working with sources of UV-A radiation.

Envelope action spectrum

The development of UVR exposure limits must consider two risks: the risks of acute and chronic injury to both the eye and skin. The literature indicates that thresholds for injury vary significantly with wavelengths for each effect. In the UV-B and UV-C regions, an action spectrum curve can be drawn which envelopes the threshold data for exposure doses in the range of reciprocity (Schmidt, 1964; Zuclich, 1980) for acute effects obtained from studies of minimal erythema and keratoconjunctvitis. Reciprocity means that irradiance E and exposure duration t have a reciprocal relation, and a constant product of E and t (i.e. exposure) results in a given effect. This EL curve does not differ significantly from the collective threshold data considering measurement errors and variations in individual response (Sliney and Wolbarsht, 1980). Although the safety factor is minimal for minimally detectable keratitis it is believed to range from 1.5 to 2.0 for acute keratitis. The curve is also well below the acute UV-B cataractogenic thresholds (Pitts et al., 1977; Sliney and Wolbarsht, 1980). Repeated exposure of the eye to potentially hazardous levels of UV is not believed to increase the protective capability of the cornea as does skin tanning and thickening of the stratum corneum. Thus, this EL is more readily applicable to the eye and must be considered a limiting value for that organ (Sliney, 1972). Any accumulation of UV-B and UV-C exposures causing photokeratitis is limited to about 48 hours since the outer corneal epithelial layers are replaced in about 48 hours by the normal repair process of this tissue. Some slight additivity of UV-A exposure exists beyond 48 hours because of the deeper penetration of UV-A rays (Zuclich, 1980).

Because of the wide variations in threshold values and exposure history (conditioning) among individuals, these guidelines should only be used as a starting point for evaluating skin hazards (Sliney and Wolbarsht, 1980). The envelope guideline has some margin of safety to protect all but the most sensitive individuals. An exact value for this margin cannot be given, but for lightly pigmented persons, it varies from about 3 to 20 depending on the spectral composition of the radiation. Since there may be more than one erythemal mechanism and, therefore, more than one erythemal action spectrum, the effect of radiations of two widely differing wavelengths in the 180 nm to 315 nm range may not be simple additive. The EL should be used with caution in evaluating sources such as the sun and fluorescent lamps, having a rapidly increasing spectral irradiance in the 300-315 nm range. Large errors can arise because of the difficulty in making accurate spectral measurements of such sources in this region.

The EL may not provide adequate protection for photosensitive individuals or for normal individuals exposed concomitantly to chemical, pharmaceutical, or phyto-photosensitizers, and special precautions must be taken for such cases (Bernstein et al., 1970).

The EL should reduce the risk of occurrence of chronic skin effects by preventing acute effects and limiting life-long UV exposure. It is assumed now that the action spectrum for UV skin carcinogenesis is similar to the erythemal action spectrum so that the spectral erythema curve can be used as a measure for the carcinogenic risk.

In addition to the UV hazard, very intense UV-C sources may also produce hazardous concentrations of ozone and nitrogen oxides from the air and of phosgene gas in the presence of degreasers.

Finally, it should be emphasized that the EL are below the new CIE spectrum in the actinic range and between 325 and 400 nm (Mc Kinley and Diffey, 1987). The CIE curve was intended for spectroradiometric measurements of sources to evaluate erythemal effectiveness. Furthermore, the three-straight-line CIE action spectra of McKinley and Diffey was not intended for simple survey instruments designed to imitate a hazard action spectrum. The IRPA/INIRC EL are more restrictive since it must also protect against ocular injury.

REFERENCES

Bernstein H.N., Curtis J., Earl F.L. and Kuwabara T., 1970, Phototoxic corneal and lens opacities, Arch. Ophthalmol., 83:336-348.

Cutchis P., 1974, Stratospheric ozone depletion and solar ultraviolet radiation on earth, Science, 184:13-19.

Everett M.A., Olsen R.L. and Sayer R.M., 1965, Ultraviolet erythema, Arch. Dermatol., 92:713-729.

Freeman R.G., Owens D.W., Knox J.M. and Hudson H.T., 1966, Relative energy requirements for an erythemal response of skin to monochromatic wavelengths of ultraviolet present in the solar spectrum, J.Invest. Dermatol., 47:586-592.

Hausser K.W., 1928, Influence of wavelength in radiation biology, Strahlentherapie, 28:25-44 (in German).
Ham W.T. Jr., Mueller H.A., Ruffolo J.J. Jr., Guerry D. III and Guerry R.K., 1982, Action spectrum for retinal injury from near-ultraviolet radiation in the aphakic monkey, Amer. J. Ophthalmol., 93:229-306.
IRPA, 1985, The International Non-Ionizing Radiation Committee of the International Radiation Protection Association: Guidelines on limits of exposure to ultraviolet radiation of wavelengths between 180 nm and 400 nm (Incoherent Optical Radiation) Health Physics 49:(2), 331-340.
IRPA, 1989, Change to the IRPA 1985 Guidelines on Limits of Exposure to Ultraviolet Radiation, Health Physics, 56:(6), 971-972.
McKinley, A.F., and Diffey, B.L., 1987, A reference action spectrum for ultraviolet induced erythema in human skin. Seminar on Human Exposure to Ultraviolet Radiation: Risks and Regulations, Amsterdam, 23-25 March 1987, ICS 744, (Edited by W. Passchier), Elsevier Science Publishers, Amsterdam.
Parrish J.A., Jaenicke K.F. and Anderson R.R., 1982, Erythema and melanogenesis action spectra of normal human skin, Photochem. Photobiol., 36:(2), 187-191.
Paul B. and Parrish J., 1982, The interaction of UV-A and UV-B in the production of threshold erythema, J. Invest. Derm., 78:371-374.
Pitts D.G. and Tredici T.J., 1971, The effects of ultraviolet on the eye, Amer. Ind. Hyg. Assoc. J., 32:(4) 235-246.
Pitts D.G., 1974, The human ultraviolet action spectrum, Amer. J. Optom. and Physiol. Optics, 51:(12), 946-960.
Pitts D.G., Cullen A.P. and Hacker P.D., 1977, Ocular Ultraviolet Effects from 295 nm to 400 nm in the Rabbit Eye, National Institute for Occupational Safety and Health, Cincinatti, OH, Contract CDC-99-74-12, DHEW (NIOSH) Publication No. 77:175.
Schmidt K., 1964, On the skin erythema effect of UV flashes, Strahlentherapie 124:127-136.
Sliney D.H., 1972, The merits of an envelope action spectrum for ultraviolet exposure criteria, Am. Ind. Hyg. Assoc. J. 33:644-653.
Sliney D.H. and Wolbarsht M.L., 1980, Safety with Lasers and other Optical Sources, A comprehensive Handbook (New York: Plenum Press).
UNEP, 1979, United Nations Environmental Programme/World Health Organization/International Radiation Protection Association, 1979, Environmental Health Criteria 14, Ultraviolet Radiation (Geneva: WHO).
Urbach F., Epstein J.H. and Forbes P.D., 1974, UV carcinogenesis, pp. 259-283, in: Sunlight and Man (Edited by T.B. Fitzpatrick, M.A. Pathak, L.C. Harber, M. Seiji and A. Kutika) (Tokyo: University of Tokyo Press).
Urbach F., and Gange, R.W. (eds.), 1986, The Biological Effects of UVA Radiation, Praeger Publications, New York.
Van der Leun J.C. and Stoop T., 1969, in: The Biological Effects of UV Radiation (Edited by F. Urbach), pp. 251-254 (Oxford: Pergamon Press).
Van der Leun J.C., 1984, UV carcinogenesis, Photochem. Photobiol. 39:(6), 861-868.
Willis I., Kligman A. and Epstein J., 1972, Effects of long ultraviolet rays on human skin: photoprotective or photoaugmentative, J. Invest. Dermatol. 59:416-420.
Zuclich J.A., 1980, Cumulative effects of near-UV induced corneal damage, Health Phys. 38:833-838.

LASER SAFETY STANDARDS: HISTORICAL DEVELOPMENT AND RATIONALE

David H. Sliney

Laser Microwave Division, US Army Environmental Hygiene Agency, Aberdeen Proving Ground, MD 21010-5422 USA

INTRODUCTION

As early as 1965, there had already been several attempts to develop some initial EL's, and these were generally two or three values: One for q-switched ruby laser irradiation, and one for non-q-switched ruby laser irradiation. It became clear quickly that one or only a few EL values are not feasible to control hazards to health from laser radiation [1]. Radiometric factors, such as: wavelength, exposure duration, and laser pulse characteristics, and biological factors related to the target biological structure/organ (e.g., the retinal image size), influence the limits. Hence, the sliding scales of values as a function of exposure duration and wavelength are more akin to a collection of a large number of limits for different chemical agents, than a few values for a single substance. It became clear by 1971, after the first decade of laser use, that detailed hazard evaluation of each laser environment was too complex for most users, and a scheme of hazard classification evolved. Today, most countries follow a scheme of four major hazard classifications as defined in Document WS 825 of the International Electrotechnical Commission (IEC). The classifications and the associated accessible emission limits (AEL's) were based upon the EL's. The EL and AEL values today are in surprisingly good agreement worldwide. There exists a greater range of safety requirements for the user for each class of laser.

Limits for laser radiation have been promulgated by both government and non-governmental organizations. The International Non-Ionizing Radiation Committee (INIRC) of the International Radiation Protection Association (IRPA) has recommend laser EL's [2] and the IEC has recommended a standard laser hazard classification with AEL's for each class and the same EL's as IRPA/INIRC [3]. In addition to the IEC, the American Conference of Governmental Industrial Hygienists (ACGIH) [4], the American National Standards Institute (ANSI) [5], the British Standards Institution (BSI) [6], the Health Council of the Netherlands [7], the Deutsche Institut fur Normung (DIN) [8], the Soviet Ministry of Health [9], NATO (OTAN) and several other national standards institutions and governmental agencies have proposed limits which are generally very similar. These limits are based largely upon biological studies of acute effects [2, 10-11]. Considerations of the adverse effects of chronic

Light, Lasers, and Synchrotron Radiation
Edited by M. Grandolfo *et al.*, Plenum Press, New York, 1990

exposure were based upon studies of the ambient optical radiation environment and theoretical studies of the interaction of ultraviolet, visible and infrared radiation with tissue [10-13].

There exists an almost world-wide consensus for the ACGIH/ANSI/IEC/IRPA EL's for laser radiation. A true consensus does not presently exist, since the Soviet Union has promulgated a differing set of EL's (e.g., see Figure 3). The Soviet Ministry of Health Temporary Sanitary Norm [9] follows many of the concepts found in the other sets of EL's for differing viewing conditions, but adds a complex formula for altering the ocular EL's with variations in ambient lighting, i.e., with changes in pupillary diameter. It is interesting to note that the Soviet limits are less conservative for short-pulse laser radiation, but more conservative for lengthy exposures lasting for seconds to minutes. It would be nice to obtain a better understanding of the rationale for these differences. The explanations given in the available literature are not very detailed. It may well be that these differences may eventually lead to revised international laser EL's.

DERIVATION OF EL's

The principal problem areas which arose during the development of these exposure limits relate to: the degree of extrapolation of biological data obtained for one laser to other pulse durations and wavelengths, the degree of sophistication (i.e., should only a few limits be given to favor simplicity, even if less accuracy is achieved), and the degree of added safety to be incorporated into the limits because of uncertainty in the data, extrapolation from animal to man and the potential for delayed effects [13]. These factors are discussed in this presentation.

As with most chemical and physical agents in our environment, current scientific knowledge defines levels of laser radiation exposure that are clearly hazardous and, to a lesser extent, those levels which are clearly beneficial or harmless. The intermediate zone between clearly safe and clearly hazardous--if it exists--may be an exposure realm where virtually no human exposure occurs (as with many laser operations), or this grey zone may be within the normal physiological range, i.e., where the majority of the population is exposed (as with solar ultraviolet radiation). Exposure limits (EL's) set to preclude acute injury are clearly possible and have been promulgated. Limits set to absolutely preclude any risk from chronic exposure to ultraviolet (UV) radiation and short-wavelength (blue-violet light) are probably not feasible, and practical judgements based upon the environmental outdoor exposure of mankind have been applied to set recommended limits in this part of the spectrum [13-15]. Considering the enormous range of genetically determined individual and ethnic susceptibility to both UV induced skin cancer [16-17], and to the accelerated onset of cataract and skin ageing, it is easy to see that realistic guidelines for chronic exposure present a challenge.

Lasers emit optical radiation in the ultraviolet, visible (light), and infrared (IR) regions of the electromagnetic spectrum. Lasers are unique sources of optical radiation having extremely high brightness. Although thousands of different types of lasers have been developed in the laboratory since the first laser was invented in 1960, only a handful have enjoyed widespread industrial and commercial use. Lasers emit wavelengths in all parts of the spectrum. The most commonly encountered laser wavelengths across the spectrum account for well over 98% of laser operations: Xenon-chloride (308 nm), Nitrogen (334 nm), He-Cd (325 and 441.6 nm), argon (488 and 514.5 nm), SHG Nd:YAG (532 nm), He-Ne (632.8 nm), krypton (647 nm, etc.), ruby (694.3 nm), Ga-As (905 nm), Nd:YAG (1064 nm) and CO-2 (10.6 um).

CURRENT STANDARDS

It is interesting to note that laser EL's were developed rapidly within a decade, despite the fact that virtually no EL's had previously existed for incoherent optical radiation [13]. It is the monochromatic nature of a laser, coupled with its ability to produce a nearly point image on the retina that has made possible relatively "simple" laser EL's--when compared to the complexity of limits for broad-band incoherent optical sources. Furthermore, the laser has often been associated with "ray guns" of the future, which led to early demands for laser regulations.

ACGIH Threshold Limit Values (TLV's) for occupational exposure to laser radiation first appeared in 1968 (based upon US Army laser occupational EL's) have been since revised several times[4]. The TLV's and the ANSI EL's published in the ANSI Standard Z136.1, Safe Use of Lasers, in successive editions in 1973, 1976, 1980 and 1986[5], have served as the basis for virtually all other laser safety standards in the USA and influenced standards elsewhere in the world. The IEC Committee on lasers, TC-76, followed the ANSI EL's and the US Federal Laser Product Performance Standard (21CFR1040) AEL's in drafting the IEC laser standard WS-825[3]. There is no major controversy as to what should be reasonable EL's for laser radiation, although the temporary norms in the USSR vary as shown in Fig. 1[9]. To understand the basis of these standards, it is necessary to briefly review the biologic effects of optical radiation upon the eye and skin.

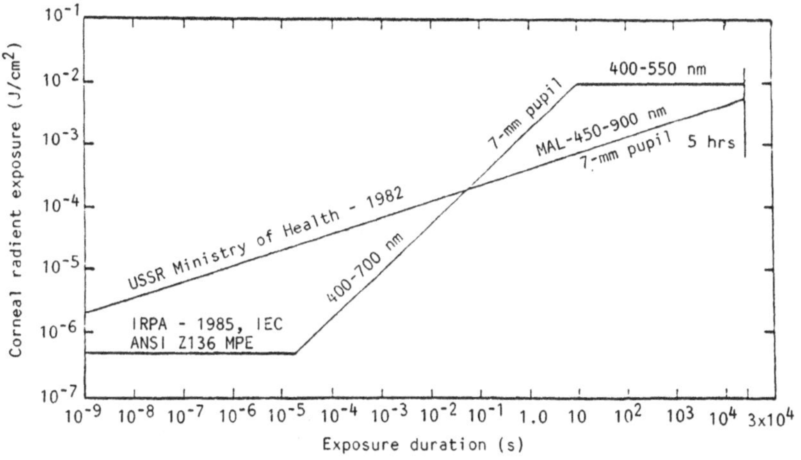

Figure 1. Selected comparison of ACGIH/ANSI/IEC/IRPA laser exposure limits [2-5] with the temporary sanitary norms of the Ministry of Health of the USSR[9]. The comparison is made for intrabeam viewing of a laser beam ("point-source" condition) for visible wavelengths and for a 7-mm pupil, corresponding to a low ambient lighting level.

BIOLOGICAL EFFECTS

Laser radiation is considered "non-ionizing radiation" because photon energies for UV wavelengths greater than about 200 nm are insufficient to individually ionize atoms found in important biological molecules. Unlike the stochastic biological effects of ionizing radiation (e.g., X rays), the acute biological effects of optical radiation are considered non-stochastic, i.e., a threshold appears to exist for each biological effect. In the ultraviolet and visible regions of the spectrum, photochemical damage mechanisms are demonstrable [12-22]. Thermal injury mechanisms dominate for most pulsed laser exposures and infrared radiation exposures. Chronic exposure of UVR produces skin cancer, and this may be a stochastic process and UV laser EL's for chronic exposure of the skin must be treated differently from the other laser EL's[16].

An understanding of both the physical and biological aspects of light interaction with structures of both the skin and the eye permit a realistic assessment of the risks of acute exposure. There remain, however, problem areas when attempting to fully answer some questions relating to chronic exposure. These problems will be addressed following a brief review of the important specific biological effects upon which EL's must be based.

There are at least five separate types of hazards to the eye and skin from lasers and other more conventional optical sources:

(a) Ultraviolet photochemical injury to the skin (erythema and carcinogenic effects), and to the cornea (photokeratitis) and lens (cataract) of the eye (200 nm to 400 nm) [21-22].

(b) Thermal injury to the retina of the eye (400 nm to 1400 nm).

(c) Blue-light photochemical injury to the retina of the eye (principally 400 nm to 550 nm) [2-].

(d) Near-infrared thermal hazards to the lens (approximately 800 nm to 3000 nm).

(e) Thermal injury (burns) of the skin (approximately 400 nm to 1 nm) and of the cornea of the eye (approximately 1400 nm to 1 mm).

THE CONCEPT OF THRESHOLD

No single concept is more critical to the development of exposure limits than is the concept of a threshold, below which an adverse effect does not occur (Figure 2). All of the aforementioned biological effects, except for carcinogenesis, are acute affects and are also generally considered to have a threshold. An understanding of the interaction of optical radiation which biological systems at the molecular level suggests, however, that the so-called threshold has meaning only on the macroscopic level. In effect, the apparent threshold for injury appears only when the normal repair processes which take place at the cellular level are overwhelmed by denaturation of proteins and similar biologically important molecules by optical radiation. Large molecules are being denatured all the time as the result of the thermal energy present at normal body temperature [16]. Individual molecules are damaged by ambient, low-level UVR all the time, but the natural on-going processes of renewal at the cellular level maintain an equilibrium condition with these processes [17-19]. Thus, deficiencies in an individual's natural repair mechanisms may render the individual more susceptible to injury from optical radiation--especially in the UV spectrum. Such increased sensitivities are referred to as photosensitive reactions. Other photosensitive reactions can occur as a result of exposure to a photosensitizing chemical. For the most part, the laser EL's are insufficient to provide protection against many photosensitizing reactions [11-13].

Individual factors can play an important role in determining susceptibility to tissue injury by exposure to laser radiation. While skin pigmentation and ambient tissue temperature will both affect the threshold for thermal injury, these factors play a relatively minor role. Thermal injury will occur when tissue such as the skin or cornea is raised by about 10 - 20 degrees Celsius. Greater temperature elevations are required to produce coagulation (a thermal burn) when the duration of exposure is shortest. For this reason, thermal injury is referred to as a rate process; the time- temperature history of the exposed tissue determines whether macroscopically visible injury occurs. Thermal injury in man's natural environment and his work environment is actually rarely the result of exposure to optical radiation. One's natural aversion response to high heat normally occurs sufficiently fast that one is not burned. Only with pulsed lasers is this exposure situation likely to occur. Recognition of the behavioral limits on exposure was not made in setting the EL's; but, behavioral responses to bright light and temperature sensations were the bases for establishing some AEL's which define hazard classification limits [13].

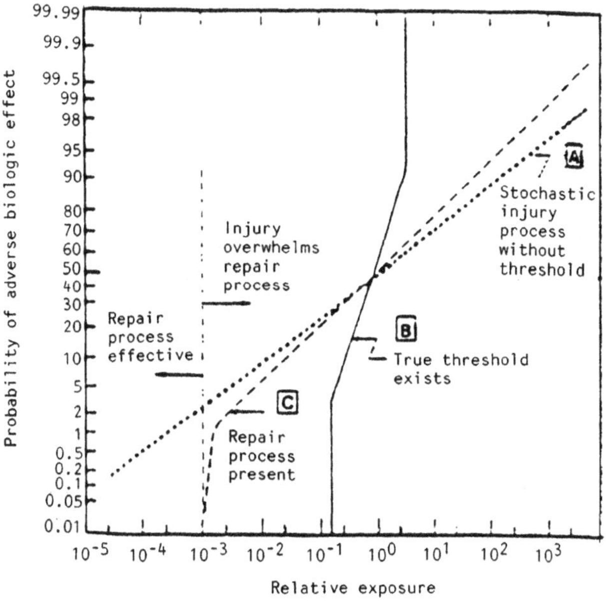

Figure 2. Probit curve used in estimating the probability of a biological effect. For acute thresholds, individual variations in susceptibility are normally within one order of magnitude (B), but many delayed effects are believed to be the result of stochastic processes, which in the simplest model (line A - the linear hypothesis), any exposure can be regarded as additive to the risk. A model similar to line C is probably more realistic [adapted from 13].

The susceptibility to UVR photochemical injury of the skin varies widely with racial factors, thickness of the stratum corneum and degree of pigmentation (i.e., presence of melanin pigment). The threshold for erythema from UV-B radiation (280-315-nm) may vary by a factor of one hundred or more. Disregarding hypersensitive reactions brought on by photosensitization of tissue, it is possible to draw a line below the thresholds for acute UVR injury of the skin and eye and set that as an EL for occupational exposure to UVR, and this has been done [13, 22]. However, no such line is likely to be possible to eliminate all risk of delayed effects from UVR. While an apparent "threshold" may exist for carcinogenesis or accelerated skin aging, as shown in Figure 2, this is still debatable and it is difficult to argue that some exposure to even low levels of UVR is "safe".

Studies of UVR carcinogenesis continue. Epidemiological evidence for UVR carcinogenesis from a lifetime exposure to the UVR in sunlight indicates that levels far above those found in indoor environments should be necessary. Laboratory studies of skin carcinogenesis in the hairless mouse [16] give only the most circumstantial clues as to the dose response curve that may be operative for skin carcinogenesis. All of the scientific studies suggest quite clearly that exposure to normal sunlight during the middle of the day can be quite dangerous for the lightly pigmented individual -- particularly the individual with a celtic heritage [12, 16]. Thus, to propose an exposure limit which would indicate that almost any out-of-doors activity for many individuals would be "hazardous," would be clearly unacceptable. For this reason, the ACGIH limits for both laser and non-laser UVR concentrates on the elimination of risk from acute effects, with the assumption that chronic exposure at the EL would still only contribute very slightly to the risk of UVR carcinogenesis. The ACGIH and IRPA UV laser EL's explicitly state that they do not preclude skin or corneal injury if the individual has been made photosensitive [2,4].

THE GEOMETRY OF EXPOSURE

The more one studies the biological effects of laser radiation and conventional optical radiation--especially UVR--upon the skin and eyes, one is struck by the enormous importance of the geometry of exposure. This is critical to the estimation of chronic exposure levels from sunlight which plays a major role in establishing limits for chronic exposure to laser radiation. Mankind appears to be remarkably well adapted to our natural sunlight environment, and customs of dress play an added importance in protecting both the skin and eye in arctic and desert environments [14]. When the sun is directly above us at noontime, the spectral irradiance at 300 nm is ten times greater than three hours prior to or following noontime. Dermatologists warn sunbathers to remain out of the sun during midday hours. Unfortunately, I do not see too many beach devotees following this advice. Natives of tropical climes often do follow this dictum and go indoors for a siesta during this period. Since any light-skinned individual who attempts to get a suntan is almost surely to exceed the present EL's for UVR, one must take a philosophical attitude toward proposals for safety standard EL's to protect against these delayed effects of UVR exposure. This would not prevent solar exposure anyway.

EL for Repetitive Pulses

Although EL's for single-pulsed exposure have not changed much since 1976, there have been improvements in the procedures to determine the ELs for repetitive pulse laser exposure to the retina. This is of particular importance in the point-of-sale products such as supermarket checkout counter laser scanners, where an individual may occasionally but not intentionally view directly the scanned beam from such a device and receive a train of laser pulses. Prior to the 1986 revision of ANSI Z-136.1, repetitive pulse exposure limits were based upon studies performed in 1971, which suggested that two mechanisms were involved, and the criteria for determining the repetitive pulse additivity were different for short-pulse exposures and long-pulse exposures. However, by 1984, it had become clear that the available data showed that small spot-sized retinal laser injury from repetitive pulse trains could be expressed as a simple additivity expression both for very short-pulses of the order of a few nanoseconds and for much longer pulses of the order of milliseconds. This empirical rule indicated that the threshold for injury of a train of N pulses could be accurately predicted by taking the threshold of injury for a single pulse exposure and multiplying it by a correction factor of N-1/4, where N is the total number of pulses in the train. This relation was published by: Hemstreet (1974) [23], Stuck et al (1978) [25], and Marshall (1978) [24].

Initially, there was some concern about adapting this simplified method, because there was no data to indicate that very lengthy exposure would not be additive. One could not state when the pulse train additivity would cease. This type of rule seemed to be unduly restrictive in such applications where exposure could be occasional during a work day, but where one could assume extremely long exposure durations as a limit.

Studies were then performed by Ham, et al. [26], using a range of pulse repetition frequencies (PRF) ranging from 100 to 1600 for exposure durations lasting from 1 to 1,000 seconds (16.7 minutes). The total number of pulses ranged up to 1.6 million pulses. The result surprised many: the N-1/4 rule predicted the experimentally determined thresholds in all cases. This held true for both blue light (488) and red light (647 nm), until photochemical injury threshold radiant exposures were exceeded. The results suggested that for intentional exposure to a point-source train of laser pulses, the N-1/4 rule probably lasted for many hours of exposure, if indeed, such an exposure could be realistic.

We had hoped that the biological studies would have provided us with a clean answer which would simplify decision making for long exposures durations. For example, had the results indicated that all pulse trains over a few thousand pulses or 10 to 100 seconds were not longer additive, it would have simplified the determination of a maximum exposure duration. Without a biologically determined endpoint, we were forced to develop a set of rules based upon practical exposure periods within certain physiological limits.

We determined from studies of individuals exposed to bright light sources and the physiological literature, that the maximum exposure duration for a bright point (less than a min) source of visible light consistent with the rest of the Standard would be 0.25 second. Only purposeful (and painful) prolonged staring at such a source would give exposures to a longer train. The 0.25-s aversion response has been used since the earliest version of Z-136.1, 1973, and it followed from the laboratory experiments described in Gerathewohl, S.J. and Strughold, H. L. (1953) [29] along with their bibliography and from Section 7.2.3.3. in Sliney and Wolbarsht (1980) [13].

As for infrared lasers, in which there could be no reason for attentive staring into the source since it would be invisible, we set a worst case stare time of 10 seconds. These values are supported by both practical experience and the scientific studies of glare, visual fixation, and eye movement [13, 27-29]. Near-infrared (IR-A) lasers are, at best, only weakly visible and may be invisible; therefore, the 0.25-s visible light aversion response will not furnish a limiting exposure duration. The 10-second limit for viewing in the infrared is based on the amplitude and frequency of unintentional eye movements or saccades which occur even when gazing at a fixed position. These move the small retinal image around, and over a ten second period of fixation would cover a 200 um diameter area [13]. This is discussed in Fender, D. H. (1964)[28] which is a very good review with references which in turn have rather extensive bibliographies, including the original work of Yarbus (1966) [27].

Two exceptions could be determined for these maximal durations--0.25 s in the visible and 10 s in the infrared. The first would be in a situation where a person intentionally stared into a light source; apparently an extremely rare event as most would agree. Secondly, in the case of an infrared source, it is conceivable that an ophthalmic instrument or visual science instrument could be developed (such as a laser eye tracker) where individuals would be fixed into a position where the same retinal area would be illuminated for periods exceeding 10 seconds. Hence, such exposures are excluded from the exposure limitation used in the ANSI Standard. The simplified N-1/4 method was later adopted by IEC TC76 for the revision of IEC 825 scheduled for publication in 1989.

EL Guidelines for Pulse Durations Less Than 1 ns

Strictly speaking, the ANSI Z-136.1, 1986, does not present EL's for exposure durations less than 1 ns; however, the guideline of IEC Standard (1984) is recommended. IEC Standard W-825 (1984) suggests limiting peak irradiances to the exposure limit expressed as a peak irradiance for a 1 ns pulse. The biological support for this approach has been reviewed by Sliney (1982) [30], and the data is shown in Figure 3 taken from this article. This shows that the biological data in the sub-ns range follows a constant peak irradiance or power rather than a constant energy.

FUTURE DEVELOPMENTS

There has been some consideration given recently toward revising the EL's that relate to extended-source (i.e., large-area retinal) exposures. Most of the recommended EL's consider two sets of values for the eye: for intrabeam (minimal image size) and extended source (large retinal image) exposures. In practical experience, we have found that the large-image EL's are seldom used and hazardous extended sources seldom exist except for the unexpanded beams of q-switched ruby and neodymium lasers. Hence, we can expect to see proposals to simplify the ocular EL's. For example, one could list only intrabeam EL's and have an angular dependent correction factor for those few cases where an extended source EL is needed.

CONCLUSIONS

Although exposures to optical radiation in the physiological range have generally been accepted as safe, greater knowledge today suggests that the absence of acute effects does not mean that some damage has not occurred at the molecular level. Delayed effects appear to be possible from UVR exposure to the skin and eye. Furthermore, growing evidence

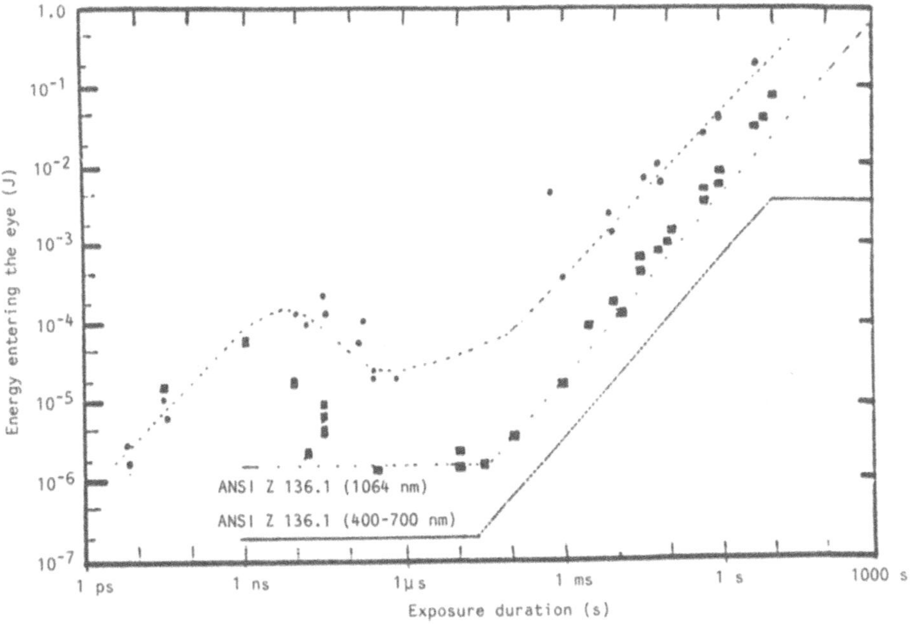

Figure 3. The Dependence of Retinal Injury Threshold Upon Pulse Duration. Injury threshold in the rhesus are plotted as a function of total laser energy entering the eye vs exposure duration. The dashed line shows the EL'S for 1014 nm; the sold line, the EL's for visible wavelengths. The dotted line runs through threshold ED-50 points for 1064 nm. Bold square data points are for green wavelengths[30].

suggested that retinal degradation may also occur from occur from chronic exposure to short wavelength light [18-20]. The present laser EL exposure limits probably reduce the risk from all of these delayed effects quite significantly, but one cannot argue that the risk is zero. As with almost all physical and chemical agents, one cannot speak of "zero risk" -- it is scientifically specious.

REFERENCES

1. Sliney, D. H., and Palmisano, W. A., The evaluation of laser hazards, Am Industr Hyg Assn J. 29: 325-431 (1968).
2. IRPA, International Non-Ionizing Radiation Committee (1985), Guidelines for Limits of Human Exposure to Laser Radiation, Health Physics, 49(5):341-359.
3. International Electrotechnical Commission, Radiation Safety of Laser Products, Equipment Classification, and User's Guide, Document WS 825, IEC, Geneva, 1984.
4. ACGIH (1988), TLV's, Threshold Limit Values and Biological Exposure Indices for 1988-1989, American Conference of Governmental Industrial Hygienists, Cincinnati, OH.
5. ANSI (1986), Safe Use of Lasers, Standard Z-136.1-1986, American National Standards Institute, Laser Institute, Laser Institute of America, Toledo, OH.
6. British Standards Organisation (1984), Radiation Safety of Laser Products and Systems, Standard BS4803, London, BSI.
7. Health Council of the Netherlands (1979), Acceptable Levels for Micrometer Radiation, Rijswijk, Gezondheidsraad.
8. Deutsche Institut fur Normung, Radiation Safety of Laser Products, Standard VDE 0837, Berlin, DIN/VDE (1984).
9. Ministry of Health, USSR (1982), Sanitarniya Normi i Pravila Ustroistva i Ekspluatatsii Lazerov, No. 2392-81, Moscow, Ministry of Health.
10. World Health Organization [WHO], (1982), Environmental Health Criteria No. 23, Lasers and Optical Radiation, joint publication of the United Nations Environmental Program, the International Radiation Protection Association and the World Health Organization, Geneva.
11. ACGIH (1980), Documentation for the Threshold Limit Values, 4th Edn., American Conference of Governmental Industrial Hygienists, Cincinnati, OH.
12. World Health Organization [WHO], (1980), Environmental Health Criteria No. 14, Ultraviolet Radiation, joint publication of the United Nations Environmental Program, the International Radiation Protection Association and the World Health Organization, Geneva.
13. Sliney, D. H., and Wolbarsht, M. L., (1980), Safety with Lasers and Other Optical Sources, Plenum Publishing Corp., New York.
14. Sliney, D. H., (1983), Eye protective techniques for bright light, Ophthalmology, 90:937-944.
15. Sliney, D. H., (1986), Physical factors in cataractogenesis: ambient ultraviolet radiation and temperature, Invest. Ophthalmol. Vis Sci., 27(5): 781-790.
16. Forbes, P. D., and Davies, P. D., (1982), Factors that Influence Photo- carcinogenesis, in: (J.A. Parrish, M. L. Kripke, and W.L. Morison, Eds.) Chapter 7, "Photoimmunology," Plenum Publishing Corp., New York.
17. Hillenkamp, F. (1980), Laser Interactions with Biological Tissue, in: (F. Hillenkamp, C. A. Sacchi, and T. Arrechi, Eds.), Lasers in Biology and Medicine, Plenum Press, New York.
18. Young, R. W. (1988), Solar radiation and age-related macular degeneration, Surv. Ophthalmol., 32(4):252-269.
19. Young, R. W. (1982), Biological renewal. applications to the eye, Trans. Ophthalmol. Soc. U. K., 102(1):42-75.

20. Ham, W. T., Jr., Ruffolo, J. J., Jr., Mueller, H. A., and Guerry, D., III, (1980) The nature of retinal radiation damage, dependence on wavelength,
power level and exposure time, Vision Res. 20:1105-1111.
21. Pitts, D. G., Cullen, A. P., and Hacker, P. D., (1977), Ocular effects of ultraviolet radiation from 295 to 365 nm, Invest. Ophthal. Vis. Sci., 16(10):932-939.
22. Sliney, D. H., (1972), The merits of an envelope action spectrum for ultraviolet radiation exposure criteria, Am. Indstr. Hyg. Assn. J., 33(9):644-653.
23. Hemstreet, H.W., Bruce, W. R., Altobelli, K. K., Stevens, C. C., and Connolly, J. S., 1974, Ocular Hazards of Picosecond and Repetitive Pulse Argon Laser Exposures, First Annual Report, February 1973-February 1974, USAF Contract for School of Aerospace Medicine, Brooks AFB, TX, Technology Inc., San Antonio, TX.
24. Marshall, W. J., 1978, A Proposal for a New Method to Determine EL Values for Repetitive Pulse Trains, US Army Environmental Hygiene Agency, Aberdeen Proving Ground, MD, June 1978.
25. Stuck, B. E., Lund, D. J., and Beatrice, E. S., 1978. Repetitive Pulse Laser Data and Permissible Exposure Limits Institute Report No. 58, Letterman Army Institute of Research, Division of Non-Ionizing Radiation, Changes in the Maximum Permissible Exposures Francisco, Presidio of San Francisco (April 1978).
26. Ham, W. T., Jr., Mueller, H. A., Wolbarsht, M. L., and Sliney, D. H., (1988), Evaluation of retinal exposures from repetitively pulsed and scanning lasers, Health Physics, 54(3): 337-344.
27. Yarbus, A. L., 1967, Eye Movements During Fixation on Stationery Objects, Plenum Press, NY.
28. Fender, D. H., 1964, Control of the Eye, Science Amer., Vol. 211, pp. 24-33.
29. Gerathewohl, S. J., and Strughold, H., 1953. Motoric Response of the Eyes When Exposed To Light Flashes of High Intensities and Short Durations, Journal of Aviation Medicine, Vol. 24, pp. 200-207.
30. Sliney, D. H., Interaction Mechanisms of Laser Radiation with Ocular Tissues, 1982. In: "Laser Induced Damage in Optical Materials", pp. 355-367, (H.E. Bennett, A.E. Gunther, D. Milam, and B. E. Newman, eds), NBS Publication, SP 669, January 1984. [Updated and published in French CEN publication, 1988].

THE IRPA/INIRC GUIDELINES ON LIMITS OF EXPOSURE TO LASER RADIATION

David H. Sliney

Laser Microwave Division
US Army Environmental Hygiene Agency
Aberdeen Proving Ground, MD 21010--5422 USA

BACKGROUND

The International Non-ionizing Radiation Committee (INIRC) of the International Radiation Protection Association (IRPA) first published Guidelines on Limits of Exposure to Laser Radiation in 1985 [1]. Since that time, as a result of an expanded laser database, the Guidelines were revised in 1988 [2]. Prior to presenting the guidelines on laser exposure, it is worthwhile to explain the process by which IRPA/INIRC develops guidelines on non-ionizing radiation (NIR).

IRPA began its activities relating to the entire subject of NIR in 1973 with a session devoted to this topic at its Third International Congress in Washington, DC. The International Non-Ionizing Radiation Committee (INIRC) of IRPA was established in 1977. The primary objectives of INIRC were to present the general principles of non-ionizing radiation protection, to determine the most appropriate exposure limits (EL's) for each type of NIR and to explore with other international organizations ways of furthering protection in this field.

Over the last decade the Division of Environmental Health of the World Health Organization and IRPA/INIRC have cooperated in the preparation of environmental criteria documents relating to NIR with the financial support of the United Nations Environmental Programme (UNEP). These criteria then become the scientific data base for the development of exposure limits and codes of practice. From the criteria documents, the INIRC has then developed guidelines on EL's for the different NIR, including laser radiation (1). Each of the guidelines for EL's have been published in Health Physics.

Each set of guidelines must, of course, be subjected to periodic revision to be kept in line with advances in knowledge of the relevant biological effects. The purpose of the guidelines is to provide guidance to international and national bodies or individual experts who are responsible for the drafting of regulations, recommendations or technical advice to protect the workers and the general public from the potentially adverse effects of any non-ionizing radiation.

Light, Lasers, and Synchrotron Radiation
Edited by M. Grandolfo *et al.*, Plenum Press, New York, 1990

The laser criteria document, like all criteria documents, was published under the joint sponsorship of UNEP, WHO and IRPA, includes a description of the physical characteristics of the radiation concerned, an overview of the sources, applications and exposure levels, measurement methods and instrumentation, a review of the data on biological effects gained from animal experimentation and observations in humans, an evaluation of the health risk of human exposure and a survey of the existing protection standards.

As an aside, I should note that the IRPA/INIRC has also collaborated with the International Labour Office by drafting codes of safe practice which are intended to inform workers of the hazards and the appropriate precautions that need to be taken to minimize NIR exposure.

EXPOSURE LIMITS

The development of EL's requires a valid assessment of the possible biological effects and of the exposure-response relationship as well as appropriate information on the various sources in use, the resulting levels of exposure and the people at risk. As noted above, the basic resource document used for the development of the laser EL's was the Environmental Health Criteria Document [3]. The data base of laser biological effects upon the human eye and skin reviewed in the criteria document was the same as that used by other national and international organizations to derive EL's, and not surprisingly, the IRPA EL's are generally in agreement with those of the International Electrotechnical Commission [4], the American Conference of Governmental Industrial Hygienists, [5] the American National Standards Institute [6], the British Standard [7], the Health Council of the Netherlands, [8] the Deutsche Institut fur Normung, [9] and others.

There is no point in simply repeating the IRPA EL's here. Each reader really should have the complete guidelines which can be found in Health Physics [1-2]. IRPA will also soon publish a book with the complete set of guidelines for all of the NIR. In Table 1, I have selected some representative EL's for the most commonly available lasers. One should also remember that the way most of the laser safety standards are organized, one makes use of a laser's hazard classification (initially based upon EL's) to perform a hazard evaluation, rather than to perform measurements of beam exposure for comparison with EL's [3,4,6,7,10]. The EL's are actually used only in special instances where human exposure is intended and the laser beam irradiance or radiant exposure may actually be measured or calculated to determine if the EL will be exceeded.

The 1988 Revisions

In 1988, IRPA published revisions to its 1985 guidelines [2]. None of these revisions affected the basic limits as outlined in Table 1. The most significant change related to the method for determining the applicable exposure duration for CW lasers and for repetitive-pulse laser exposures. The previous method was simplified and clarified. Several previous rules for determining pulse additivity were reduced to one expression. All tables of ELs remained unchanged.

A correction for large-area illumination of the skin was also introduced. Among other things, that change minimized the transition between the guidelines for limits of exposure to microwave radiation at 300 GHz and the guidelines for infrared laser exposure at 1.0 mm (i.e., 300 GHz).

Table 1. Selected Occupational Exposure Limits (EL's) for Some Common Lasers

Laser	Wavelength	Exposure Limit
Argon-flouride laser	193 nm*	3.0 mJ/cm^2 over 8 hrs.
Xenon-chloride laser	308 nm	40 mJ/cm^2 over 8 hrs.
Argon ion laser	488, 514.5 nm	3.2 mW/cm^2 for 0.1 s
		2.5 mW/cm^2 for 0.25 s
Helium-neon laser	632.8 nm	1.8 mW/cm^2 for 1.0 s
		1.0 mW/cm^2 for 10 s
Krypton ion laser	568, 647 nm	
Helium-neon laser	632.8 nm	17 μW/cm^2 for 8 hrs
Neodymium-YAG laser	1064 nm	5.0 μJ/cm^2 for 1 ns to
	1334 nm	100 μs
		No EL for t < 1 ns
		5 mW/cm^2 for 10 s
Erbium glass laser	1540 nm	1.0 J/cm^2 for 1-1000 ns
Erbium:YAG	2940 nm	10 mJ/cm^2 for 1-100 ns
Hydrogen-fluoride laser	2.7-3.1 μm	10 mJ/cm^2 for 1-100 ns
Carbon-dioxide laser	10.6 μm	100 mW/cm^2 for 10 s to 8 hrs, limited area
		10 mW/cm^2 for >10 s

*Not all standards/guidelines have EL's below 200 nm.
Source: IRPA/INIRC Guidelines [1,2]
Note: to convert EL's in mW/cm^2 to mJ/cm^2, multiply by exposure time t in seconds; e.g., the He-Ne or Argon EL at 0.1 s is 0.32 mJ/cm^2.

Infrared Laser EL's

It may seem remarkable that with the variation of the spectral absorption properties of water in the IR-B and IR-C (wavelengths between 1400 nm and 1 mm), that the EL's are all constant with wavelength with the exception of short-pulse, submicrosecond exposures to 1.54 um laser radiation. However, the available data are insufficient to define additional wavelength corrections (relative to the extensive data base at 10.6 um) over the entire infrared range (1.4 um to 1 mm). At 1.54 um, the infrared EL's are increased by a factor of 100 for exposure durations shorter than 1 μs. No further extrapolation to other wavelengths is justified on the basis of presently available information.

A change in the EL's for the skin was made in 1988. For beam cross-sectional areas between 100 cm^2 and 1000 cm^2, the EL for exposure durations exceeding 10 s is 10,000/A_s mW/cm^2, where A_s is the area of the exposed skin in cm^2. For exposed skin areas exceeding 1000 cm^2, the EL is 10 mW/cm^2 [Note: the basis for this change is supported by reference Ro74]

Exposure Duration

A significant clarification in the 1988 revision related to guidance for selecting the appropriate exposure duration to attribute to a CW laser exposure. Determining the applicable EL for a specific CW or repetitively pulsed laser exposure requires a determination of the exposure duration. For a single pulse exposure, this duration is obvious; however, the following criteria should be followed where repeated exposures or lengthy exposures occur.

For any single-pulse laser exposure, the exposure duration is the pulse duration, t, defined at its half-power points. For all skin ELs and for ocular exposure to non-visible wavelengths (less than 400 nm or greater than 700 nm), the cw exposure duration is the maximum time, T, of anticipated direct exposure. For exposure of the eye to any cw laser, the exposure duration is the maximum time of anticipated direct viewing. However, if purposeful staring into a visible (400- to 700-nm) beam is not intended or anticipated, then the aversion response time, 0.25 s, should be used. For ocular exposures in the near-infrared (700 to 1400 nm), a maximum exposure duration of 10 s provides an adequate hazard criterion for either unintended or purposeful staring conditions. In this case, eye movements will provide a natural exposure limitation and thereby eliminate the need for consideration of exposure durations greater than 10 s, except for unusual conditions. In special applications, such as intentional exposure from diagnostic medical instrumentation, even longer exposure durations may apply.

Exposure limits for pulse durations less than 1 ns were not provided by IRPA/INIRC because of a lack of biological data. However, a conservative interim guideline would be to limit peak irradiances to the exposure limit applicable to nanosecond pulses at the wavelength of interest.

Repetitive Laser Exposures

Repeated exposure within any one day to laser radiation can occur from multiple repeated exposures to a cw beam, or from exposures to repetitively pulsed lasers and some scanning beam lasers. Scanning beams create repetitive-pulse exposures to the eye in the retinal hazard region (400-1400 nm). Both the individual pulse duration and the total cumulative exposure duration must be determined. In this case, the total exposure duration of the train of pulses is determined in the same manner as is used for cw laser exposures, i.e., the time T elapsed from the beginning of the exposure (first pulse) to the end (last pulse). The method for determining the ELs for repetitive laser exposures are as follows:

1. Repeated Exposures, Ultraviolet (315 to 400 nm) Laser Radiation. For repeated exposures, the exposure dose is additive over a 24-hour period, regardless of the repetition rate. This rule is consistent with all experience with UVR biological effects [10-15]. The EL for any 24-hour period should be reduced by a factor of 2.5 times relative to the single-pulse EL if exposures on succeeding days are expected (Zuchlich, 1980) [14].

2. Repeated Ocular Exposures, Visible (400 to 700 nm) and Infrared (greater than 700 nm up to 1 mm). Both scanned CW lasers and repetitively pulsed lasers can create a series (train) of laser pulses entering the eye. The EL per pulse for repetitively pulsed intrabeam viewing is $N^{-1/4}$ times the EL for a single pulse of the same duration (t) where N is the number of pulses found from the product of the PRF and the total exposure duration (T). The duration T is determined in the same manner as for a cw laser of the same wavelength (see above). This EL applies to all wavelengths greater than 700 nm where thermal injury predominates. For wavelengths less than 700 nm where photochemical damage mechanisms may also apply, the EL as calculated on the basis of $N^{-1/4}$ also must not exceed the EL calculated for Nt seconds when t is the duration of a single pulse in the train and Nt is greater than 10 s. For pulse repetition frequencies greater than 15 kHz, the average irradiance or radiant exposure (radiance or integrated radiance) of the pulse train shall not exceed the EL for a cw exposure for the viewing duration T. Average irradiance is the total radiant exposure delivered during time T, divided by T. The empirical relation of N-1/4 was based upon several studies [16-20]. The basis of the 0.25-s and 10-s viewing criteria are studies of eye movements [21-22].

3. Repeated Exposure of the Skin. For repetitive-pulsed laser exposure of the skin, the EL based upon a single-pulse exposure shall not be exceeded and the average irradiance of the pulse train shall not exceed the EL applicable for the total duration T of the pulse train.

FUTURE OUTLOOK

The only area of EL guidelines being considered for possible revision relates to extended-source limits. This change would still not affect many applications, as one normally uses the more conservative intrabeam ("point-source") EL's in practice [23].

REFERENCES

1. IRPA, International Non-Ionizing Radiation Committee (1985), Guidelines for Limits of Human Exposure to Laser Radiation, Health Physics, 49(5):341- 359.
2. IRPA, International Non-Ionizing Radiation Committee (1988), Recommendations for minor updates to the IRPA 1985 Guidelines for Limits of Exposure to Laser Radiation, Health Physics, 54(5):573-575.
3. World Health Organization [WHO], (1982), Environmental Health Criteria No. 23, Lasers and Optical Radiation, joint publication of the United Nations Environmental Program, the International Radiation

Protection Association and the World Health Organization, Geneva.
4. International Electrotechnical Commission, Radiation Safety of Laser Products, Equipment Classification, and User's Guide, Document WS 825, IEC, Geneva, 1984.
5. ACGIH (1988), TLV's, Threshold Limit Values and Biological Exposure Indices for 1988-1989, American Conference of Governmental Industrial Hygienists, Cincinnati, OH.
6. ANSI (1986), Safe Use of Lasers, Standard Z-136.1-1986, American National Standards Institute, Laser Institute of America, Toledo, Ohio.
7. British Standards Organisation (1984), Radiation Safety of Laser Products and Systems, Standard BS4803, London, BSI.
8. Health Council of the Netherlands (1979), Acceptable Levels for Micrometer Radiation, Rijswijk, Gezondheidsraad.
9. Deutsche Institut fur Normung, Radiation Safety of Laser Products, Standard VDE 0837, Berlin, DIN/VDE (1984).
10. ACGIH (1980), Documentation for the Threshold Limit Values, 4th Edn., American Conference of Governmental Industrial Hygienists, Cincinnati, OH.
11. Sliney, D. H., and Wolbarsht, M. L., (1980), Safety with Lasers and Other Optical Sources, Plenum Publishing Corp., New York.
12. Ham, W. T., Jr., Ruffolo, J.J., Jr., Mueller, H. A., and Guerry, D., III, (1980) The nature of retinal radiation damage, dependence on wavelength, power level and exposure time, Vision Res. 20:1105-1111.
13. Pitts, D. G., Cullen, A.P., and Hacker, P. D., (1977), Ocular effects of ultraviolet radiation from 295 to 365 nm, Invest. Ophthal. Vis. Sci., 16(10):932-939.
14. Zuclich, J. A. Cumulative effects of near-UV induced corneal damage, Health Phys., 38:833-838, 1980.
15. Sliney, D. H. (1972), The merits of an envelope action spectrum for ultraviolet radiation exposure criteria, Am. Industr. Hyg. Assn. J., 33(9):644-653.
16. Hemstreet, H. W., Bruce, W. R, Altobelli, K.K., Stevens, C. C. and Connolly, J.S., 1974. Ocular Hazards of Picosecond and Repetitive Pulse Argon Laser Exposures, First Annual Report, February 1973-February 1974, USAF Contract for School of Aerospace Medicine, Brooks AFB, TX, Technology Inc., San Antonio, TX.
17. Marshall, W.J., 1978. A Proposal for a New Method to Determine EL Values for Repetitive Pulse Trains, US Army Environmental Hygiene Agency, Aberdeen Proving Ground, MD, June 1978.
18. Stuck, B.E. Lund, D.J., and Beatrice, E. S., 1978. Repetitive Pulse Laser Data and Permissible Exposure Limits Institute Report No. 58, Letterman Army Institute of Research, Division of Non-Ionizing Radiation, Changes in the Maximum Permissible Exposures, Presidio of San Francisco, (April 1978).
19. Ham, W.T., Jr., Mueller, H.A., Wolbarsht, M.L., and Sliney, D.H. (1988) Evaluation of retinal exposures from repetitively pulsed and scanning lasers, Health Physics, 54(3): 337-344.
20. Yarbus, A. L., 1967, Eye Movements During Fixation on Stationery Objects, Plenum Press, NY.
21. Fender, D.H., 1964, Control of the Eye, Science Amer., Vol. 211, pp. 24-33.
22. Gerathewohl, S.J. and Strughold, H., 1953. Motoric Response of the Eyes When Exposed To Light Flashes of High Intensities and Short Durations, Journal of Aviation Medicine, Vol. 24, pp. 200-207.
23. Sliney, D. H. Interaction Mechanisms of Laser Radiation with Ocular Tissues, 1982. In: "Laser Induced Damage in Optical Materials", pp. 355-367, (H.E. Bennett, A.E. Gunther, D. Milam, and B. E. Newman, eds), NBS Publication, SP 669, January 1984. [Updated and published in French CEN publication, 1988].

MEDICAL SURVEILLANCE IN LASER WORKING SITES

Per G. Söderberg, D. Sliney*

Dept. of Ophthalmology (Head: Björn Tengroth) and
Dept. of Medical Biophysics (Head: Rudolf Rigler),
Karolinska Institutet, Box 60 400, S-104 01 Stockholm
Sweden

*U.S. Army Environmental Hygiene Agency,
Aberdeen Proving Ground, Maryland, USA

INTRODUCTION

The discovery of occupational diseases associated with prolonged exposures to toxic agents has triggered a demand for monitoring of potentially hazardous effects in the working environment. Furthermore, the establishment of an insurance system for occupational health has created a need for a basis for an objective decision on whether a health problem is associated to work activity or not. The medical activity devoted to these areas is denoted "Medical surveillance".

OBJECTIVES FOR MEDICAL SURVEILLANCE

The main purposes for programs on medical surveillance are to protect against threats to the health of the employee elicited by occupational activity and to avoid injury to other persons caused by health problems in professionals. The tool to obtain these goals is biological monitoring. Biological monitoring includes measures to detect cumulative damage not recognized by the worker, regular checking of crucial physiological functions and periodic monitoring of physiological functions at threat.

If the occupational environment exposes the professional to toxic chemical agents and/or hazardous physical stress, it is essential to avoid cumulative effects. For this reason, e.g. dentists have been regularly monitored for mercury in the blood. Accumulating lesions in the periphery of the retina has been considered a potential longterm hazard of laser beams. However, existing medical surveillance programs for laser workers have examined thousands of employees without identifying any retinal lesions that can be exclusively correlated to laser damage (Hathaway et al. 1977).

If faulty performance of professional activity is a potential hazard to other persons, it is vital that the skill of the employee remain unaltered. Thus, it is necessary to regularly check visual ability in pilots. Laser installations are either for scientific use or equipment for manufacture. Neither of these activities constitutes a major threat to second person.

Sometimes an occupational exposure induces clearly noticable damage to physiological functions. For example, workers in the steel industry have been found to suffer from hearing problems after life long exposures to noise. Yet, the thousands of examinations of laser workers, so far having been carried out, have not discerned any accumulative injury to the eye (Hathaway et al. 1977).

In some countries from a medical-legal viewpoint it may be in the interest of the employer or an insurance company to have a firm basis for rejection of inadequate claims for compensation for problems stated to be related to occupational activity. Conversely, it is crucial for the employee to have an objective backup when demanding compensation for work related health problems.

SURVEILLANCE REGIMEN

The surveillance may include a pre-placement examination, periodic examinations, and an end-of-job examination.

The pre-placement examination serves to assure that the physical and mental capability of an applicant matches the demand (Alexander et al. 1975). It is also in the interest of the employer to prevent hiring of applicants who carry above average medical risks. Furthermore, it is advantageous from the legal viewpoint to have a base line recording for the health condition of the worker at the onset of the employment. The claim for capability of a worker is applicable for all professions. The power of pre-employment medical evaluations, in filtering out above average absence risks is limited (Alexander et al. 1975). The importance of having base line information on the visual function and the structure of the eye of laser workers is undebatable.

A periodic examination of the eyes of laser workers would aim at monitoring alterations associated with the occupational exposure to lasers. Such a regimen requires examination methods sensitive enough to discriminate laser induced eye damage specifically. The experience of trials of periodic examinations since the 60's demonstrate that if eye damage is at all evoked by laser work, methods to detect the damage are not currently available (Hathaway et al. 1977). Therefore, periodic examination of laser workers is at present inadequate.

The end of job examination provides the employer with documentation of the health condition at the termination of the employment. This avoids financial claims for health problems occuring after the employment. Concurrently, the end-of-job examination assists the employee in demanding compensation for work related injuries. Both of these aspects are significant for laser workers. Therefore, an end-of-job examination should be included in a surveillance program for many laser workers.

RELEVANT EYE EXAMINATION

The selection of examinations in a surveillance program for laser workers depends on the potential hazard of the laser and the risk of being exposed.

Laser hazard evaluation

According to the four step classification system available for hazard evaluation of lasers (Sliney and Wolbarsht 1980) a Class I laser can be stared into without any eye hazard, a Class II (or IIIa) laser is a laser that emitts visible light and which has a power output allowing an accidental exposure to the bare eye, a Class IIIb laser has a power output high enough for any direct exposure to cause ocular hazard, and a Class IV laser emitts such high powers that even diffuse reflec- tions may pose an ocular hazard. Thus, with regard to this classification system no surveillance is required for Class I, Class II, and Class IIIa lasers.

Risk groups

Employees in laser working sites are classified "incidental personnel" and "laser personnel", thereby denoting the risk for being exposed. "Incidental personnel" includes employees that are working in a laser environment but who are not working directly with the laser, indicating that no accidental exposure is expected. "Laser personnel" comprises all employees routinely working with lasers. In this category accidental exposures to the skin are expected and incidental hits against the eye cannot be excluded.

Medical history

The principal goals for the medical history in a pre-employment examination for laser workers are to, before the initiation of the laser work, discern ophthalmological and dermatological problems, and to disclose risk factors such as medication with photosensitizing drugs. Therefore, the medical history is relevant for "laser personnel" but is of doubtful value for "incidental personnel".

Physical examination

The ophthalmic examination should be restricted to procedures that with a high specificity are likely to identify eye injury related to laser damage. Consequently, the specific effects in the eye, caused by radiation in different wavelength regions, must be considered. Radiation of wavelengths shorter than 290 nm is completely absorbed in the cornea (Maher 1978, Boettner and Wolter 1962). At wavelengths between 290 nm and 400 nm the lens is the main absorber of radiation. The iris almost completely absorbs all optical radiation. In the visual waveband, 400-760 nm, and in the near infrared wavelength region, 760-1400 nm, a considerable amount of the incident radiation energy is transmitted to the retinal-choroid layer. At longer wavelengths, 1400-2600 nm the corneais again the primary absorber, and at very long wavelengths all optical radiation hitting the eye is absorbed in the cornea.

Functional examination of the eye

Visual function includes; refraction in the cornea, in the aqueous humour, in the lens, and in the vitreous; phototransduction in the retina; and image processing in the retina and the brain.

The clinically most useful compound test for these functions is the test of discrimination of letters at varying visual angles, the visual acuity test, with and without corrective spectacle glasses. Yet, this test is insensitive and non specific. Recognition of gratings on a television screen as a function of the contrast between the stripes, contrast sensitivity testing, is a more sophisticated method for objective analysis of the complete visual function. However, even this method suffers from being too insensitive to detect small laser lesions.

There is no method available to functionally evaluate the refractive media of the eye, isolated. The phototransduction can be evaluated specifically by electrical recording of retinal potentials induced by light, electroretinogram (ERG), but the routine ERG is to insensitive to provide useful information of retinal laser damage.

The phototransduction and image processing can be analyzed by recording potentials in the brain cortex, evoked by visual stimulus, visually evoked potentials (VEP), but this method is not sensitive enough to detect small laser induced lesions in the retina.

The retinal image processing can efficiently be evaluated by asking the patient to reproduce a matrix that is presented, i.e., Amsler grid testing. Central retinal lesions are detected as deviations from exact reproduction of the matrix. Yet, peripheral lesions are not possible to detect.

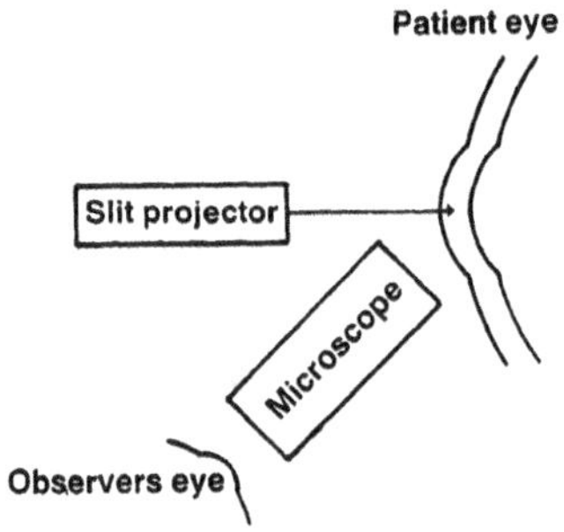

Fig. 1. Schematic drawing of slit lamp microscopy.

Intraocular pressure is not related to chronic exposure of the eye to laser light.

For pre-placement and end-of-work examinations, only the aforementioned functional tests are really needed. However, if the functional tests reveal abnormalities, or if laser injuries are suspected, structural examination is needed.

Structural examination of the eye

The conjunctiva, the cornea, the aqueous humour, the iris, and the lens may be examined by observing a slit of light that is projected into the eye, slit lamp microscopy (Fig. 1). Any pathology inducing a shift of refractive index, e.g. edema, disturbs the projected light and gives rise to light dissemination. Slit lamp microscopy discloses very small structural alterations but the diagnostic decision have a considerable intra- and inter- observer variability.

The vitreous and the retina can be visualized by projecting an image to the observers retina while illuminating, ophthalmoscopy (Fig. 2). Even small retinal lesions from laser light are easily detected over the entire retina. Yet, the apperance of the lesions is not specific for laser damage.

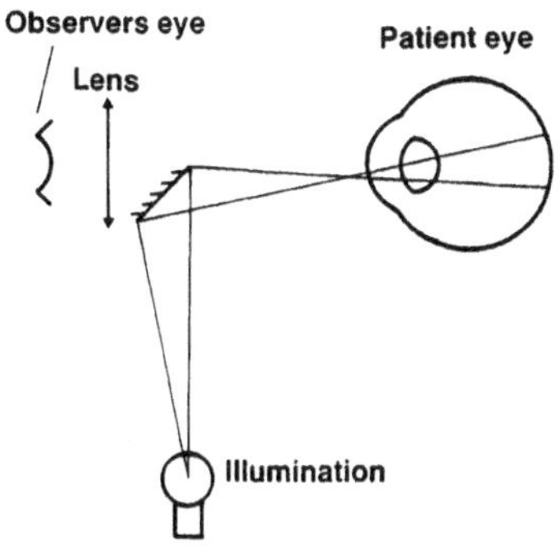

Fig. 2. Schematic drawing of ophthalmoscopy.

Suitable physical examination

Considering the potential risk for different cathegories of employees in laser working sites and the efficiency of the different techniques available for eye examination, the examination of "incidental personnel" should be limited to visual acuity testing. For "laser personnel" only visual acuity testing, and Amsler grid tests are needed unless vision is not normal or an injury is suspected. Then, slit lamp microscopy and ophthalmoscopy are relevant to include in the eye examination.

RESPONSIBILITY FOR MEDICAL SURVEILLANCE

In order to fulfill a medical surveillance program in laser working sites the responsibility for the program must be defined. The ANSI Z-136.1 (1986) laser safety standard proposes that the individual in charge of safety aspects at the laser installation should be the "laser safety officer".

ACUTE EYE DAMAGE FROM LASERS

Once laser equipment is installed, the national ophthalmic center with special interest for laser damage should be established. The general ophthalmologist normally does not have the special knowledge to effectively examine and treat laser damages. Furthermore, the incidence of laser injuries to the eye is so small that international collaboration between groups with special knowledge is required to improve protection and treatment.

If an eye is accidentally hit by a laser beam the victim should immediately be referred to an ophthalmologist with a special interest in this problem. In addition, the accident should be reported to the national occupational health organ for the initiation of an investigation of the cause of the injury.

ACKNOWLEDGEMENTS

We are indebted to Björn Tengroth for reviewing the manuscript and to Lotta Vig for help with the preparation of the manuscript.

REFERENCES

Alexander, R. W., Maida, A. S., and Walker, R. J., 1975, The validity of preemployment medical evaluations, J. O. M., 17:687.

ANSI Z-136.1, 1986, Safe use of lasers, American National Standards Institute, New York.

Boettner, E. A., Wolter, J. R., 1962, Transmission of the ocular media, Invest. Ophthalmol., 1:776.

Hathaway, J. A., Stern, N., Soles, E. M., and Leighton, E., 1977, Ocular medical surveillance on microwave and laser workers, J. O. M., 19:683.

Maher, E. F., 1978, Transmission and absorbtion coefficients for ocular media of the rhesus monkey, Report SAM-TR-78-32, USAF School of Aerospace Medicine, Aerospace Medical Division (AFSC).

Sliney, D., and Wolbarsht, M., 1985, 1.4. Laser classification, in: Safety with lasers and other optical sources, p. 7, Plenum Press, New York.

MEDICAL LASERS AND BIOLOGICAL CRITERIA AND LIMITS OF THEIR THERAPEUTIC EFFECTS

Louis A. Court, Daniel Courant,
and Dominique Dormont

Centre de Recherches du Service de Sante des Armées
Commissariat a l' Energie Atomique
Grenoble and Fontenay aux Roses France

I INTRODUCTION

The use of lasers in medicine has undergone and still undergoes considerable developments without the man in the street, the patient and even the practitioner really knowing the efficacy of these news means of diagnosis or treatment, their eventual contraindications, and harmful or beneficial effects.

Scientific publications are numerous but their contents resemble more a list of the benefits rather than an objective analysis of the possibilities and qualities of the different lasers.

There is a real interest in these new techniques in many professional categories :

. the clinician seeking new therapies
. the physicist or engineer looking to develop new techniques which the industry does not as yet possess
. the director of the clinical or health care etablishment desiring greater frequentation of his institution and thus substantial profits.

These hopes are however moderatedby the control exercised by the officials of the agencies who have to foot the bill !

At the moment, purveyous of alternative medicine and beauty treatment are triumphant in the treatment of functional disorders by taking advantage of the scientific character of the tool, its esoteric aspect and its real and supposed properties every time that conventional medicine is too often ineffective through lack of time, reflexion or confidence in the power of the fantastic.

The laser is the promised hope for the correction of disfiguration, the effects of aging or the results of nutritional excess.

Taking together these different orientations, a thematic and common reflexion is applicable in the search for a new mean of action.

The problematic aspects arise in the definition of objective criteria of laser efficacy. This search of efficacy depends upon the perfect and detailed knowledge of the biological effects of the differents beams and in particular their hazardous properties.

Light, Lasers, and Synchrotron Radiation
Edited by M. Grandolfo *et al.*, Plenum Press, New York, 1990

The etablishment of precise rules of utilisation will obligatorily have to preceed all research into efficacy in its own right.

II NATURE OF MEDICAL AND PARAMEDICAL LASERS

They are multiple, continuous or pulsed emitting from ultra violet to deep infra red. Their areas of applications are extremely far reaching.

MEDICAL LASERS

II-1 OPHTALMOLOGY

. PHOTO COAGULATION	Krypton
. MICRO SURGERY	Argon
- PARS ANTERIOR	He-Ne
. CORNEA KERATOTOMY	Excimer CW or Q Switched Laser
. GLAUCOMA	Nd : Yag
. CAPSULECTOMY	Dye Laser
. MEMBRANOTOMY	
. SYNECHOLYSIS	
- PARS POSTERIOR : RETINA	
. PROPHYLAXY AND SURGERY OF RETINAL DETACHMENT DIABETES RETINOPATHY MACULOPATHY	

II.2 GENERAL SURGERY

. THERMO COAGULATION
. PHOTO ABRASION
. CUTTING
. VAPORIZATION
- TUMORAL TREATMENT
- ENDOSCOPY
- VASCULAR SUTURE

1. NEURO SURGERY	
MENINGIAL TUMOR	CO_2
MENINGIOMA	Nd : Yag
EPENDYMOMA	
ACOUSTIC NEURINOMA	
ANGIOMA	
CRANIO PHARYNGIOMA	
GLIOMA OF CHIASMA AND OPTICAL NERVUS	
BRAIN STEM TUMOR	
2. OTO RHINO LARYNGOLOGY	
TONSILLECTOMY	CO_2
LARYNGEAL, PHARYNGEAL TUMOR	Nd : Yag
3. THORACIC SURGERY	
BRONCHIAL AND LUNG TUMOR	CO_2
	Nd : Yag
4. STOMATOLOGY	
ANGIOMA	
LEUKO KERATOSIS AND LEUKOPLASIA	Argon
GINGIVAL HYPERPLASIA	CO_2
PERIONDOTAL DISEASES	Nd : Yag
LICHEN PLANUS	

BENIGN AND MALIGN TUMOR
TECHNICAL FIXATION OF DENTAL
PROTHESIS OR REPLANTATION/IMPLANTATION

5. GYNECOLOGY - LASER LAPAROSCOPY AND ADHESIOTYSE - TUBAR SURGERY AND SALPINGOSTOMY - ENDOMETRIOSIS	CO_2 Nd : Yag
6. VASCULAR SURGERY . VASCULAR SUTURE . MICRO ANASTOMOSIS ATHEROSCLEROSIS ANGIOPLASTY	CO_2 Nd : Yag Holmium Excimer Argon
7. GASTRO ENTEROLOGY BENIGN AND MALIGN STENOSIS HEPATIC TUMORS AND PARTIAL HEPATECTOMY HEMORRHOIDECTOMY AND ANAL AND ANO RECTAL SURGERY	CO_2 Nd : Yag
8. UROLOGY URETERAL LESIONS TUMORS OF KIDNEY AND BLADDER VASO-VASOSTOMY	CO_2 Nd : Yag
9. PEDIATRIC SURGERY OESOPHAGEAL AND ANO-RECTAL FISTULAE TRACHEO-OESOPHAGEAL FISTULAE SLEEN'S RESECTION HEMI HEPATECTOMY	CO_2 Nd : Yag
10. REPAIR SURGERY BURNS AND SEQUELLAE TUMORS	CO_2 Nd : Yag
II.3. DERMATOLOGY TUMORS ANGIOMA TATTOOS REMOVAL	CO_2 Nd : Yag He-Ne Ar Kr Ruby
II.4. ACUPUNCTURE	He-Ne Ruby
II.5. APPLICATION OF PHOTO SENSITIZERS DIAGNOSTIC AND THERAPEUTIC PREOCCUPATIONS SENSITIZERS . FLUORESCENT DYE ACRIDINORANGE ACRIDINORANGE EOSINE FLUORESCEIN	Argon Dye laser

. PORPHYRINE
HEMATOPORHYRINE
- DIHEMATO PORPHYRINESTER
I-ETHER
. CHLORINE
. PHTALO CYANINES

II.6. LITHIASIS

LITHOTRYPSY	Nd : Yag pulsed
- BILIARY TRACTUS	8 ns - 100 mJ - 50 Hz
- PANCREATIC TRACTUS	12 ns - 45 mJ - 20 Hz
- UROGENITAL TRACTUS	10 ns - 800 mJ - 1P Hz

SOFT MILD AND PARA MEDICAL LASERS

ANTI INFLAMATORY AND ANTI-OEDEMATOUS ACTION ANTALGIC ACTION BIOSTIMULATION OR STIMULATING "EFFECT OF LOW-POWER LASER RAYS ON BIOLOGICAL SYSTEMS" ANTI-CELLULITIS EFFECTS IMPROVING EFFECT OF MICROCIRCULATION OR LYMPHATIC DRAINAGE	Ruby He-Ne Argon Nd : Yag Diode	0,05 - 20 m $W.cm^2$

III EXPERIMENTAL APPROACH OF MEDICAL LASER EFFECTS

Numerous parameters are to be considered; the wavelength and the exposure time (continuous or pulsed lasers where the dominant factors are the impulse duration and repetition frequency and the diameter of the focussed beam (reduced to a single point 0.1 m for sectioning or divergent the widest 1-204 m for vaporising. Figures 1, 2 and 3 borrowed from Sliney show the different effects and interaction mechanisms as a function of the length of exposure and the beam profile.

In effect, the appearance of mechanical effects then (for ultrashort impulses) plasma formation and thermoacoustic wave production lead to mechanical ruptures between cells and also destruction of lithiasic concretions.

It is evident that the use of pulsed lasers, particularly those of very high frequency may pose problems due to the intensity of their effects.

Two rules should be imperatively respected :

1 for the development of a laser technique necessarily implying the study of biological effects only these criteria should be used in the design and the construction of the apparatus and not the adaptation of an industrial laser to the demands of the operative medicine.

2 the criteria which permit the evaluation of laser efficacy should be of the same nature as those need in therapeutics in general and in pharmacology in particular. Any efficacy test should be performed double blind on a sufficent number of patients to allow statistical interpretation of the results and in comparison with groups of patients treated by conventional means.

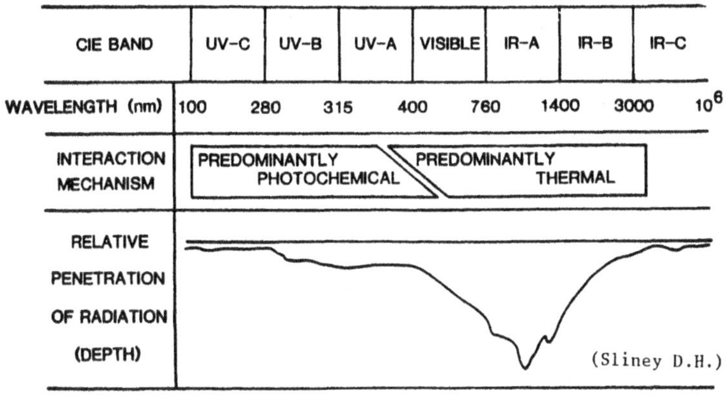

Figure 1. Interaction mechanisms. Tissue interaction mechanisms are predominantly thermal for long wavelengths and photochemical for shorter wavelenghts. Attenuation is the combined effects of scattering and absorption and defines the penetration depth.

Safety with Lasers and Other Optical Sources, New York, Plenum Press, 1980.

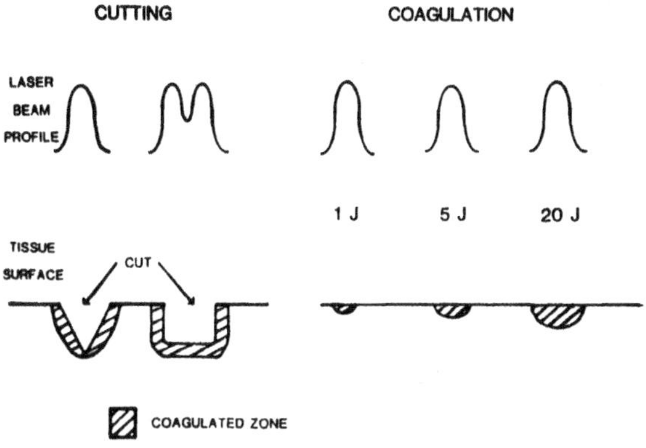

(Sliney D.H.)

Figure 2. Beam profiles and tissue heating patterns. Because most laser beams are intended to be gaussian (bell-shaped), heat tissue is greatest near the beam center. An irregular beam could, in theory, produce a different cut except for heat conduction. Dwell time at a tissue site increases the energy (in J) delivered as well as the depth and width of the zone of damaged tissue. In this example, constant power (10 watts) and different exposure times (0.1, 0.5 and 2 seconds) produce different tissue response in shape and depth.

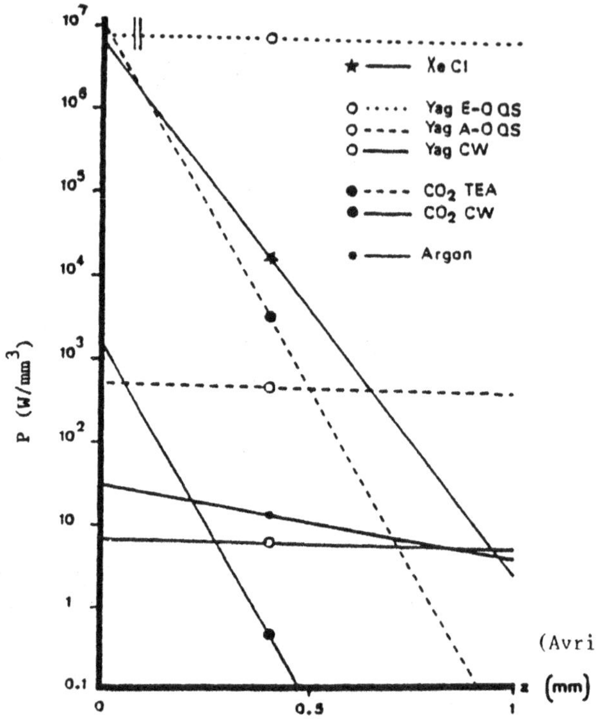

(Avrillier et. Al)

Figure 3. Laser-tissue interactions. Biologic effects depend on exposure duration. The tissue temperature for these effects vary, and the values and dividing lines shown are very approximate.

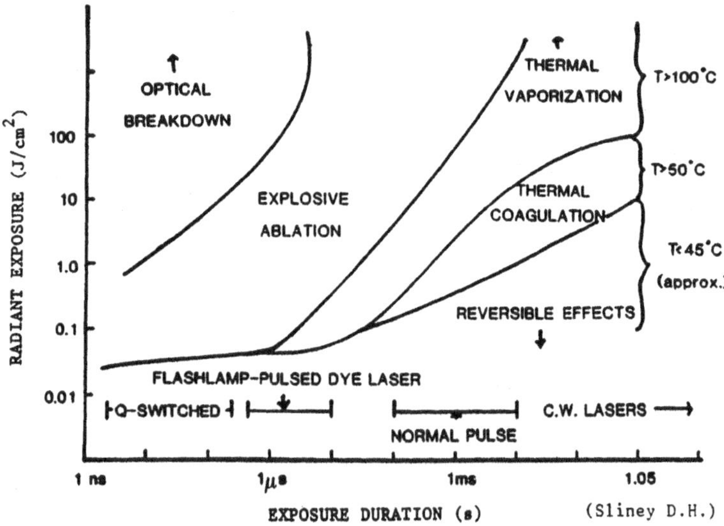

Figure 4. Calculated variations of the instantaneous volumetric power density P(0,z) supplied in tissues by the indicated lasers.

Finally, short and long term post operative surveillance will be instituted to select eventual side effects and to compare, for example, the evolution of scar formation, the presence and rize of cheloids.

III-1 Use of a fiberoptic XeCl excimer laser system for angioplasty

Avrillier S. et al.(2) and Mohr F.W. et al.(3) compare main characteristics of lasers for which disintegration of atherosclerosis plaque has already been obtained in vitro or in vivo studies. The figure 4 exhibit calculated variations of the instantaneous volumetric power density P (0,z) supplied in the tissue by several types of lasers.

Two groups of lasers can be easily distinguished :

1 the first one is the Ar and Nd:Yag laser for which the deacrease of laser energy deposition with penetration depth is small or very small. This fact explain the difficulty to control precisely the boundary of the tissue injury and the risk of perforation or dissection of adjacent tissue.

2 the second group with CO_2 and excimer laser for which the volumetric power density is large near the surface and decreases very rapidly inside the time to become negligible for the depth z ≠ 0.5

The CO_2 pulsed laser is interesting with special characteristics as a very small light penetration (≠ 50 m), a time of vaporization smaller than the pulse duration and much smaller than the characteristic thermal relaxation time, but unfortunately there is not really far IR optical fiber commercially available. The Q switched Nd:Yag laser induce electro mechanical or photodisruptive processes which may be dangerous for the tissues.

The ablative effects of different wavelengths 308, 248 and 193 nm, with slightly focussed beam and the pulse energies range from 60-115 mJ with 10-2500 pulses have been studied. These effects were almost similar with the different wavelengths and were calculated to be between 0.3 and 0.7 depending on the levels of the pulse energies.Highly calcified tissue could only ablated using 193 nm. Very clear cuts of fibrotic cardiac valves could be observed by histological examinations. A tissue ablation was performed with pulse energies from 60-115 mJ (0-5) at fluence of 50 $J.cm^{-2}$. The figure 5 give the variations of energy absorbed for a pulse as a function of the distance from the surface with XeCl laser. Temperatures of the tissue after ablation are indicated on the right hand vertical axis. Comparison of wavelengths give best results with shorter wavelenghts; 308 nm caused a certain amont of thermal energy.

The clinical application of excimer lasers for the treatment of cardiovascular disease is possible but long term studies are necessary to prove that there are not risks by embolic complications due to gaz or debris or by preparation of smooth surface created by UV ablation and by carcino genetic effects after far UV laser radiation.

III-2 Laser induced shock wave lithotrypsy in animal experiments and patient treatment. (Reichel et al.) (4)

Laser induced shock waves can be released in narrow cavities immediately in front of the surface of the stone. The duration of these shock waves is very short and no fixation of the stone is necessary. The laser is generally Q-switched Nd:Yag laser with the following characteristics; pulsewidth 8 ns, pulsenergy 20-800 mJ, repetition rate 10, 50, and 100 Hz.

The pulses are transduced by a flexible quartz fiber which are focussed as the tip of the fiber. Figures 6, 7 and 8 give the laser induced shock wave pressure versus time, the maximal amplitude of pressure of laser induced shock waves as a function of pulse energy, laser shock wave pressure versus distance from laser induced breakdown.

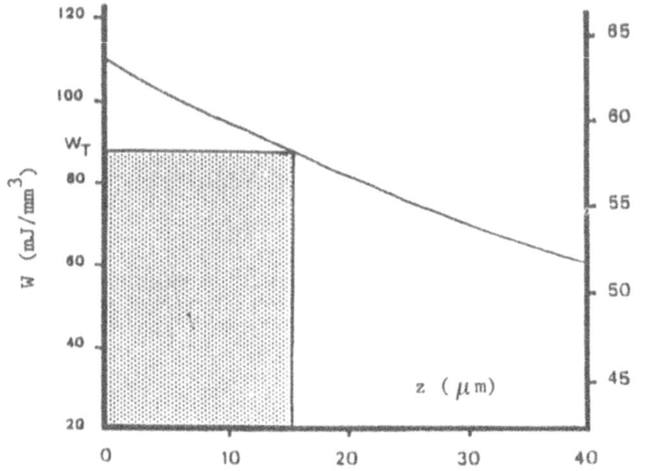

(Avrillier et. Al)

Figure 5. Variations of the energy absorbed per pulse as a function of the distance from the surface. The XeCl laser beam threshold energy for photoablation of vascular tissue corresponds to a etch depth of 15 m. Temperatures of the tissue after ablation are indicated on the right hand vertical axis.

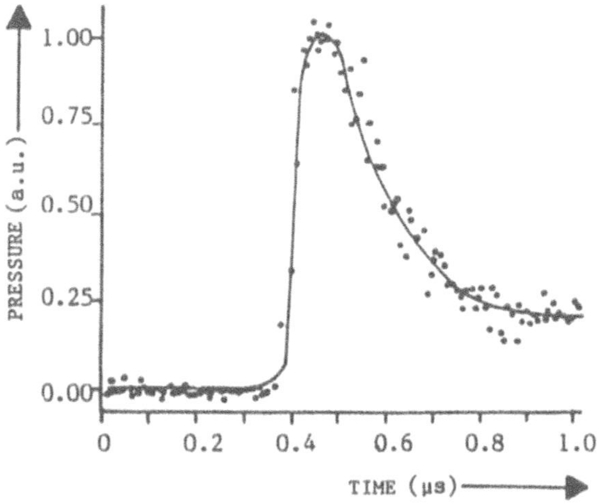

(Avriller et. Al)

Figure 6. Laser induced shock wave : pressure versus time.

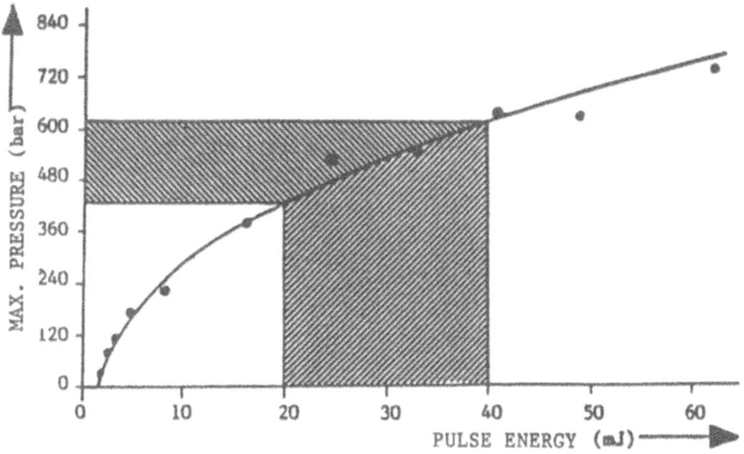

(Reichel et. Al)

Figure 7. Maximal amplitude of pressure of laser-induced shock waves as a function of pulse energy fed to the laser-induced plasma break down.

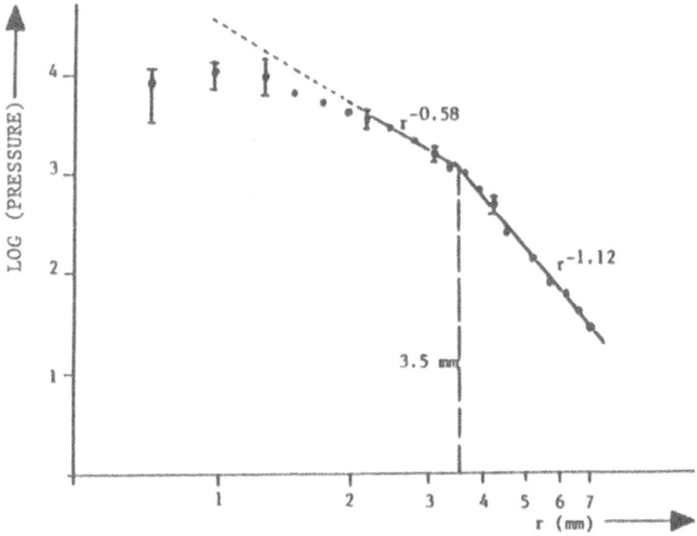

Figure 8. Laser-induced shock wave : pressure versus distance from laser-induced break down.

(Reichel et. Al)

The first clinical application of these shock waves give good results. The application of 1800 shock waves on the rabbit bladder did not cause severe lesions of urothelium and without damage of the muscularis or perforation. The destruction of urinary calculi - 80 mJ,13 ns,25 Hz-,(Schmell et al) (5) or pancreatic concrements in the pancreato-biliary tract -45 mJ,

12 ns, 20 Hz-are easy in human patients. (Wenck et al) (6).

III-3 The long term biological effects of medical lasers : the carcinogenetic effect

The study of long term effects is more difficult, since there exists pratically no means of appreciation of carcinogenicity. This observation is even more serious as :

- the use of the laser is not palliative
- the subject is younger
- the treatment concerns tissues where the presence of oncoviruses is frequent (urinary tract and genital tract surgery particularly in women in the latter instance).

There exists moreover a contradiction between recognising a stimulating of numerous laser radiations and admitting that this stimulant effect may be accompagnied by the expression of growth factors or pro-oncogenes.

It is a very controversial aspect of medical and surgical lasers. Our results, (Court, Courant, Dormont et al) (7) obtained with the pulsed Nd:Yag laser, bring the proof that it is possible to induce in the culture cell (human leucocyte and human fibroblast) for low level doses the following events :

. chromosomic aberrations

. decrease of cell's growth

. and simultaneously appearance of oncogenic factors

The hazards of new techniques laser (Q switched and excimer laser with specially in UV the absorption band of DNA) have never been really studied. The processes are probably very dependant of dose rate wavelength and cell's metabolism.

IV EVALUATION OF MEDICAL LASERS EFFICIENCY AND CLINICAL USE

Two operative procedures are used :

. with focussed mode; the spot is very small (0.1 mm) and muscular section is of excellent quality, at least as good as that of a scalpel but with the following advantages, a total absence of bleeding if no vessels of more than 0.5 mm diameter are severed and without stimulation of the muscle fibers, which remain perfectly inert

. with defocussed mode, 1 to 4 mm with the diverging beam, for coagulation, hemostasis or vaporization of the abnormal tissue.

These characteristics make it possible to perform a surgery with excellent quality and precision and with the minimal damages to adjacent tissue, with the adjustment of the size of the irradiated area and the power output, it is possible to destroy level by level, layer by layer, the pathologic tissues with perfect visual control. The haemostatic properties of laser beam are remarkable since many angiomatous lesions can be vaporized without severe bleeding. Often, with the particular properties of laser beam, the post operative comfort of patient is also better than with the normal procedures. Generally, the appearance of optical fibers fabricated from silver crystals (eg $AgCl_x$; Br_{1-x}) highly transparent in the IR spectrum between 3000 and 15000 nm and the special sapphire probe for Nd:Yag laser, attached as a light scalpel, extends considerably the application's field of near and far IR laser. The optical fibers can transmit CW laser beams of power density higher than 3 x 10^3 W.cm^{-2} or short pulses of power density 5x 10^8 W.cm^2 and the napphire probe with Nd:Yag laser in direct touch with the tissue permit an excellent cutting effect with only minimal damage to the surrounding tissue. There are some examples of a good appreciation of new surgical laser techniques; without real advantages, with restrictions or with better and very interesting results.

IV-1 Without real advantages

1. Study of conventional versus laser tonsillectomy in adults (Iversen P.B. et al) (8)

   Tonsillectomy was carried out in 20 young adults by the same experienced surgeous. One tonsil was remoded in the conventional way and the other by laser : 10 W CW-CO_2 laser incision of anterior pillar, after retraction of tonsil, ligature great vessels, vaporization and complete haemostasis with laser. The patients were followed regularly for one month. The only significant difference between the two procedures was less pain during the immediately post operative course on the laser side. But the length of the operation, the post operative bleeding, the last post operative pain and scar building were without significant difference for conventional and laser procedures and the healing tendancy seemed a little delayed on the laser side. Using the CO_2 laser in tonsillectomy as a routine is not recommanded.

2. In dermatology. The removal therapy of Tattoos (Seipp W. et al) (9)

   In laser treatment of 182 patients which decorative tattoos -excepted for tattoos which consist of single lines or thoses which are located in area with thin skin like ie back of the hand- there is not marked superiority of laser therapy against conventional methods of tattoo removal.

IV-2 With restrictions. CO_2 laser in gynecology

1. CO_2 laser in gynecology (Bruhat et al) (10)

   The efficiency of CO_2 laser compared to conventional techniques is very difficult to etablish, without long series with long term post operative observations. The use of surgical laser after laparatomy must be considered of poor value, but procedures which require a laparatomy with conventional techniques ie difficult adhesiolysis, endometriosis and salpingostomy can be performed by laparascopy with the CO_2 laser. CO_2 laser is clearly the ideal instrument to destroy endometriosis. (200 patients)

IV-3 With better and very interesting results

1. In neurosurgery (Beck et al) (11)

   In neurosurgery the use of laser is justified by :

   - the superior coagulation effect of the 1.06 nm Nd:Yag laser

. the fact that circumscribed areas such as a tumor residuals can be destroyed by means of the defocussed laser beam under accurate control and with predictable depth of penetration.

In laser operated patients compared to conventionally operated patients, there is an evident reduction of the reccurence rate between 10 and 30 % during the same long period of follow up versus 7.8 % with laser. With both Nd:Yag 1.06 nm and 1.032 nm, about the same results be obtained regardling the radicality of tumor exeresis (254 and 89 patients respectively) These two factors are very important in the treatment of the benign and malign meningial tumors.

2. In urology
(Hofstetter et al) (12)

In treatment of bladder tumors, randomized prospective studies comparing the efficiency of trans-urethral resection (TUR) and laser treatment, indicate the superiority of Nd:Yag laser application. The tumors have the same histological type and the treatment associates or no the chimiotherapy with mitomycine. The recurrence rate is give by the table I

TABLE I

	LASER AND CHIMIOTH.		LASER ONLY		TUR AND CHIMIOTH.		TUR ONLY	
	✱	0	✱	0	✱	0	✱	0
Nb of patients	15	15	9	5	13	2	3	4
Nb of patients with recurrence	1	2	1	3	6	2	3	4
Recurrence rate	0.17	0.4	0.27	2.2	1.32	7.14	7.89	11.1

✱ Primary tumor 0 Relapse tumor TUR

TUR : Trans urethral resection

CHIMIOTH.: Chimiotherapy

Recurrence rate : $\frac{\text{Nb recurrences x 100}}{\text{Month follow up}}$

V THE SOFT OR MILD LASER

This problem is the interpretation of growing number of clinical reports claiming to provide evidence in relation to the beneficial effects of irradiation of tissues by extremely low levels of optical radiation .

These effects are generally the promotion, the acceleration and the growth increase of cells or the regeneration of the tissue and have been described by Reed R.D. et al (13) or biostimulative effects. A report of Mester E. (14) collect many data out.

It is very difficult to have a scientific opinion on these experimental data. The most part have been collected coming from Eastern Block and are confined to measurements of DNA, RNA and protein synthesis in micro organisms or clinical effects without real statistical analysis. The results are very difficult to interpret.

1 in culture cell of two fibroblast cell lines, - human embryonic foreskin fibroblasts and human skin fibroblasts Bolton and Marshall (15) confirm biostimulation after 1 mW He-Ne laser (633 nm) at an irradiance of 100 $W.cm^{-2}$

. cell counts of both irradiated cultures exhibited a significant increase in the number of cells in comparison with their respective non irradiated control The effect is observed 24 h following a single laser irradiation of 15 mn. This effect became even more pronounced when the cells were examined 48 h following a single irradiation

- laser irradiation stimulate the attachment of cells to plastic substrate

- no significant change is observed in cell counts between irradiated and control cultures in experimental studies with the incoherent source

- the measurements of thermal load show no elevation of temperature

. Fava et al (16) on tumor cells in culture find that the continuous irradiation by the He-Ne laser at dose rate of 10^{-1} and 1 $W.m^{-2}$ did not induce a marked effect on cells reproduction, but the DNA synthesis increases significanttly at 10 $J.cm^{-2}$. The effect is found to be dose rate dependant

. Yew et al (17) analyse the low dose He-Ne CW laser (0.5 mW - 8.3 $mW.cm^{-2}$) effects in chorioallantoic cultures of retinal pigment cells. The cell sizes decreases and the number of mitosis increases in the experimental irradiated culture ; a higher proportion of detached cells was observed with two mechanisms, induction of defects in cell-cell junction or an induction of cell shapes transformation.

. However Strube et al (18) have examined with a radiodermitis 45 Gy model of guinea pig the influence of He-Ne CW laser (irradiance 10 $mW.cm^{-2}$, doses 2 $J.cm^{-2}$). The laser irradiation effect, appreciated during 40 days have not stimulated the healing process.

Mechanisms :

The interpretation of most experimental data concerning biostimulatory effect is very difficult

1 The low level irradiation, in visible or near IR spectrum does not induce thermal process 1 °C

- Boulton and Marshall (15) did not observe thermal increasing in culture cell
- Haina D. et al (19) measure the temperature of the skin during application of soft laser (He-Ne with 2mm beam diameter and power densities up to 600 $mW.cm^{-2}$)
  The maximum in temperature arise 1°C .
  cf figures 9, 10 and 11

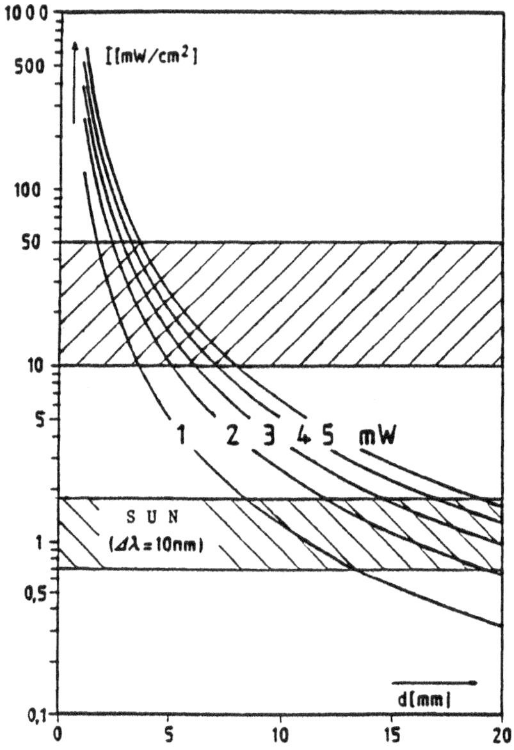

(Haina et. Al)

Figure 9 . Power density I as a function of beam diameter d for laser output powers from 1 mW to 5 mW.

2 there is no empirical evidence in relation to the molecular basis or absorption of light

3 there is no really photochemical processes

4 there is an hypothetical mechanism :
- the changes by He-Ne or IR radiation of the optical and biochemical properties of the coenzyme nicotamide adenine deshydrogenase (NADH) and a two-proton absorption process by ruby light
- the increase in transmembrane electrical potential in mitochondria resulting in a proton force which drives ATP synthesis

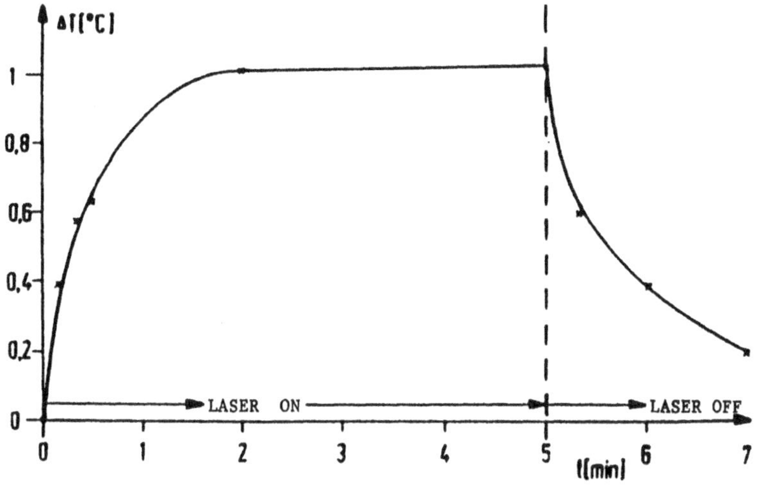

(Haina et. Al)

Figure 10. Temporal change in temperature of human skin during irradiation with a HeNe laser (power 18.8 mW; beam diameter 2 mm). After 5 mn the laser was switched off.

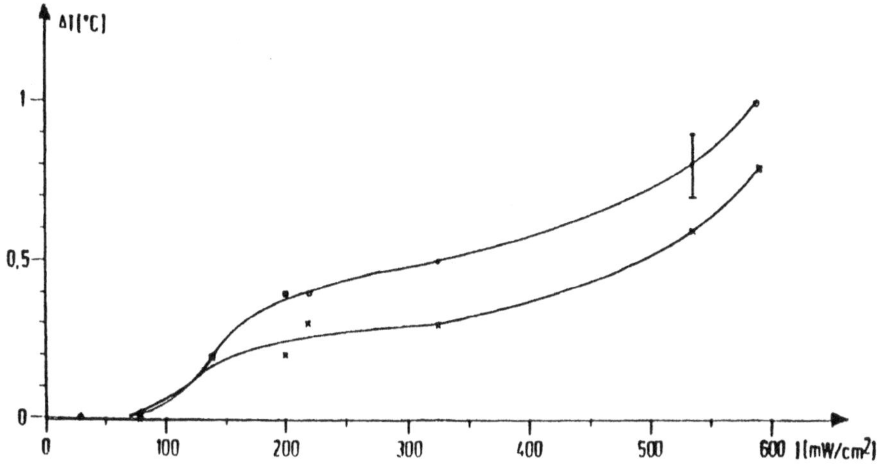

(Haina et. Al)

Figure 11. Temperature increase in dependance of power density, measured on a test person after a 2 min. HeNe laser irradiation.

- Iwase T. et al (20) using an in vivo system, study the 5 mW He-Ne laser effect on the Jaw -Opening Rabbit Reflex and confirm after 5 mn exposure the elevation of threshold and an analgesic effect of the laser on the dental pain induced by an electrical stimulation. (figure 12)

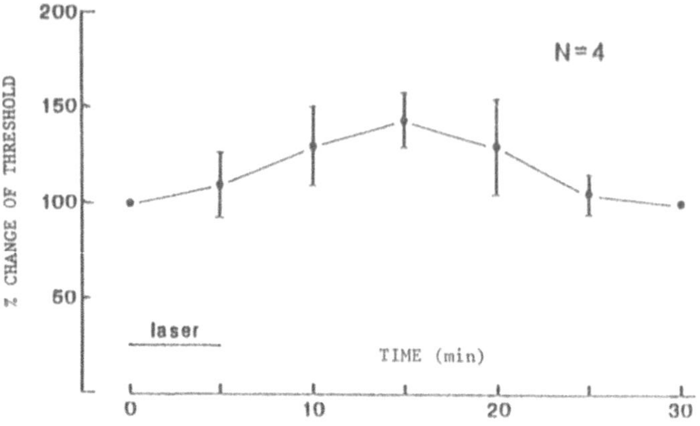

(Iwase et. Al)

Figure 12. Effects of HeNe laser on dental pulp stimulated electrically The graph indicates percent of threshold to control. The teeth exposed to HeNe laser for 5 minutes.

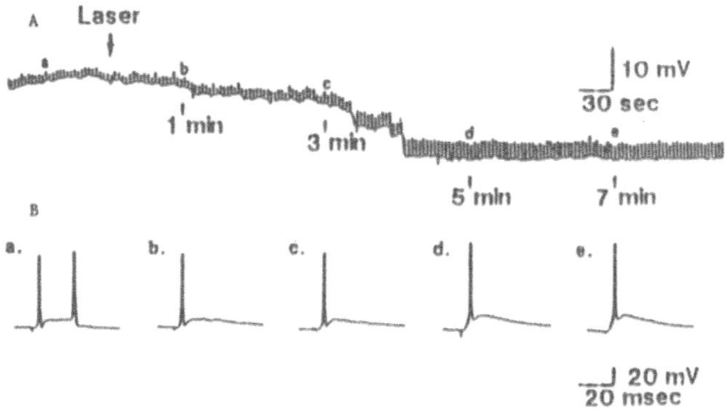

(Iwase et. Al)

Figure 13. Effects of HeNe laser on membrane potential (A), and EPSPs of action potential (B) recorded from the pyramidal cells.

In vitro on rat brain slices preparation, the laser irradiation change the membrane potential of pyramidal cells which is hyperpolarized (figure 13). The pyramidal cells which had lower membrane potential (about 50 mV) are more sensitive to the laser than the cells whose membrane potentials were higher (about 70 mV). As the membrane potential is going deeper and deeper, the excitability of the pyramidal cell is decreasing and the excitatory post synaptic potential is increasing. This phenomenon is probably related to the met metabolic system in the cell. (figures 14 and 15)

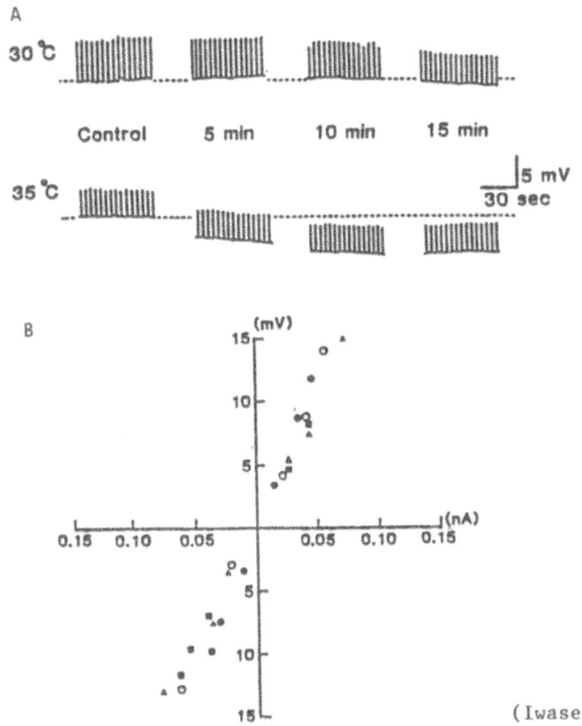

(Iwase et. Al)

Figure 14. Effects of the temperature tissue exposed to laser. A : the change of the membrane potentials by HeNe laser at 30 °C and 35 °C, B : laser effect on the input impedance at 30 °C, Control: (filled circles), 5 minutes (filled squares), 10 minutes (open circles), 25 minutes (filled triangles.

VI CONCLUSION

The development of medical lasers will only be harmonious and beneficial if two problem areas are respected.

- the one based upon the precise knowledge of the biological effects of lasers (in the diversity of their characteristics) permitting the development of laser systems specific for the medical use for which they will be constructed.

- the other being clinical trials of these prototypes. Owe the the construction, the technical development and the efficacy tests are performed it will be necessary to envisage a precise training programme for the users, comprising the dangers and limits of their use and also the mechanisms of interactions and the energy exchange processes whose inconveniences may outweigh their efficacy.

There is no room for complacency. Under no circumstances should be the use of this systems, which are both complex and are capable of doing great harm if mis-used, for purposes other than those described above be supportable by badly conceived scientific analysis and by pseudo science.

" Primum non nocere sed etiam in veritate curare "

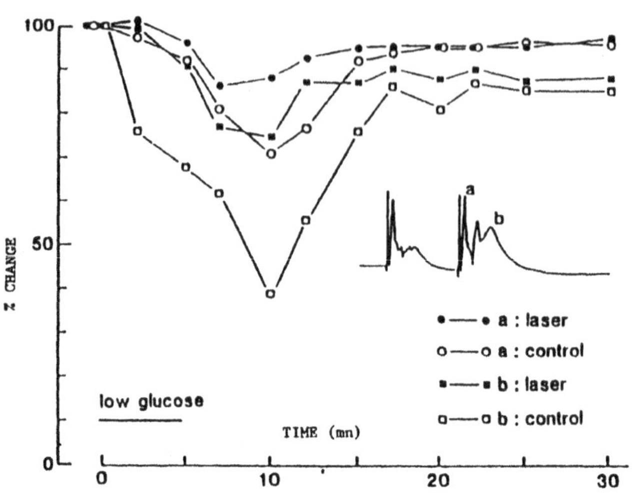

(Iwase et. Al)

Figure 15. Effects of HeNe laser on the amplitude of indicated positions (a, b) in field potentials recorded from the pyramidal cells under the condition of low glucose (2.5 mM). The tissue was continuously exposed to HeNe laser for 10 minutes. The low glucose ringer was perfused the later 5 minutes instead of normal ringer solution.

REFERENCES

1. Sliney D.H.
Laser tissue interactions
Clinics in Chest Medicine 6, 2:203-208,(1985).
2. Avrillier S., Ollivier J.P., Berthier J.P., Reguer M. and Raynal E.
Simplified physical considerations about laser angioplasty and preliminary experimental results with a fiberoptic XeCl excimer laser system.
in procedings of First International Symposium on Laser Biological Effect and Exposure Limits, L.A. Court et Al Eds.(1988)
3. Mohr F.W., Lenz W., Kusserow S., Greulich O., Weller R., Wolfrum J. and Kirchhof P.J.
Excimer Laser for angioplasty and cardiac valve repair
Laser in medicine and surgery, 3:93-97, (1987)
4. Reichel E., Schmidt-Kloiber H., Schöffmann H., Dohr G., Hoffmann R., and Hartung R.
Physical mechanisms of the laser induced shockwave lithotrysy
Laser in medicine and surgery 3:117-183, (1987)
5. Schmeller M.G., Hofstetter A.G., Pensel J., Thomas S., Franck F. and Wondrazek F.
Laser induced shock wave lithotrypsy (USL)
Laser in medicine and surgery 3:184-193, (1987)
6. Wenck H., Lange F., Schildberg F.W. and Hofstetter A.
Pancreato-biliary system with laser induced shock wave lithotrypsy (USL)
Laser in medicine and surgery 3:194-200, (1987)
7. Court L.A., Courant D. and Dormont D.
Biological effects of coherent and non coherent IR radiation
in Optical sources, lasers and synchrotron radiation- Biological coherent and hazard potential.(this book)

8. Iversen Palle B. and Jörgen W. Paulsen
Conventional versus laser tonsillectomy in adults
Laser in medicine and surgery 1:35-37, (1986)
9. Seipp W., Haina D., Seipp V. and Waidelich W.
Is laser therapy suitable for removal or tattoos ?
Laser in medicine and surgery 4:188-195 (1986)
10. Bruhat M.A., Mage G., Canis M., Pouly J.L., Manhes H. and Wattiez A.
CO_2 laser laparascopy : a ten years experience.
Laser in medicine and surgery 4:149-151, (1988)
11. Beck O.J., Franck F. and Schönberger J.L
Clinical evaluation of the 1,320 nm Nd:Yag laser in neurosurgery
Laser in medicine and surgery 3:110-113, (1987)
12. Hofstetter A.,
Neodymium Yag laser treatment of bladder tumors
Laser in medicine and surgery 2:148-150, (1987)
13. Reed R.D.,
Low power laser alterations of physiological processes
in Report SAM-TR-79-29 USAF School of aerospace medicine,Brooks,
Air force base,Tx 78235, (1979)
14. Mester E.,
The stimulating effect of low power laser rays on biologicalsystems
Biol. Eng. 8:2-430, (1970)
15. Boulton M. and Marshall J.,
HeNe laser stimulation of human fibroblast proliferation and
attachment in vitro
Lasers in the life sciences 1 (2):125-134, (1986)
16. Fava G., Marchesini R., Melloni E.,Milani M. and Schiroli A.
Effect of low energy irradiation by HeNe laser on mitosis rate of
HT-29 tumor cells in culture
Laser in medicine and surgery 1(2):133-141, (1986)
17. Yew D.T., Lam S.T.L. and Chan Y.W.
Effect of low dose laser on the chorioallantoic culture of retinal
pigment cells
Acta radiologica oncology, 21,6:433-438, (1982)
18. Strube D., Haina D., Landthaler M., Braun-Falco O. and Waidelich W.
Stimulation of wound healing. Disturbing factors in animal experiments
Laser in medicine and surgery 4:15-20, (1988)
19. Haina D., Spahn J., Landthaler M., Braun-Falco O and Waidelich W.
Temperature of the skin during application of softlaser
Laser in medicine and surgery 1:26-29, (1988)
20. Iwase T., Hori T. and Morioca T.
Possible mechanisms of the HeNe laser effects on the cell membrane
characteristics
Laser in medicine and surgery 4:166-171, (1988)

CURRENT AND FUTURE APPLICATIONS OF LASERS IN MEDICINE

M.H. Repacholi, G. McLennan, *A. Pugatschew and R. Hancock

Royal Adelaide Hospital, Adelaide, South Australia

*School of Applied Physics, South Australian

Institute of Technology, Adelaide, South Australia

INTRODUCTION

Using Bohr's theory of optical resonators, Einstein (1917) predicted that atoms could be stimulated to emit light. But it was not until 40 years later that Towns and Schawlow (1958) conceived of the basic principles of operation of the maser (microwave amplification by stimulated emission of radiation) that Maiman (1960) was able to construct a device to emit radiation in the visible region of the spectrum. Within one year of the visible laser being developed, the first medical application was reported in ophthalmology for treatment of an eye tumour (Meyer-Schwickerath, 1961). Since this time application of lasers in medicine has progressed rapidly.

It is the purpose of this review paper to give a brief introduction to lasers in medicine, identify some of the key lasers and their applications, include a summary of the different laser delivery systems and conclude with descriptions of a few promising future applications. Although the speciality most associated with laser use is ophthalmology, this review will concentrate mainly on surgical applications.

Table 1 gives a brief chronology of events leading to early use of lasers in medicine. Since these early days laser application in medicine has diversified and challenged conventional techniques. It is now clear that, with the tremendous advances occurring in laser research, lasers in medicine are at the stage of development that computers were for business some 20 years ago. Further, it is apparent that they have the same potential to transform many clinical specialties in terms of treatment modalities available to patients.

Table 2 provides some examples of lasers currently used in medicine or that are being researched for possible clinical application.

Table 1. Early History of Laser Development in Medicine

Year	Scientist (Ref)	Discovery
1917	Einstein (1917)	Theory enumerating concepts of simulated and spontaneous emission
1955	Gordon et al (1955)	Stimulated emission of microwaves
1958	Schawlow and Townes (1958)	Described physical principles of maser
1960	Maiman (1960)	Observed stimulated emission in ruby (ruby laser)
1961	Javan et al (1961)	Developed first gas laser (HeNe)
1961	Johnson (1961)	Nd:YAG laser developed
1961	Meyer-Schwickerath (1961)	Treatment of eye tumor (ruby laser)
1962	Bennett et al (1962)	Argon laser developed
1963	Campbell et al (1963)	Ruby laser used as retinal photocoagulator
1964	Patel (1964)	Developed CO_2 laser
1965	Zweng and Flocks (1965)	Expanded uses of lasers in ophthalmology
1965	Stellar (1965)	CO_2 laser in neurosurgery
1966	Yahr & Strully (1966)	CO_2 laser cutting of dog tissue
1967	Goldman and Blaney (1967)	Treated malanomas with ruby Laser

Table 2. Examples of Lasers used in Medicine

Type	Wavelength (micrometres)	Application
Carbon Dioxide	10.6	General surgery, gynae., ophthal., ENT, dermatol., thoracic etc.
Neodymium-YAG	1.06	Urology, ophthal., dermatol., gynae., general surgery, angioplasty, any endoscopic application.
Gallium Arsenide	0.840	Biostimulation
Ruby	0.694	Ophthalmology
Helium Neon	0.633	Aiming beams, wound healing, biostimulation, acupuncture.
Gold vapour	0.633	Photoradiation therapy (PRT).
Krypton	0.647 0.568 0.530	Ophthalmology
Copper vapour	0.578 0.510	Ophthalmology, dermatology
Argon	0.514	Ophthalmology, angioplasty, endoscopy
Excimer		
Xe-F	0.351	Dermatology
Ar-F	0.193	Ophthalmology
Dye Lasers	0.35-1.0	Dermatology, PRT, ophthalmology

There are many reasons why such intensive research and development is being conducted on lasers. Lasers offer treatments for diseases or conditions where no previous therapies existed. Certain lasers can be transmitted into natural body cavities or through small incisions, and are able to be absorbed in tissue components to produce excellent cutting or coagulation so that the following benefits are derived:

Patients

Due to the characteristics of the laser there is minimal peripheral tissue damage, with less inflammation possibly resulting in reduced postoperative pain, more rapid healing and less scar tissue. In most cases patients spend less time in hospital and less time off work, lowering the cost to them for the procedure.

Clinicians

Lasers permit surgery in many instances not otherwise possible. Improved patient care using non-touch, sterile techniques that result in a reduced risk of infection and blood loss, due to the sealing of small blood vessels and lymphatics during surgical procedures. The capability of transmitting laser energy down flexible fibre optic channels enables access to difficult sites within the body.

Hospitals

Laser treatments involve advanced and developing technology that can be very cost-effective, resulting in reduced patient stays in hospital and so shorter lists of patients waiting for treatment. Overall, laser surgery appears to reduce morbidity and mortality. Higher patient turnover could be expected as patients with "marginal" conditions would be attracted to non traumatic laser techniques.

LASER INTERACTIONS WITH TISSUE

Offering the unique characteristic of monochromaticity, and the ensuing spatial and temporal coherence of emission, lasers are able to produce a range of biological responses in tissue determined by the various processes of energy conversion within biomolecules. Laser parameters important in determining it's interaction with tissue are wavelength, power density and pulse duration. Figure 1 (adopted from Boulnois, 1986) indicates the four basic types of tissue interaction that can occur: Electromechanical, photoablative, thermal and photochemical.

Electromechanical (or photomechanical) - This interaction requires extremely high power density (10^9 - 10^{12} W/cm^2) beams in extremely short pulses (ns) to deliver fluences of about 100 J/cm^2 and very high electric fields (10^6 - 10^7 V/cm) which are comparable to the average atomic or intermolecular electric field. This induces dielectric breakdown in tissue resulting in a microplasma or ionized volume with a very large number of electrons. A localised mechanical rupture of tissue occurs due to the shock wave associated with the plasma expansion. This may be seen as a local plume of tissue at the point of interaction.

Such interactions are currently being investigated for laser assisted shock-wave lithotripsy for calculi in the biliary tree. Techniques involving this mechanism are routinely used in ophthalmology for photodisruption of the posterior capsule when the posterior membrane becomes opaque.

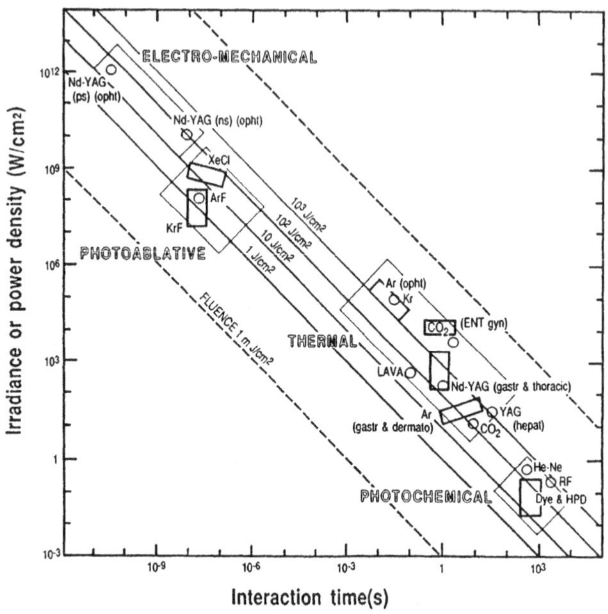

Figure 1. Four basic types of laser-tissue interactions (from Boulnois 1986).

Photoablation - This involves photodissociation or direct breaking of intramolecular bonds in biopolymers, caused by absorption of incident photons and subsequent release of biological material. Molecules of collagen for example may dissociate by absorption of single photons in the energy range 5 - 7 eV. Excimer lasers at several ultraviolet wavelengths (ArF 193nm/6.4eV; KrF 248 nm/5eV; XeCl 308 nm/4eV) with nanosecond pulses focused on tissue at power densities of about 10^8 W/cm^2 can produce this photoablative effect.

Ultraviolet radiation is extremely strongly absorbed by biomolecules and so absorption depths are small, of the order of a few micrometres. These interactions offer selective advantages in medicine and are currently being investigated in ophthalmology for radial keratotomy (sculpturing the cornea as an alternative to eye glasses) and orthopaedic surgery for treatment of bone surface.

Thermal - The thermal mechanism involves conversion of laser energy into heat. With the laser's ability to be focused to points of a few micrometres or millimetres in diameter, high power densities can be spatially confined to heat target tissues. Depth of penetration into the tissue varies with wavelength of the incident radiation, determining the amount of tissue removal and bleeding control (haemostasis).

The photothermal process occurs first with the absorption of photon energy producing a vibrational excited state in molecules and then inelastic scattering with neighbouring molecules increasing their kinetic energy, resulting in a temperature rise. Under normal conditions the kinetic energy per molecule (kT) is about 0.025eV. Thermal lasers commonly used in surgery have corresponding photon energies much larger (CO_2 = 10600nm, e = 0.12 eV; Nd:YAG = 1064nm, e = 1.17 eV; Ar = 514nm, e = 2.4 eV).

Heating effects are largely controlled by molecular target absorption such as free water, haemoproteins, melanin, and other macromolecules such as nucleic acids. Figure 2 shows the variation in absorption with wavelength for water, haemoglobin (HbO_2), melanin and haematoporphyrin derivative (HPD). The wavelength of

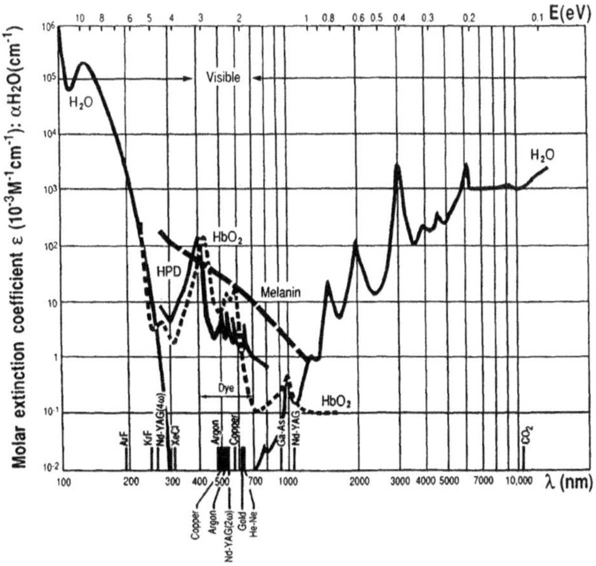

Figure 2. Absorption of various tissue components (water, haemoglobin (HbO_2) and melanin) and haematoporphorin derivatives (HPD) - after Boulnois (1986).

some medical lasers is shown to identify their major absorbing components in tissue. The figure (adapted from Boulnois 1986) indicates particularly the very strong absorption of the 10600 nm wavelength of the CO_2 laser and the high absorption in melanin and haemoglobin of the argon laser. Knowing the distinctive absorption characteristics of a given wavelength can lead to selective advantages of a laser in medicine.

Tissue heating produces different effects depending on the temperature. Table 3 shows these effects.

Table 3. Temperature Effects on Tissue/Cells

Temperature °C	Tissue Effect
< 43	No effect
43 - 45	Cell mortality if sustained Conformational changes in cells
50	Reduced enzyme activity
52	Edema, enzyme changes
60	Protein denaturation, coagulation
80	Cell membranes become permeable Collagen denatures, tissue carbonization
> 100	Vaporization and ablation

Photochemical - This interaction involves uptake by target cells of chemicals (chromophores) capable of causing light-induced reactions in molecules to produce a cytotoxic effect. If chromophores, sensitive to a narrow wavelength band, are introduced and selectively retained in cancer cells at specific sites, subsequent exposure to laser energy in this waveband, especially if delivered to deep target cells by penetrating wavelengths, can offer therapeutic advantage. PUVA (Psoralen Ultra-Violet-A) therapy has long been used in the treatment of psoriasis. In this case, the drug psoralen is taken by the patient and subsequent exposure to a UVA lamp produces a photochemical modification of nucleic acids in rapidly dividing cells to reduce the rate of cell turnover.

A similar principle has been used to treat malignant tumours where haematoporphyrin derivative (HPD) is taken up predominantly in rapidly dividing cells. Exposure to laser energy at the appropriate wavelength produces resonant excitation in the electronic states of the HPD molecules, causing them to undergo a series of decays which result in intramolecular transfer reactions. This culminates in the release of highly reactive cytotoxic species, causing irreversible oxidation of some essential cellular components, and destroying the affected host tissue. Readers are referred to Boulnois (1986) for an excellent description of events in this photodynamic reaction.

Another "photochemical reaction" which is subject to some controversy is the low intensity laser (up to about 5 mW) and the suggestion that it produces a stimulatory effect in tissue. Biostimulation lasers normally operate in the red (HeNe) or infrared (GaAs) region with energy densities in the range 0.5 to 10 J/cm^2, and are reported to stimulate DNA synthesis, activate enzyme - substrate complexes, transform prostoglandins and produce microcirculatory effects. One of the leading investigators of the biological effects of biostimulation lasers has been Karu (1989). She has concluded that:

a) Laser biostimulation is a photobiological phenomenon. Both incoherent and coherent red light seem equally effective for stimulatory treatment. Irradiation in the red and far red regions can enhance metabolic processes in the cell and activate proliferation.

b) Primary photoacceptors are components of the respiratory chain. Absorption of light by the respiratory chain components (flavine, dehydrogenases, cytochromes and cytochrome oxidase) cause a short-term activation of the respiratory chain and oxidation of the NADH pool. This leads to changes in the redox status of mitochondria and cytoplasm. An increase in intracellular hydrogen ion and ATP concentrations in eukaryotic cells causes activation of membrane ion carriers as well as Na^+, K^+ and ATPase, which influence cell proliferation. This explains the universality of low power laser effects.

c) The respiratory chain components can be the photoacceptors in the case of cellular metabolism stimulation, as well as inhibition depending on the dose of light. At low doses, irradiation causes redox regulation of cellular metabolism; at high doses photodynamic damage prevails.

d) Light quantum is only a trigger for cellular metabolism regulation. This explains the low doses and intensities needed.

e) The magnitude of the biostimulation effect depends on the physiological state of the cell before irradiation. This explains why the biostimulation effect is not always possible.

f) The therapeutic effects of low power laser irradiation can be explained by an increase of proliferation of G_0 and G_1 cells or by changes in the physiological activity of excitable cells.

One can conclude from this general review of mechanisms of laser-tissue interactions that greatly different effects can be produced with the same laser by varying the power density, using continuous or pulsed beams and altering the pulse duration. Once the beam characteristics have been established, clinical effects can be varied using different beam delivery systems.

DELIVERY SYSTEMS

The lasers commonly used in medicine today are shown in figure 3. Each has a wavelength producing unique absorption characteristics in tissue components.

The wavelength determines the type of delivery system which can be used. Three types of delivery systems are in common use today:

. Direct Delivery
. Articulated Arm
. Fibre Optic

Direct Delivery

With the advent of the laser diode and small gas lasers such as the HeNe, biostimulation lasers, laser pointers and similar devices have been developed typically with power output limited to a level where no mandatory requirement for a safety shutter to interrupt or control the laser beam is required. Laser energy is delivered directly from the emitting aperture to tissue. Output may be controlled by switching on or off via a press button or timer. Users should be aware of the power output and the possible need for safety glasses.

Articulated Arm

Since the carbon dioxide laser is absorbed by glass, it cannot be delivered through a glass fibre or make use of conventional glass lenses to focus the beam. A special articulated arm has been developed as shown in figure 4.

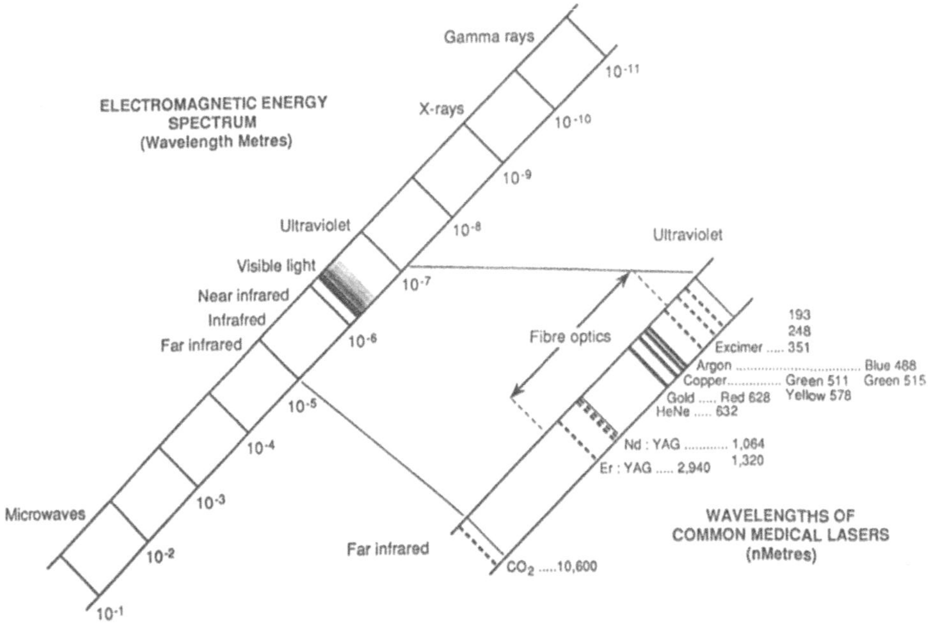

Figure 3. Electromagnetic spectrum indicating laser wavelengths used in medicine and those that can be transmitted through glass fibres.

Because radiation from the carbon dixoide laser is invisible (far infrared) a low power, visible, (typically HeNe) laser is used to designate target tissue and indicate spot size. Both the carbon dioxide and HeNe lasers are optically combined to be coincident at the laser source and propogate through the hollow arm, reflecting from special front surfaced mirrors at each joint to emerge at the distal end as a coincident, collimated beam. The articulated arm may couple to surgical accessories such as a handpiece, micromanipulator (microscope attachment), rigid fibre delivery system, waveguide or rigid endoscope. A lens in the accessory may focus the beam.

The micromanipulator and endoscopes both make use of a joystick to control a mirror and direct laser energy to the target tissue.

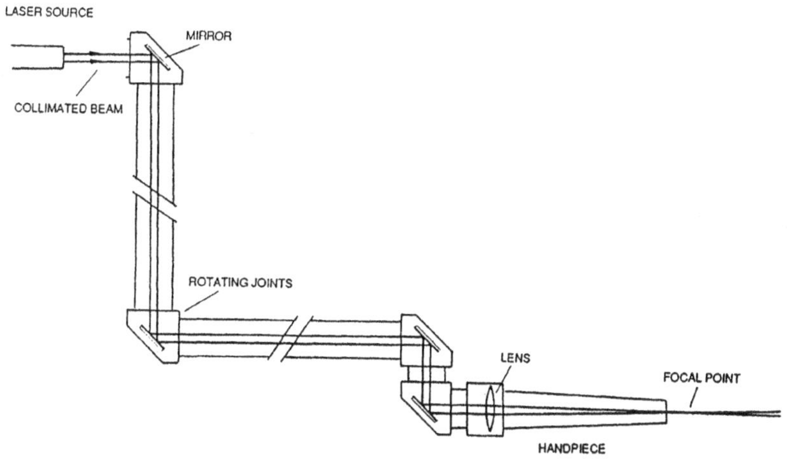

Figure 4. Articulated arm laser delivery system.

Limitations of an articulated arm include:

a) Laser energy delivery is restricted to "line of sight" or via straight segments of a delivery path.
b) The arm is prone to shock (mechanical bumping), so the aiming beam and CO_2 beam must always be checked for coincidence and spot shape before and during use.
c) Manufacturers sterilization procedures must be rigidly observed, otherwise expensive lens/mirror coatings may be damaged.
d) If handled, grease from the hands can damage the optics of mirrors or lenses.
e) The articulated arm transmits a collimated beam which is potentially hazardous, particularly if a handpiece or lens is not fitted. The collimated beam diameter changes little over distances of some metres and the power density may be high enough to cause injury.
f) The articulated arm should be safely stored both for transport and when not in use. Dust must be kept from the optics.

Fibre Optic

Fibre optic delivery systems have been developed in the communications industry. One fibre can simultaneously carry thousands of telephone channels interference free.

In medicine the flexible endoscope makes use of this fibre optic technology for both vision and light source input. These same fibres can be used to deliver a number of different laser wavelengths (refer figure 3). Argon and Nd:YAG lasers commonly use this technology in their delivery system.

As shown in figure 5, laser energy is focused by a lens into a glass fibre and is propagated through it by internal reflection off the fibre-cladding interface to emerge at the distal end as a divergent beam.

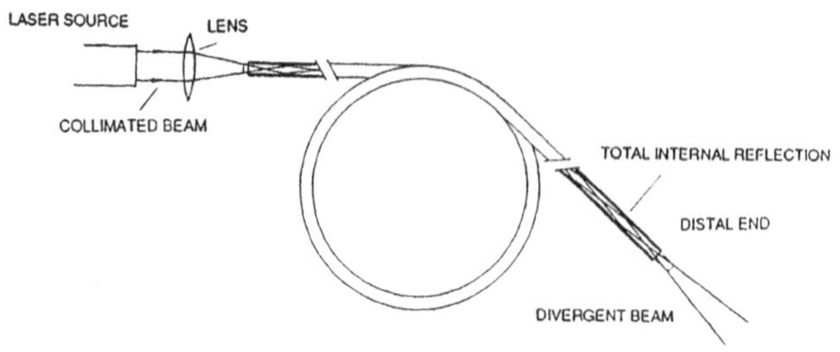

Figure 5. Fibre optic delivery system.

In the case of the near infrared Nd:YAG laser a HeNe aiming beam is combined with the Nd:YAG beam at the source to produce a coincident beam. The Argon laser is visible and can be heavily attenuated to provide a visible aiming beam for the surgeon, before the higher power surgical beam is used for treatment.

Typically the fibre delivery system is used in conjunction with a rigid or flexible endoscope. A special tip (as shown in figure 6) is often used to provide insufflation (flow of gas or fluid) to the surgical site which cools the end of the delivery system, clears the surgical area and reduces the risk of fibre tip contamination. This system can only be used in a non-contact mode. The Argon laser fibre can be used directly onto tissue without tips.

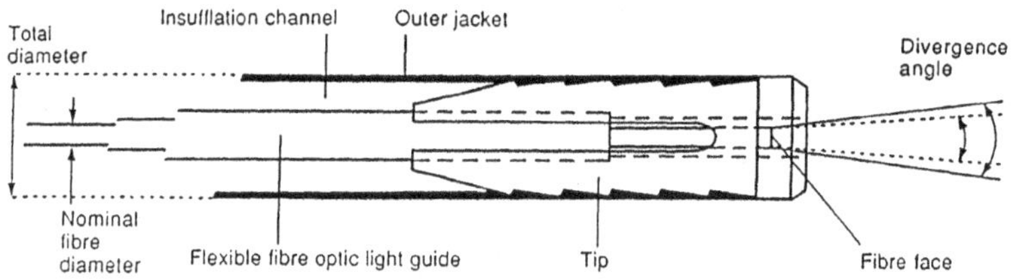

Figure 6. Fibre tip designed for non-contact surgery.

Sapphire Tips

In an attempt to improve the cutting characteristics, control depth of penetration and allow tissue contact, artificial sapphire tips have been developed for medical lasers, predominantly for the Nd:YAG. These tips act as a lens on the end of the fibre delivery system allowing a reduction in laser energy required (maximum typically 25-30 watts). They improve cutting characteristics by modifying the energy distribution to tissue, and allowing greater control of the depth of penetration of the Nd:YAG laser, as shown in figure 7. They also show advantages in coagulation as they allow direct contact with tissue.

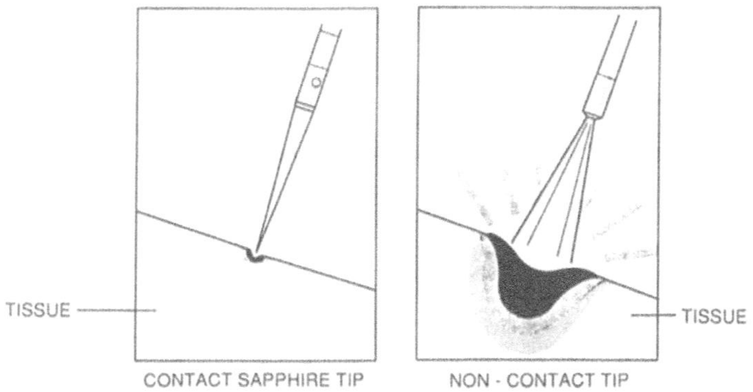

Figure 7. Distribution of heating tissue using the direct beam and sapphire tip.

Figure 8 shows the simple coupling that can be used to screw on different shape sapphire tips.

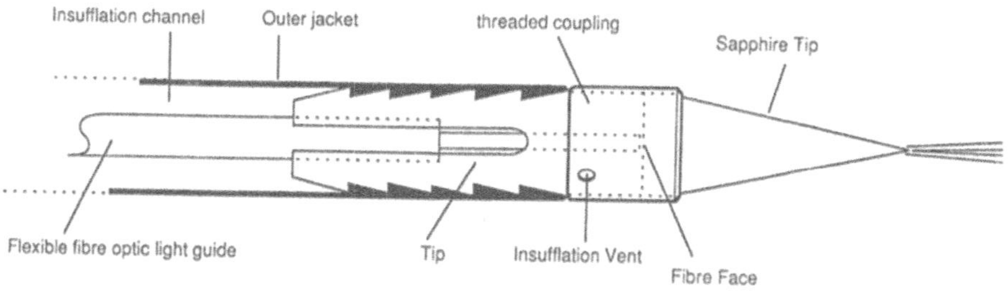

Figure 8. Special fibre coupling for different shape sapphire tips.

Metallic/Ceramic Tips

Laser energy can be used to heat a metallic or ceramic tip (as shown in figure 9) to a temperature where the tip can be used as a reconnulisation probe. A guide wire to allow the tip to track the lumen of a blood vessel may be provided. This technology is still under development.

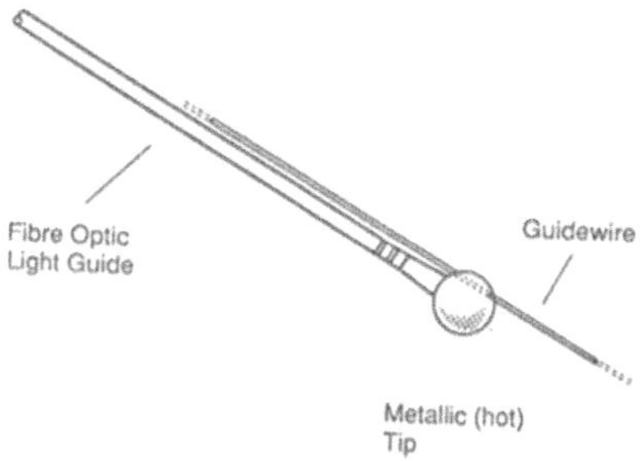

Figure 9. Metal tip with guide wire

Diffusers/Photodynamic Therapy Probes

In order to take advantage of the photochemical properties of certain laser wavelengths a fibre optic delivery system with a diffuser fitted to the distal end has been developed as shown in figure 10. The diffuser shape determines the energy distribution to target tissue.

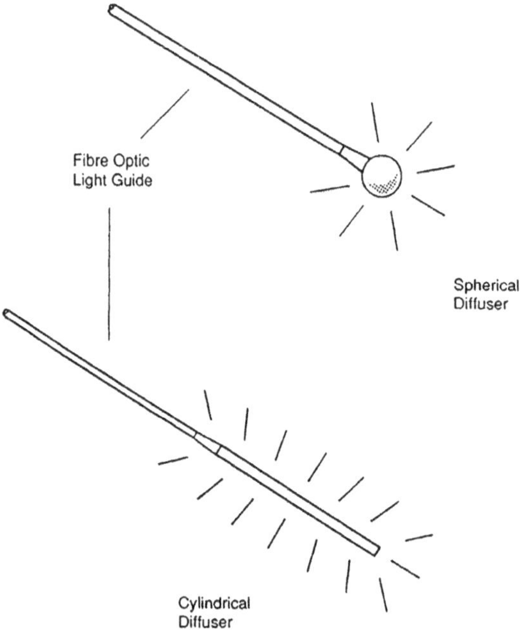

Figure 10. Photodynamic therapy probes.

Limitations with fibre delivery systems include:

a) Non contact fibre tips are prone to contamination. If contaminated the tip may overheat and separate from the fibre. The first indication of this is deterioration of the aiming beam shape.
b) The fibre needs to be inspected along its entire length for damage before use. Crimps in the fibre can lead to breakage and the laser beam being transmitted uncontrolled into the theatre.
c) The end of the fibre needs to be visualised through the endoscope before firing the laser, otherwise the laser may be fired while inside the scope causing severe damage to it.
d) Sapphire tips, metal tips or the laser fibre should not be overdriven with power levels higher than recommended by the manufacturer. Fibre destruction or tip separation can result. Typically small diameter fibres handle less power than larger fibres.
e) In the case of the Nd:YAG laser, the aiming beam and invisible infrared beam diverge at different angles from the tip of the non-contact delivery system. (See figure 6).
f) The manufacturers recommended sterilization procedures need to be followed for the fibre delivery system and tips.
g) Glass fibres have limitations on flexibility, which need to be appreciated. Some endoscope deflection mechanisms not designed for laser work can break the fibre inside the scope.

h) The metal coupling to the laser head may cause a serious burn if caution is not used when disconnecting the fibre delivery system.
i) The endoscope to be used with the laser must be appropriate, remembering the viewing optics offers a path for laser "flash back" for both the Nd:YAG and Argon laser. A suitable filter, safety glass, safety shutter or television system should be used to protect the users eyes. The Argon laser has the potential to damage some video systems so a safety shutter may be needed.

Users must understand the characteristics and limitations of the laser delivery system to obtain optimum performance. Any change of laser effect on tissue for a given power output should be questioned. Safety is paramount.

Other Developments

Smart Lasers

One of the current limitations with laser treatment is knowing the exact nature of the target tissue and what adjacent organs or blood vessels may be damaged if the laser is used.

The Smart Laser (shown in figure 11) makes use of a fibre bundle, where target tissue is irradiated with a low power diagnostic laser through one or more fibres. Reflected energy is analysed spectroscopically from another fibre to determine the nature of the target tissue. Once the target is identified the therapeutic laser is used to treat through a third fibre. Such a device offers great potential in laser angioplasty for the removal of plaque in blood vessels.

Endoscopes

New endoscopic developments which can only enhance the use of the surgical laser include integration of small high resolution television cameras into the endoscope tip and the use of ultrasonic transducers at the tip to determine the nature of the surrounding body organs/tissue.

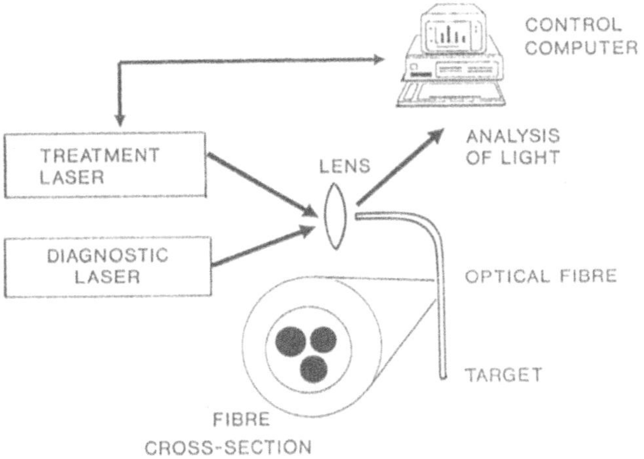

Figure 11. Schematic diagram of a Smart Laser

CURRENT APPLICATIONS

Surgical techniques using scalpels and other cutting instruments have changed little since the days of the "barber-surgeon". These techniques involve removal, repair or replacement of a diseased or damaged organ (or part thereof) and make use of sharp or blunt instruments and bleeding control by ligature or tourniquet.

Major advances in surgery have been the development of anaesthetic techniques, control and treatment of infections and post operative care improvements. Advances in the surgery itself have been relatively minor and include such things as electrocautery and stapling guns in lieu of sutures.

Problems which have not been overcome (or overcome with difficulty), in terms of surgical techniques include:

1. Surgery to sites of difficult access (ie, central trachea).

2. Surgery to very delicate tissues (ie, retina).

3. Surgery to very vascular areas (ie, liver).

4. Tumour surgery where
   (i) there are multiple areas of disease (ie, bladder cancer).
   (ii) the tumour has an ill defined tissue invasion (ie, carcinoma of the lung).

Lasers offer the potential and in many instances, a practical solution to these problem areas. It MUST be remembered that lasers are relatively new in medicine/surgery and that their use in many areas is not generally accepted. However, this is changing as the cost of lasers reduce, and as appropriate staff become more familiar with them. In addition, delivery systems have improved recently allowing easier use of the laser.

Using a laser is very different to using a scalpel or diathermy, a new set of skills is required. This may be likened to an experienced trumpet player being asked to play the piano - the music is similar but the instrument is quite different and therefore a period of retraining and practice is required.

The most common lasers used in surgery today are the carbon dioxide, argon and Nd:YAG lasers. The carbon dioxide laser with a wavelength of 10600nm is heavily absorbed in water and so is very good for precise cutting and vapourisation of tissue. It is the most widely used surgical laser for treatment of superficial lesions or removing small volumes of tissue.

The argon laser with a wavelengths of between 458 - 515nm is heavily absorbed in haemaglobin and thus is especially useful for non bleeding vascular lesions requiring a minimum depth of penetration of about 1 mm. While heavily absorbed by blood, the argon laser energy is readily transmitted through water, gastric fluids and urine.

The Nd:YAG with a wavelength of 1064nm is specifically absorbed in tissue protein and has a lower absorption in water and haemoglobin than either the carbon dioxide or argon lasers. The depth of thermal injury from this laser is in excess of 3 mm in most tissues, this being very useful for coagulation of large volumes of tissue such as tumour masses, with consequent effective haemostasis during ablative surgical procedures.

In general the current clinical applications of lasers can be divided into two main areas. The first deals with conventional incision to provide surgical access for laser techniques and the second deals with the rapidly growing area of endoscopic access for laser surgical techniques. The latter provides minimumly invasive techniques for surgery.

Fibre delivery systems which allow laser beams to be transmitted into the body through natural body cavities has meant remarkable advances in surgery. With endoscopic devices providing excellent visualisation of tumour tissue or other luminal obstructions and the use of operative channels for the introduction of the optical glass fibres and subsequent treatment by lasers has meant dramatic improvements in the care of patients. One improvement that is intending to benefit the health care system remarkably is the conversion of many conventional procedures requiring long stays in hospital into day surgery procedures. One of the recently developed clinical applications which has the potential to alter current management markedly is the application of Nd:YAG for endometrial ablation as an alternative to conventional hysterectomy. Endoscopic laser treatments have now expanded to such an extent that they have become well established in many surgical specialties.

Gastroenterology

A summary of the available literature indicates that the Nd:YAG laser used down an endoscope can stop acute gastrointestinal bleeding in 80 - 90% of patients with a re bleeding rate of 10% for ulcers and 30% for gastrointestinal haemorrhage. In controlled trials it has been found that Nd:YAG laser therapy has a role in the management of acute gastrointestinal haemorrhage, and should be available as one method to be used. In particular, in patients with actively bleeding vessels in an ulcer crater, or non-bleeding visible vessels within an ulcer crater. Lasers are not of any particular value in bleeding oesophageal varices or diffuse gastritis bleeding. Although relatively rare causes of gastrointestinal bleeding, vascular malformations in the gut are easily and effectively ablated with laser techniques, with perhaps the Argon laser being slightly more effective and safer than the Nd:YAG laser; and both modalities compare favourably with alternatives such as electrocautery probes.

Intestinal polyps can easily and successfully be ablated with Argon laser therapy.

Treatment of gastrointestinal malignancy has been a very useful application of lasers. The bowel can be cleared by using lasers if larger malignant tumours are present, and the laser procedure is immediately effective with minimal risks or post operative problems, compared to conventional radiotherapy and/or surgery; in addition, laser treatment can be repeated regularly if required, frequently under local anaesthesia as day or outpatient procedures. Clinically, improvement occurs in approximately 90% of patients lasting from three to six months. The Nd:YAG laser can be used for these debulking procedures. New laser treatments are also being applied not only to obstructing bowel lesions but also to biliary duct tumours.

There have been case reports of opening a pancreatic pseudocyst into the stomach using an endoscopically directed laser, and fragmenting of bile duct stones using fibreoptics and lasers has also been achieved.

Ear, Nose & Throat

Ear applications including stapedectomy (Argon laser) and acoustic neuroma resection (Argon laser) are only being performed in experimental situations at this stage, and advantages over conventional methods have yet to be demonstrated, although theoretically advantages occur.

Nasal surgery includes a variety of potential applications including rhinophyma removal using carbon dioxide laser.

Tumours and capillary angiomas of the mouth, pharynx and larynx are treatable with lasers, either carbon dioxide or argon.

Whilst routine tonxillectomies can be performed using the carbon dioxide laser or Nd:YAG laser with contact sapphire probes, advantages of laser procedures over conventional methods have yet to be identified, especially in view of the relative ease of usual surgical methods.

For ENT applications, the Argon and carbon dioxide lasers can be utilized down the operating microscope and the Argon and Nd:YAG lasers down fibreoptic fibres via endoscopes.

Dermatology and Plastic Surgery

Port wine stains can be effectively treated with the Argon laser but these should probably not be treated in children. Other pigmented skin lesions can be successfully treated using lasers, as well as tattoo removal. Newer lasers are being evaluated including the copper vapour and copper bromide lasers - this may produce more selective photo thermolysis of blood vessels than the Argon laser, but this requires confirmation.

Gynaecology

There is a major and common usage of lasers in gynaecology practice, for treatment of lesions of the lower genital tract.

Cervical intra-epithelial neoplasia (CIN) occurs frequently in the younger population, ages 20-30; usual treatment has been conization of the cervix - this having several problems.

Laser treatment, usually carbon dioxide, has advantages, these being;

1. able to be performed as outpatient.
2. only local anaesthesia required.
3. increased speed with less bleeding.

Human papilloma virus infections with condylomata acuminata are treated better by laser than other therapies; and the carbon dioxide laser is generally considered the treatment of choice. There may be a question regarding the transmission of HPV to the operator by the smoke plume. Vaginal intra-epithelial neoplasia (VAIN) is also well treated using the carbon dioxide laser.

Laparoscopic laser applications have been developed for the last 10 years; and has been successful for endometriosis, benign ovarian cysts etc. Long term follow up of treated patients is yet to be reported. Recently, endometrial ablation has been successfully introduced for post menopausal bleeding using the Nd:YAG laser.

Urology

Whilst urethra strictures can be treated with lasers, there appears no advantage over simple dilatation in either short term or long term results; the same is the case also with bladder neck obstruction from benign prostatic hypertrophy.

The main area of laser treatment in urology is in the treatment of superficial and invasive bladder cancer. For superficial bladder cancer, it is likely that the Nd:YAG laser treatment gives similar results to routine electrocautery treatment. Large controlled trials are underway to assess whether there is less local recurrence after Nd:YAG laser vs. electrocautery as early studies seem to suggest. For invasive bladder cancer, Nd:YAG laser treatment offers advantages over routine surgical resection and extensive electrocautery. The technique allows for tumour debulking with relative ease and may save patient radical bladder resection.

Carcinoma involving ureters is easily treated with the Nd:YAG laser with very good early results; and long term follow up studies are currently being performed. Nephrectomy can also be performed using the Nd:YAG laser in combination with an ultrasonic surgical aspirator. However, this approach is not ideal. Currently Nd:YAG laser application using sapphire tip technology is being evaluated for partial nephrectomy.

Neurosurgery

Brain tumours (malignant astrocytoma) can be surgically treated using carbon dioxide laser, but has no survival advantages over regular surgery. Follow up treatment with photosensitizers (ie, HPD) may extend long term remission and is being evaluated.

In general, tumours that are very fibrous or very vascular may be operated upon with laser, without mechanical disruption of normal brain and therefore lasers have a definite role in neurosurgery. However, this may be to only a limited patient group. This is also true for spinal cord tumours especially intra-medullary tumours where electrical and mechanical stimulation is kept to a minimum. Brain tumour surgery in children has especially benefited from the use of lasers; particularly the carbon dioxide laser for microsurgery applications.

Cardio-Vascular Surgery

All of the laser use in this specialty is currently still investigational. Uses include laser endarterectomy of coronary arteries (which may or may not be in conjunction with angioscopy), and myocardial incisions for treatment of arrythmias.

The development of the so-called "smart laser" for vascular work may lead to a reasonable technique for atheroma removal. However, there are only a few centres experimenting with these techniques and it is unlikely to be fully evaluated for a few more years. In the meantime, other non laser methods of atheroma removal are being evaluated and appear to have equal promise at this early stage.

Thoracic Medicine

The Nd:YAG laser can be used effectively for debulking large intra-luminal bronchial or tracheal tumours. This procedure improves survival. Importantly, this procedure cannot be performed by other methods in the majority of patients, without putting these patients at undue risk. Secondary tumours of the airway can also be successfully debulked using the Nd:YAG laser. Benign strictures of the airway can also be effectively treated.

Photodynamic therapy using haematoporphyrin derivatives excited with a laser beam is also used to advantage for the treatment of carcinoma-in-situ in the airway and is a curative procedure in those patients with this tumour characteristic, identified at an early stage.

Photodynamic Therapy

This treatment has already been described. At this stage the therapeutic ratio (ie, cancer cells killed vs. normal cells killed) is not very great (about 3:1), the light penetration is not very deep (about 5-7mm), and the patient develops generalised photosensitivity for several weeks.

However, HPD treatment is useful for some applications. The problems are being overcome by using newer photosensitizers (ie, Photofrin II etc.), which allow a greater therapeutic ratio and less toxicity, and by using light delivery systems to put light into tumours. Results are still being evaluated.

Conclusions

According to Dixon (1988), certain observations can be made about current laser surgical procedures.

1. For general surgery, laser applications have achieved limited but defensible indications for certain selected procedures.

2. Endoscopic laser surgery techniques have impacted on conventional incisional procedures to the point that in endoscopic surgery future general surgeons will need skills in at least upper GI endoscopy, colonoscopy and choledochoscopy to utilise minimumly invasive laser techniques.

3. Photodynamic therapy techniques open up very interesting possibilities for primary management or adjunctive treatments in association with conventional surgery.

We agree with these conclusions. In addition, it is our view that laser techniques and technology are at a very interesting stage of development. With a large number of promising applications already demonstrated. It is imperative that laser techniques are reviewed critically by the "main stream" surgical and physician specialties, and that serious evaluation of various procedures be undertaken to demonstrate if advantages for lasers exist or not. There is a tendancy for laser procedures to be diminished by some groups. However, there is little doubt that the defensible applications of lasers will increase rapidly over the next 5-10 years.

FUTURE OF LASERS IN MEDICINE AND SURGERY

There is an established role for laser treatments in ophthalmology; in gastroenterology, thoracic medicine, urology, ENT, the treatment of some malignancies and some benign conditions.

In all areas of surgery, lasers are being evaluated in clinical trials for non-tumour surgery with the likelihood of advantages in some conditions over conventional surgical methods. Such areas as endometrial ablation using laser is becoming more widely accepted as more clinical trials appear; as is laser assisted angioplasty in large arteries. Other techniques such as cholecystectomy using Nd:YAG laser with sapphire tips via the laparoscope have been successfully performed in a couple of centres in 1989; with the early promise of significant advantages over usual surgery.

Lithotripsy (breaking renal or gall stones) using lasers is in an advanced clinical investigational stage with early promise of success.

New lasers or new models of old lasers and new delivery systems make the following prospects likely:

1. Lasers will become much cheaper and more reliable.
2. Lasers will have "bio feedback" mechanisms.
3. Lasers will be produced with "specific" tissue interactions.
4. Lasers will find genuine application in more areas.

If only a small proportion of the laser trials currently underway show a benefit to laser treatments in new areas, then lasers will become very common in medicine and surgery as the "treatment of choice".

A frequent question is "are lasers better than conventional methods?" In many instances one can say that laser treatment is probably as good as conventional methods, but it will be some time before adequate clinical trials are conducted to say that lasers offer a true advantage in many of the newer applications.

In examining this question, one must be aware of total costs ie, laser cost, plus patient time away from work and home etc, to gain a realistic cost-risk-benefit appraisal. It is very important that this type of evaluation is conducted quickly and properly to gain a true perspective of the role of lasers in medicine and surgery.

A few examples of promising laser applications likely to become established in the future are given below:

Laser Angioplasty

Lasers are used in some hospitals for angioplasty of the femoral arteries and veins. These use argon, Nd:YAG, excimer and dye lasers, usually in conjunction with a specialised tip which may be metal, or a combination of metal with window in the end to allow a directional laser beam or a sapphire tip. Open beams are not generally preferable because of the inability to steer the fibre through the artery without perforating the wall. Metal tips offer the possibility of heating to a given temperature and penetration through blockages in the vein. Problems are encountered in the penetration of calcified plaque in the vessels. The successful application of excimer laser to coronary artery disease has already been reported (Karsch, 1989).

Photodynamic Therapy

Investigations are currently under way to determine the efficacy of new photosensitising agents with preference for those agents which are photosensitive at a wavelength in the 700-800nm. This wavelength range provides maximal penetration into tissue. If such photo-sensitisers are found, treatment at greater depths into solid tissue masses would occur.

Anastomosis

The mechanism by which a laser can weld tissue is not completely understood, however, it is likely that the heat generated causes protein denaturation allowing the sealing of the wound's edges. This laser assisted microvascular anastomosis have the potential advantages of reducing operational time, less trauma and elimination of foreign body response to suture material. The tensile strength of laser welds apparently compare to conventional sutured anastomosis. Many carbon dioxide lasers offer a low power setting (less than 1W) to allow tissue welding to be conducted.

REFERENCES

Bennett, W.R., Faust, W.L., McFarlane, R.A. et al., (1962), Dissociative excitation transfer and optical maser oscillation in NeO_2 and ArO_2 rf discharges (letter) Physiol. Rev. 8: 470-473.

Boulnois, J.L., (1986), Photophysical processes in recent medical laser developments: A review. Lasers in Medical Science 1: 47-66.

Campbell, C.J., Rittler, M.C and Koester, C.J., (1963), The optical maser as a retinal coagulator : An evaluation. Trans. Amer. Acad. Ophthalmol. Otolaryngol. 67: 58.

Dixon, J.A., (1988), Whats new in general surgery: Current laser applications in general surgery. Anals of Surgery 207(4): 355-372.

Einstein, A., (1917), The quantum theory of radiation, Physikalische Zeitschrift 18:121-128.

Goldman, L., Blaney, D., (1967), Laser therapy of melanomas. Surg Gynecol Obstet 124:49.

Gordon, J.P., Zeigler, H.J and Townes, C.H., (1955), The maser - new type of amplifier, frequency standard and spectrometer. Physiol. Rev. 99: 1264-1274.

Javan, A., Bennett, W.R. and Herriott, D.R., (1961), Population inversion and continuous optical maser oscillation in a gas discharge containing a HeNe mixture. (letter) Physiol. Rev. 6: 106-110.

Johnson, L.F., (1961), Optical maser characteristics of rare-earth ions in cyrstals. J. App. Physiol. 34: 897-909.

Karsch, K.R., Haase, K.K., Mauser, M., Ickrath, O. et al., (1989), Percutaneous coronary excimer laser angioplasty: Initial clinical results. Lancet 8664; 647-650; Sept. 1989.

Karu, T., (1989), Photobiology of low-power laser effects, Health Physics 56(5): 691-704.

Maiman, T.H., (1960), Stimulated optical radiation in ruby. Nature 187:493.

Meyer - Schwickerath, G., (1961), The preservation of vision by treatment of intraocular tumors with light coagulation. Arch. Ophthalmol. 66: 458.

Patel, C.K.N., McFarlane, R.A. and Faust, W.L., (1964), Selective excitation through vibrational energy transfer and optical maser action in N_2 - CO_2. Physical. Rev. 13: 617-619.

Schawlow, A.L. Townes, C.H., (1958), Infrared and optical masers, Phys. Rev. 112:1940.

Stellar, S., (1965), Effects of laser energy on brain and nerve tissues. Laser Focus, August P 3.

Yahr, W.Z. and Strully, J.K., (1966), Blood vessel anastomosis by laser and other biomedical applications, JAAMI 1: 1.

Zweng, H.C., and Flocks, M., (1965), Clinical experiences with laser photocoagulation. Fed. Proc. 24(1) part 3: 5-65.

TOWARDS A SOLUTION OF THE OZONE LAYER DEPLETION PROBLEM

Branko F.M. Bosnjakovic

Radiation Protection Directorate
Ministry of Housing, Planning and Environment
Leidschendam, The Netherlands

INTRODUCTION

The sun is the most ubiquitous and most important source of UV radiation in our environment. The ozone layer is the region in the stratosphere where ozone makes up a greater proportion of the air than at any height in the earth's atmosphere. The presence of ozone is vital because it absorbs UV and prevents too much of it from reaching the earth's surface.

The potential effect on ozone of nitrogen components released by fleets of high altitude supersonic aircraft gave rise to concern in 1971. A few years later a paper by Rowland and Molina postulated that long - lived man - made compounds containing chlorine would be transported to the stratosphere, decomposed into free chlorine which would decrease the concentration of ozone. After setting up a Coordinating Committee on the Ozone Layer in 1977, UNEP (United Nations Environmental Programme) decided in 1981 to organise a Convention for the protection of the ozone layer.

In the meantime research has led to a wealth of information on atmospheric processes and on the possible effects of ozone depletion. A convention was signed in Vienna in 1985, and a protocol was signed in Montreal in 1987 under auspices of UNEP, aiming at substantial reduction of the production of ozone-layer depleting substances.

The present contribution attempts to give a brief overview of the relevant facts and policy developments, highlighting the way towards solving one of the threats to the global environment.

Much of the parts on ozone chemistry, global variations of atmospheric ozone and the Antarctic ozone, is based upon the comprehensive reports by the United Kingdom Stratospheric Ozone Review Group (1987, 1988). For more detailed information, the reader is also referred to the Proceedings of the 3rd US-Dutch International Symposium on "Atmospheric ozone research and its policy implications" (Ozone Symposium, 1988) .

OZONE IN THE EARTH'S ATMOSPHERE

The atomosphere's density decreases as we ascend, so that about half of its mass is below 5 km. Because of rapid vertical mixing, the temperature usually decreases with height in the troposphere, until a height of about 10 km where it starts to increase again. In the region above the troposphere, known as stratosphere, vertical mixing (convection) is suppressed and the temperature increases with altitude. The stratosphere extends to a height of about 50 km; only about 0,1% of the mass of the atmosphere is above that altitude. The source of heat in the stratosphere is the absorption of solar ultraviolet (UV) radiation by the ozone molecules present there.

The atmosphere has evolved into a mixture of 78% nitrogen and 21% oxygen. It contains many other gases with concentration from a fraction of a percent through the ppm (10^{-6}), ppb (10^{-9}) to the ppt (10^{-12}) ranges. The mean composition of the atmosphere is as follows:

Nitrogen	78%
Oxygen	21%
Noble gases	9300 ppm
Carbon dioxide	340 ppm
Methane	1,7 ppm
Hydrogen	0,5 ppm
Ozone	0,15 ppm
Carbon monoxide	0,09 ppm
CFC11	0,22 ppb
CFC12	0,36 ppb

The biological, geochemical and geophysical importance of the "trace" gases is incommensurable with their low concentrations. This is particularly true of ozone and a number of other trace gases ("source gases") which control the abundance of ozone. Ozone is produced and destroyed at a wide range of altitudes: in the polluted boundary layer (up to a few hundred meters) and in the free troposphere as well as in the stratosphere. The production and loss mechanisms are different in the different regions. They include chemical, radioactive and dynamical processes. The resulting ozone concentration varies throughout the atmosphere.

Ozone in the atmosphere is important for several reasons. It absorbs UV radiation which may adversely effect human health, ecosystems and materials. By absorption of UV, the stratosphere is heated, and the atmospheric circulation is influenced. In the troposphere, ozone is a greenhouse gas. Like carbon dioxide and other source gases, it is transparent to most of solar radiation, but absorbs strongly in the infra-red. Finally, ozone at ground level is a noxious environmental agent.

OZONE PHOTOCHEMISTRY

Ozone is produced directly in the atmosphere by the action of UV on oxygen molecules:

$$O_2 + UV \rightarrow O + O \qquad \lambda < 243 \text{ nm}$$

$$O + O_2 \rightarrow O_3$$

Ozone is photodissociated by both UV and visible light

$$O_3 + \text{light} \rightarrow O_2 + O \qquad \lambda > 270 \text{ nm}$$

The production processes are balanced by loss through chemical reactions. Chapman proposed in 1930 the reaction

$$O + O_3 \rightarrow O_2 + O_2$$

Now it is realised that ozone in the statosphere is removed predominantly by catalytic chain reactions involving reactive molecules

$$X + O_3 \rightarrow XO + O_2$$
$$XO + O \rightarrow X + O_2$$

$$\text{net: } O + O_3 \rightarrow 2O_2$$

Hereby the catalyst X is H, OH, NO, Cl and Br. These catalyst species can initiate up to several thousand ozone-destroying cycles before they are themselves removed. The catalysts are derived from natural or man-made emission of trace constituents ("source gases"). A common feature of these "source gases" is that they are relatively inert in the troposphere, but are transported up to the stratosphere. There they are broken down to produce the reactive catalyst species. Their relative importance depends on the extremely complex chemical reactions which form, remove and interconvert the active components of each family.

At least twenty individual sources gas containing chlorine and/or fluorine have been identified positively in the atmosphere. Only three components, CH_3Cl (methyl chloride), CH_2Cl_2 and $CHCl_3$ are thought to have substantial natural sources. The major anthropogenic source gases are chlorofluorocarbons such as $CFCl_3$ (CFC 11) and CF_2Cl_2 (CFC 12) with increasing contributions from other compounds. Many bromine and iodine compounds have predominantly natural sources. However, halons, such as CF_3Br, are entirely man-made.

Chlorine and bromine are released from their source gases either by photolysis or by reaction with free radicals. The lifetimes of halogen source gases cover a very large range from a few days to well in excess of 100 years. Generally speaking, the anthropogenic compounds have longer lifetimes, because they are fully halogenated.
Nitrous oxide, the source for reactive oxides of nitrogen (NO, NO_2) is of lesser importance. Methane and carbon monoxide can indirectly affect stratospheric ozone by altering the lifetime of other source gases. In the present overview, the emphasis will be upon the halogenated source gases.

MODELS OF THE ATMOSPHERE

The total ozone column consists mainly (80%-90%) of stratospheric ozone. The seasonal transport cycle has large influence on the seasonal total ozone distribution. It appears that the largest transport in the stratosphere and mesosphere (above 50km) takes place from low latitudes in the summer hemisphere to moderate latitudes in the winter hemisphere. The maxima in the total ozone distribution will occur in winter to early spring. Other meteorology factors influencing the ozone distribution include horizontal transport and the tropopause height.

To try to assess how ozone will change over the next 50 years or more, estimates of future gas concentration are required for numerical model calculations. Large uncertainties attach to these estimates, and to the model calculations. One-dimensional (altitude only) models make predictions on the globally averaged ozone. Many aspects of the circulations of the atmosphere have been reproduced by general circulation models (GCMs)-comprehensive three dimensional numerical models of the atmosphere, with the main physical processes represented in some way. At present computer time is to expensive to include full photochemistry in GCMs. Much effort has therefore been directed into development of two-dimensional (altitude, latitude) models.

One dimensional models usually have very detailed chemical schemes and high resolution both in the vertical dimension and in time. These models are commonly used to derive parametrizations which can be used in more dimensional or more detailed models later on.

Two dimensional zonally averaged models are generally used in order to study the effects of anthropogenic and natural emissions upon the composition of the earth's atmosphere at larger temporal and spatial scales.

Recent two-dimensional models predict overall larger decreases in globally-averaged column ozone than those predicted by the one-dimensional models. Using a scenario with CFC emissions assumed to grow at 3% per year, they predict a typical global ozone depletion of 3% by the year 2000. Model calculations show that once changes in ozone have occurred it will take tens of years or more for ozone to return to its original value. Furthermore the

2-D models indicate that depletion is likely to be greater at high latitudes.

SPATIAL AND TEMPORAL VARIATIONS OF OZONE

The integrated amount of ozone in a vertical column extending from the earth's surface to the top of the atmosphere is referred to as the column ozone. This quantity is often stated in Dobson Units (DU), related to the thickness of an equivalent layer of pure ozone at standard temperature and pressure (STP). There are 1000 DU in a layer one centimetre thick at STP. Observed values range from 150 to 650 DU, with a mean value in the tropics of 220 DU.

The first measurement of the column ozone was performed by Fabry and Buisson in 1913. The Dobson spectrophotometer, first developed in the 1920's, remains the basis instrument in the global network of stations measururing column ozone which has existed since around 1930. It measures the intensity of sunlight at two different wavelengths between 305 and 340 nm. From the knowledge of the different absorption coefficient at the two wavelenghts, the ozone column density can be deduced. The distribution of the world-wide network of Dobson stations, mainly established during the International Geophysical Year 1957/1958, is quite uneven.

Estimates of the vertical ozone profile have been made since the 1930's by baloon sounds and by the ground-based "Umkehr" method. In the latter technique, the zenith clear blue sky is observed while the sun traverses a range of solar zenith angles until just after sunset. Scattering from different altitudes ("Umkehr" layers) in the atmosphere allows the height profile of ozone to be deduced.

The availability of satellites has led to the development since 1967 of several satellite-based measurement systems, such as SBUV (Solar Backscatter UV) and TOMS (Total Ozone Mapping Spectrometer).

Variations and trends in ozone concentrations have been observed in the troposphere, in the stratosphere and for the column ozone.

The seasonal variation of surface ozone concentrations at mid latitudes on the Northern hemisphere follows a general pattern (Erisman and Beck, 1988).

In winter concentrations vary around 20 ppb while summer concentrations range from 40 to 50 ppb. Daily maximum concentrations fluctuate between 45 and 100 ppb. Concentrations on the Southern hemisphere are about 20 ppb lower than the typical values on the Northern hemisphere, whereby differences may occur between the wet and dry season. The surface ozone production is a result of photochemical disintegration of the pollutant NO_x gas into NO and oxygen followed by a reaction between NO and O_2. Detailed analysis of the last decades of surface ozone records at remote European stations seem to confirm a significant increase of 1-1,5% per year.

The variations and trends in the stratospheric ozone are most important. The annual variations is such that the ozone column reaches a maximum in late winter to early spring. The seasonal variations amount to ±15% at moderate latitudes. Solar activities can influence the amount of stratospheric ozone by ±20% through variation in solar UV radiation. The solar cycle has a time scale of 11 years. At the end of the eighties the solar cycle just passed its deepest point. The quasi biennial oscillation (OBQ) is an alternating easterly and westerly zonal wind system, with a period of 26 months. The resulting oscillation of the ozone concentration is of the order of ± 2%. The amount of stratospheric ozone is also influenced by volcanic aerosols and by aperiodic climatological events such as El Niño current.

In order to establish a possible trend in total ozone and in stratospheric ozone and causes of these phenomena, NASA, WHO, NOAA, FAA and UNEP formed a trend panel in 1986. The panel has reanalysed all key data taking into amount all disturbing factors like drift in instruments, aerosols, trace gases, geophysical factors, and a summary of the report has already been published (WMO, 1988).

Ground based Dobson instruments data show measurable decreases in the annual averaged total ozone column ranging from 1,7% to 3% at latitudes between 30° and 60°N in the period from 1969 to 1986. In this analysis the effects of solar cycle and QBO were taken into account. The decreases are most pronounced during the winter months and ranged from 2,3% to 6,2%.

The magnitude of the QBO and solar cycle effect are -2,0% and +1,8% respectively, so the trends cannot be explained by these natural factors. Calculations using two-dimensional models predict that increasing atmospheric concentrations of trace gases would have caused a small decrease in ozone globally between 1969 and 1986. Predicted decreases between 30° and 60°N for this period ranged from 0,5 to 1,0% in summer and 0,8 to 2,0% in winter, where the range reflects the results from most models.

The quantity and quality of ground based observations in the tropics and the Southern hemisphere are such that the determination of trends is difficult in these latitudes. Changes can be studied since 1978 when satellite data became available. The normalized satellite data show a decrease of the total ozone column between 1978 and 1985 in all latitude zones in both the Northern (1,1 to 3,7%) and Southern (1,1 to 9,0%) hemispheres. The range of predicted decreases due to the solar cycle varies form 0,7% to 2.0% in this period. A decrease of 0,5% to 1,5% in winter and 0,2% to 0,6% in summer due to anthropogenic influences over the same time span was predicted. So according to these models the solar cycle may have had a comparable or larger impact on the total ozone column than the man-made anthropogenic gases over this period.

Theoretical calculations of the vertical profile predict that local ozone concentration near 40 km altitude should have decreased between 1979 and 1985 by 5% to 12% in response to the decrease in solar UV output and the increased atmospheric abundance of trace gases. This range is predicted for the latitude belt 30° to 60°N whereby 4%-9% is due to trace gases, and 1%-3% due to the reduced solar UV output. Analyses of satellite and ground-based data show small decreases in the same period; these decreases peak near 40 km altitude with mean values of 3% and 9%, respectively, and agree within the range of their errors.

ANTARCTIC OZONE

It is now common knowledge that in spring the ozone layer in the stratosphere above the Antartic is reduced when the sun rises during that season. What is worrying, however, is that ground-based and satellite data have shown conclusively that the spring-time Antartic column ozone decreased rapidly after the late 1970's. This phenomenon is large, sudden and unexpected. Total column ozone (at all latitudes south of 60°S) was lower in the Antartic spring-time in 1987 than in any previous year since satellite measurements began. In October 1987, the monthly zonal mean amount of total ozone at latitudes 60, 70 and 80 degrees south was about 20%, 40% and 50% lower, respectively, than in October 1979. Normalized satellite data indicate that total column ozone has decreased since 1979 by more than 5% at all latitudes south of 60°S throughout the year.

In view of the well documented year-to-year variability in the meteorological conditions and the abundance of Antarctic ozone, the observation that the 1988 Antarctic ozone levels are not as low as in 1987 is not unexpected.

It should be mentioned that the ozone data seem to have been corroborated by the increased amounts of UV flooding through the ozone hole. The measurements indicate that the UV radiation in October equalled or exceeded the levels of December 1988 (New Scientist, 1989).

During the winter the circulation over Antarctica from about 10 to 50 km altitude is dominated by a very stable polar vortex which breaks down in late October, due to stratospheric warming. The air within the vortex is not mixed up with air outside the vortex during its residence above the Pole. In stratospheric air active chlorine molecules are captured in stable species like HCl. Within the polar vortex, Polar Stratospheric Clouds (PSC) are formed and on the ice surfaces, stable chlorine species are converted into more active forms (e.g. Cl_2). These molecules can be readily photolysed in the Antarctic spring sunlight to form active atomic chlorine for O_3 depletion cycles. The observed strong anti-correlation between active chlorine and ozone from 1987 aircraft measurements indicate that these largely industrially derived chlorine species are primarily responsible for the observed decrease of ozone within the polar vortex.

Assuming our current understanding of the role of man-made chlorine in producing the Antartic ozone hole is correct then the Antarctic ozone hole will not disappear until the atmospheric abundance of chlorine is reduced to the levels of the late 1960's of about 2ppb.

Current two-dimensional models predict reduction of only a few percent at Antarctic latitudes. These models did not predict the Antarctic ozone hole because of omission of heterogeneous chemistry. The discrepancy between predicted and observed ozone reduction at polar latitudes is a matter for additional concern, both from scientific and policy point of view.

A decrease of atmospheric ozone will have effects on human health, ecosystems and economy. Some adverse effects on human health can be quantified now. Other effects, mainly those on animals and plants, and expecially the food economy are less clear. Progress in the understanding in these fields will crucially depend on the research effort in the coming years.

Human health

UVB radiation influences the human health in positive through the formation of vitamin D and the improvement of skin diseases. Many effects are damaging such as sunburn, snow blindness, cataract, ageing of the skin, immunological changes, and skin cancer. Right from the beginning of the problem of ozone depletion, the possibility of increasing the skin cancer incidence has played a central role. This is also the only adverse effect where the dose-effect relationship and the action spectrum can be quantified in a rather reliable way, at least for non-melanoma cancers.

The dose-effect relationship has been known for some time (De Gruyl et al., 1983). An experimentally determined action spectrum (wave length dependency) for UV carcinogenesis became available recently (Sterenborg and v.d. Leun, 1987 , Slaper et al., 1987).

Ozone depletion leads to a more-than-linear increase of the skin cancer incidence (Mc Donald, 1971), described by an "amplification factor". The non-linearity is caused by a non-linearity in the dose-effect relationship ("biological amplification") and the fact that the change in UV irradiance due to ozone depletion is spectrally selective ("optical amplification").

From animal experiments, it is well known that the dose-effect-relationship for the incidence I_c of skin cancers is a power function of effective irradiance E_{eff}. A large set of human data was carefully collected in the USA (Scotto et al, 1981). From these data Slaper (1987) derived the dependence of the incidence on dose. For basal call carcinoma the exponent of effective irradiance is 1.7, for squamous cell carcinoma the exponent is 2.9. Thus the average value of the biological amplification factor is 2.3.

Animal experiments have yielded (Cole 1986, Sterenborg 1987, Slaper 1987) an action spectrum for photo-carcinogenesis. The Sterenborg-Slaper action spectrum can be approximated by the action spectrum for UV-erythema. Carcinogenic effectiveness depends very steeply on wavelength just above 300 nm, precisely where a steep increase of the solar irradiance is expected due to ozone removal. Effective irradiance E_{eff} can be represented as a power function of the column ozone.

A recent preliminary computation showed that the optical amplification factor has a value of about 1.5 (Van der Leun, 1988). The biological and optical amplification factors must be multiplied in order to obtain the combined, overall amplification factor of about 3.5. This means that 1% reduction in ozone will increase the skin cancer incidence by 3% to 4%.

Ecological and economical impact

The potential effects of ozone depletion on ecosystems were summarized recently by Van der Leun (1988).

The sensitivity of aquatic ecosystems to UV-B radiation is not well mapped. Various experiment have demonstrated that UV-B causes damage e.g. to fish, shrimp and crab larvae. In natural marine plant communities a change in species composition is more likely than a decrease in net production as a result of enhanced UV-B levels.

In particular, both invertebrate zoo plankton and food-fish larvae near-surface populations are likely to be diminished. Penetration of UV-B into natural waters is a key variable in assessing the potential impact of UV on any aquatic ecosystems. Since one half of the world's protein is derived from marine species, more research is needed to assess the influence of ozone depletion on the world food system.

Out of 300 terrestric plant species screened for UV tolerance approximately two thirds were found to be sensitive. Despite uncertainties, data suggest that crop yields including forest productivity are potentially vulnerable. Crop quality may be also impaired by increasing levels of UV-B radiation. In particular, disease resistance may be modified.

Finally, studies suggest that increased UV from ozone depletion may increase the rate of urban ozone formation. Further increase in ground-based ozone would adversely effect public health and welfare, crops, ecosystems and materials.

CHARACTERISTICS AND USES OF THE SOURCE GASES

The emission of halocarbons forms a grave risk for the protective ozone layer of the earth. Model studies performed in the early 1980's indicated that once changes have occurred it will take tens to hundreds of years for ozone to return to its original value. Therefore consensus emerged that it is desirable to reduce the production, use and emissions of halocarbons as far as possible. From the beginning, it was clear that a solution must have an international character. It was also clear, that technical and economical consequences must be taken into account.

Not all halocarbons have the same ozone depletion protential (ODP). For practical purposes, it is useful to define ODP as a dimensionless number, which is the predicted ozone depletion for a certain gas divided by the ozone depletion due to the same emission of CFC-11. ODP depends on variables such as the reactivity of the gas, the number of halogen atoms released per molecule of the gas, whether the release takes place in the troposphere or in the stratosphere, and the residence time in the atmosphere. It should be remarked that most halocarbons have in addition a greenhouse potential, thus contributing to the global increase of the temperature.

If the contributions of different compounds to the chlorine concentration in the atmosphere are compared it emerges that the contribution of methyl chloride is predominantly of natural origin, whereas most other compounds are mainly man-made and on the increase.

Some characteristic data on the present emission, average concentrations, annual increase, ozone depletion potential and residence time in the atmosphere for some halocarbons are summarized in the following table (De Leeuw and Beck, 1989).

Why is the use of halocarbons so popular? The well-known uses of CFC's are

Table 1

Compound	Formula	Emission (ktons/ year)	Average concentration (pptv)	Increase per year	ODP	Residence time (years)
CFC-11	$CFCl_3$	280	226	4%	1,0	75
CFC-12	CF_2Cl_2	370	392	4,3%	1,0	110
Methyl chloroform	CH_3CCl_3	470	139	4,5%	0,15	7
Carbon tetrachloride	CCl_4	70	129	1,4%	1,2	40
CFC-113	$CF_2ClCFCl_2$	140	32	11,3%	0,8	90
Halon-1211	CF_2BrCl	3	1,7	12%	3	25
Halon-1301	CF_3Br	3	2,0	15%	10	110

as blowing agent in sprays, as refrigerant in household, industrial and automobile cooling-systems, as ingredient in fabrication of insulating rigid and flexible foams, as cleaning solvent in electronics and in metal and dry cleaning. Halons are mainly used as fire extinguishant. The approximate percentages of different uses of the CFC's in the non-communist countries and the European Community, respectively, can be summarised as follows (Reijnders, 1989):

Table 2

	Blowing agent	Refrigerant	Foams fabrication	Cleaning agent
Non-communist countries	30%	30%	30%	10%
European Community	50%	10%	30%	10%

In 1986, the CFC consumption by region was as follows (Pollock Shea, 1988):

Region	Share of total (%)
United States	29
Other industrial countries	41
Soviet Union, Eastern Europe	14
Other developing countries	14
China and India	2

The present uses of halocarbons represent huge markets which are rapidly growing especially in the developing countries.

INTERNATIONAL AGREEMENTS

Bearing in mind these developments, UNEP set about organising a Convention for the Protection of the Ozone Layer which, following negotiations, was adopted at Vienna on 22 March 1985. It was followed by the Montreal Protocol on 16 September 1987.

Where the Vienna Convention was only an intentional agreement, the Montreal Protocol constitutes a commitment package of real measures. The far target of the protocol is to reduce the emission to zero of substances which deplete the ozone layer. The near target for industralized countries is a successive reduction of consumption in this century. With 1986 as reference year, production should be reduced by 20% in 1996 and by 50% in 1999. No increase in the halon production is allowed.

Developing countries are entitled to delay compliance with these controlling measure by 10 years. In fact, the Montreal Protocol left room for increase of production in development countries. Therefore a 75% level of production in 1999 in comparison with 1986 is still within the terms of the protocol.

By October 1988, the formal status of the Vienna Convention and the Montreal Protocol was as follows (UNEP, 1988).

		Number of parties
Vienna Convention:	signed	28
	: ratified	29
	: ratification in progress	9
Montreal Protocol:	signed	46
	: ratified	10
	: ratification in progress	14

In total, 53 parties (most CFC's producing states including the European Communities as separate signatory party) have joined, or are in the process of joining, the Vienna Convention or the Montreal Protocol.

SUBSTITUTES AND ALTERNATIVES

To comply with the requirements and targets set by the Montreal Protocol, environmentally safe substitutes and alternatives must be found for the present use of halocarbons. The industry is aware of the problem and is actively developing such substitutes and alternatives.

In this context some of the following questions need to be answered:

- o Is a present practice, involving halocarbons, justified
- o If so, how can it be improved from environmental point of view
- o Is a substitute for halocarbons technically available
- o If so, is the substitute safe enough with respect to flammability, toxicity etc.
- o Is the substitute clean enough, e.g. how large are the ozone depletion potential, greenhouse potential etc.
- o What are the economic consequences of substitutions or alternative practices.

All the chlorine containing substitutes (HCFC's) have ODP's substantially lower than the fully chlorinated chlorocarbons which are to be controlled under the Protocol. The estimated ODP's obtained for potential substitute species (UNEP, 1988) are shown in the table which follows.

Table 3

Name	Formula	ODP
HCFC- 22	$CHClF_2$	0,05
-123	$CHCl_2CF_3$	0,02
-124	$CHClFCF_3$	0,02
-141b	CH_3CCl_2F	0,1
-142b	CH_3CClF_2	0,06
HCF-125	CHF_2CF_3	0
-134a	CH_2FCF_3	0
-143a	CH_3CF_3	0
-152a	CH_3CHF_2	0

The greenhouse potential is in all cases an order of magnitude lower than that for CFC's. The toxicity of these substitutes is low, but in some cases the toxicologic knowledge is still incomplete. The flammability is generally low, with the exception of HCFC-141b, 142b and HCF-143a, 152a. HCFC-22, HCFC-142b and HCF-152a are existing commercial products, whereas the other compounds are still under consideration or on the verge of market introduction.

During the UNEP Scientific Review of the Depletion of the Ozone Layer, held at The Hague in October 1988, the current status of progress towards substitutes and alternatives has been summarized. The main conclusions are as follows (UNEP, 1988).

Refrigeration, air conditioning - the solutions are emission reduction in the short term and equipment redesign in the long term, using ozone-safe, health-safe substitutes without a significant climate warming effect. HCFC-123, HFC-134a, HFC-152a, and possibly HFC-125 are promising substitutes in this respect. Car air conditioners offer re-cycling possibilities as do other kinds of refrigeration and general air conditioning.

Rigid foams - these are used, for instance, for insulating houses. HCFC-123 and HCFC-141b can be substituted for CFC-11, as can a blend of water and CFC-11. Toxicology testing for the possible substitutes, HCFC-123 and HCFC-141b, is not complete. Other technologies, notably vacuum panels, are emerging. Vacuum panels have a higher insulating value while occupying much less space. There is no longer a need for using foams, blown with the controlled CFC's, in packaging.

Flexible foams - used, for example, in car seats and furniture. The technology is pending for producing all grades without CFC's, that is to say, we are in the product-testing stage.

Electronics solvent - a 50% reduction in CFC-113 could be achieved by conservation measures alone. The reduction programme is concentrating on the develepment of fluxes which do not require cleaning, on water-cleaning technologies, alternate solvents, recycling and recovery.

Metal cleaning - alkaline or acid cleaning looks the most promising.

Dry cleaning - the only promising substitute at the moment is HCFC-123.

Aerosols - there is no technically valid reason for not moving away from the use of CFC's now, with the exception of a few critical medical uses such as asthma sprays.

Halons - the scope for emission reduction is enormous for gases called

Halon-1211 and Halon-1301. Current programmes focus on eliminating unwanted discharges, alternatives to discharge testing, recovery and recycling, developing alternative agents, etc.

POLICY DEVELOPMENTS

In view of the above-mentioned technical developments, the UNEP review meeting concluded that the Montreal Protocol phase-down schedule of 20% by 1993 cannot be significantly hastened. A tremendous technical progress is anticipated for 1994. Therefore, the following 30% reduction can be achieved by mid-1998. Indeed, there now appears to be a broad agreement that phasing out between 80% and 90% of the controlled CFC's before the end of the century is an achievable goal. Concerning halons, conservation measures alone should be able to reduce global consumption by more than 80%, New substances being developed could permit their almost total elimination in five to ten years.

Since the Montreal Protocol in 1987, significant scientific progress mentioned earlier has led to even more alarming conclusions concerning the ozone depletion.

As mentioned before strong evidence emerged that man-made chlorine species are primarily responsible for the observed decrease of ozone within the polar vortex. Larger ozone reductions are predicted in mid-latitudes and polar regions than at low-latitudes. Measured column ozone reductions in the Antarctic are much more serious than predicted by any two-dimensional models so far. Current model calculations do not include the effects of heterogeneous chemistry which might result in an underestimate of the impact of CFC's on ozone.

Even on the basis of current models, the near term targets of the Montreal Protocol are not sufficient to stop the increase of CFC's and chlorine in the atmosphere. Without more stringent and faster phasing-out of the use of halocarbons, the ozone decrease and greenhouse effect will continue to increase. EPA called in October 1988 for the complete ban on using fully halogenated CFC's and halons, and a freeze on emission of methyl chloroform (EPA, 1988).

The Commission of the European Communities took the decision in March 1989 to stop the use and production by the end of the century. Before then, a 85% reduction will be achieved in the European Community as soon as possible. In the mean-time possibilities are discussed how to form a fund to help the developing countries to accelerate the phasing out of the use and production of halocarbons. This is significant in view of the fact that some very large developing countries hesitate to participate in the international agreements.

The international effort in solving the problem of the ozone layer depletion has so far involved many participants: scientists, industries, governments and pressure groups. A long and arduous way is still in front of us. There is a great hope and well founded perspective that these efforts will succesfully result in a cleaner global environment.

REFERENCES

Cole, C.A., P.D. Forbes and R.E. Davies, 1986
An action spectrum for photocarcinogenesis, Photochem. Photobiol. 43, 275-284

McDonald, J.E., 1971
Relationship of skin cancer incidence to thickness of the ozone layer, Congr. Rec., 117, 3493

Erisman, J.W. and J.P. Beck, 1988
A general description of atmospheric ozone with the accent on changes in its concentrations and possible causes
Rijksinstituut voor Volksgezondheid en Milieu (RIVM)
Bilthoven, The Netherlands.

European Communities, 1989
Statement by European Communities Commissioner Ripa di Meana
for the conference "Save the ozone layer",
London 5-7 March.

De Gruijl F.R., J.B. van der Meer, and J.C. van der Leun, 1983
Dose-time dependency of tumor formation by chronic UV-exposure, Photochem. Photobiol. 37, 53-62.

De Leeuw, A.A.M. and J.P. Beck, 1989
Afbraak "goed" ozon maakt emissiestop CFK's nodig.
Chemisch Magazine, maart 1989, 137-141.

Montreal Protocol, 1987
Montreal Protocol on the Substances that Deplete the Ozone Layer
United Nations Environmental Programme
Final Act, 1987.

New Scientist, 1989
"Radiation" flood" through ozone hole is measured", New Scientist, 15 April

Ozone Symposium, 1988
Atmospheric ozone research and its policy implications
Proceedings of the 3rd US-Dutch International Symposium,
Nijmegen, The Netherlands, May 9-13, 1988 (T. Schneider et al. eds)
Elsevier, Amsterdam-Oxford-New York-Tokyo.

Reijnders, L., 1989
Prevention of stratospheric modification
In: Atmospheric Ozone Research and its Policy Implications
(T. Schneider et al. eds.)
Elsevier Science Publishers B.V., Amsterdam.

Scotto J., T.R. Fears and J.F. Fraumeni, Jr., 1981.
Incidence of nonmelanoma cancer in the United States
National Cancer Institute
U.S. Department of Health and Human Services (NIH) 82-2433.

Slaper, H., A.A. Schothorst and J.C. van der Leun, 1987.
Risk evaluation of UV-B therapy for psoriasis,
Photodermatology 3, 271-283.

Slaper, H., 1987.
Skin cancer and UV exposure: investigations on the estimation of risks,
Ph.D. Thesis, Utrecht.

Sterenborg, H.J.C.M. and J.C. van der Leun, 1987.
Action spectra for tumorigenesis by ultraviolet radiation,
in: Passchier, W.F. and B.F.M. Bosnjakovic, (eds.),

Human exposure to ultraviolet radiation: risks and regulations, pp. 173-190, Elsevier Science Publishers, Amsterdam

UNEP, 1988.
United Nations Environment Programme (UNEP) Meeting:
Scientific Review of the Depletion of the Ozone Layer
The Hague, 17-18 October 1988.

United Kingdom Stratospheric Ozone Review Group, 1987.
Statospheric Ozone, First Report
Prepared at the request of the Department of the Environment
and the Meteorological Office
London: Her Majesty's Stationery Office

United Kingdom Stratospheric Ozone Review Group, 1987.
Statospheric Ozone, Second Report
Prepared at the request of the Department of the Environment
and the Meteorological Office
London: Her Majesty's Stationery Office

World Meteorological Organization, 1988.
Executive summary of the Ozone Trends Panel
Geneva, Switzerland.

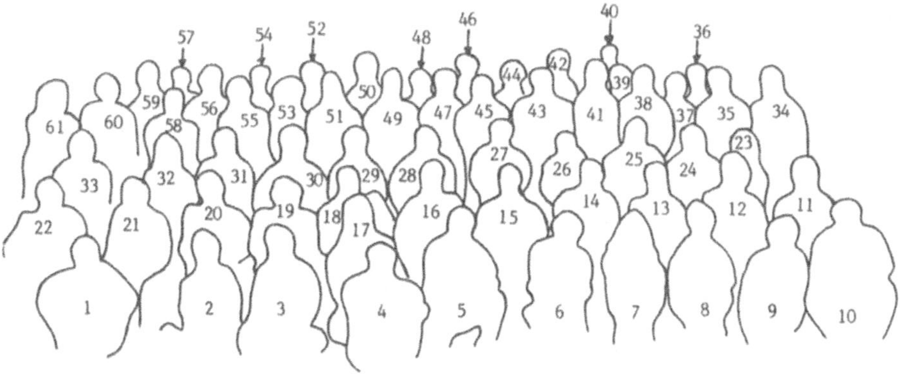

PARTICIPANTS

1 - P.VECCHIA
2 - N.BONDING
3 - H.J.HANSEN
4 - P.BALUGANI
5 - D.B.SURIJON
6 - R.E.Van LOON
7 - L.TODD
8 - R.MISURACA
9 - M.HANNEVIK
10 - M.NORSELLI
11 - W.J.MARSHALL
12 - T.C.M.WIJSMAN
13 - R.C.AVANZO
14 - G.MARIUTTI
15 - A.RINDI
16 - D.H.SLINEY
17 - R.T.MIHRAN
18 - M.GRANDOLFO
19 - O.CEREZCI
20 - P.ERIKSEN
21 - I.QUINTO
22 - B.F.M.BOSNJAKOVIC
23 - G.TROMBA
24 - J.FRANKS
25 - Z.GOURGOULIATOS
26 - J.BERNHARDT
27 - M.H.REPACHOLI
28 - C.PAPA
29 - M.HIETANEN
30 - A. ETTI
31 - S.SEKER
32 - P.K.GALOFF
33 - S.BATTISTI
34 - G.CIARROCCO
35 - B.ALTMAN
36 - A.F.McKINLAY
37 - C.BORDOLOI
38 - J.A.SILVA
39 - L.A.COURT
40 - P.G.SODERBERG
41 - J.BAKOS
42 - J.C.FEIJ
43 - H.BRUEGGEMEYER
44 - Y.ULGEN
45 - T.A.SPRINGER
46 - G.J.EGGINK
47 - M.VINCENTINI
48 - N.BIANCHI
49 - S.J.KALITA
50 - J.C.VAN DER LEUN
51 - K.BRINK
52 - R.BIANCOTTO
53 - F.GERMOLI
54 - G.D'AMORE
55 - J.C.LOPES
56 - A.LAZZARI
57 - M.CHECCUCCI
58 - A.CHECCUÇCI
59 - R.DONAHUE
60 - R.ORTEGA-MARTINEZ
61 - S.OZSAR

INDEX

Absorption, 8
Absorption coefficient, 8
Accessible emission limit, 135
ACGIH, 229
ACGIH weighted exitance, 31
Actinic radiation, 130
Acupuncture, 355
Aerosols, 19
Agricultural application, 27
ALARA principle, 301, 309
American National Standards Institute, 259, 265
American Conference of Governmental Industrial Hygienists, 293, 309, 329
American National Standards Institute, 329
Anatomy
  of the eye, 181
  of the skin, 185
Angioplasty, 359
ANSI Standard Z136.1, 331
Aphake photic hazard, 294
Aqueous humour, 181
Arc welding, *see* Welding arcs
Artificial sources
  agricultural application, 27
  arc sources, 24
  cosmetic application, 28
  fluorescent lamps, 24
  gas discharge source, 23
  germicidal application, 28
  industrial application, 29
  medical application, 28
  metal halide lamps, 24
  physical characteristics, 26
  quartz halogen lamps, 23
Atmospheric attenuation, 264
Attenuation
  coefficient, 264
  of solar radiation, 20
Axons, 183

Band gap, 102
Beam diameter, 260
Beam divergence, 260
Beam visualizers, 142
  burn papers, 145
  image converters, 145
  liquid crystal sheets, 145
  luminescent detectors, 145
Biological effects
  action spectrum, 121
  dose requirements, 121
  irradiance, 121
Blackbody radiation, 15, 277
  blue-light weighted exitance, 32
  erythemally weighted exitance, 32
  ACGIH weighted exitance, 31
Blacklight
  spectral emission, 42
Blinking reflex, 181
Bloch's theorem, 112
Blue-light
  retinal photochemical hazard, 31
  hazard, 229, 293, 294
  weighted exitance, 32
Born-Oppenheimer approximation, 114
Bowman's membrane, 181
Brewster's angle, 7
Brillouin zone, 112
British Standards Institution, 329
Broadband sources
  protection standards, 293
  spectral distribution, 121
Brows, 181

Calibration, 130, 149
Capsule of tenon, 181
Cataract, 325
Chlorophyll photosynthesis, 43
Choroid membrane, 182
Ciliary body, 182
Classical beam
  diffraction limit, 90
Classical image formation, 87
Coherence
  length, 79
  time, 79
  volume, 80

focusability, 80
spatial, 80
temporal, 79
Coherent optical sources, 77
Coherent states, 118
Colour temperature, 17
Compact lamps
arc, 54
fluorescence, 39, 42
xenon, 59
Compton scattering, 109, 113, 115
Conductivity, 93
Cones, 182
Conjunctiva, 181
Conjunctivitis, 123
Cornea, 181
Corneal irradiance, 134
Corneal radiant exposure, 134
Correspondance principle, 4
Cosine correction, 150
Cosine law, 129
Critical grazing angle, 105
Cutaneous appendages
adenomer, 188
eponichius, 188
escretory duct, 188
lunula, 188
nails, 188
pilosebaceous units, 188
sebaceous glands, 188
sebum, 188
sweat glands, 188

Dermis
collagen fibres, 187
dermal nerve network, 187
endothelial cells, 187
fibroblasts, 187
Golgi corpuscles, 187
istiocytes, 187
Krause corpuscles, 187
limphocytes cells, 187
Meissner corpuscles, 187
mucocutaneous bodies, 187
papillary layer, 187
pericytes cells, 187
reticular layer, 187
reticulin fibres, 187
Ruffini corpuscles, 187
Schwann cells, 187
Descemet's membrane, 181
Detectors
accuracy, 121
angular response, 121
bolometer, 125
pneumatic, 125
power range, 121
precision, 121
pyroelectric, 126
saturation, 149
spectral response, 121
thermal detectors, 125
thermocouple, 125
thermopile, 125
Deutsche Institut fuer Normung, 329
Diazo film, 129
Dielectric function, 97
asimptotic behaviour, 95
Diffraction, 12
Diode lasers, 88
Diopters, 183
Dipole
approximation, 111
radiation, 98
Direct fire simulators, 263
Direct reading instruments, 130
calcium phosphor, 130
calcium tungstate, 130
Dyes, 290

Effective irradiance, 321
Electroluminescent lamps, 58
Electron-phonon interaction, 114, 116
Emissivity, 16
Endothelium, 181
Envelope action spectrum, 325
Environmental Health Criteria Documents, 319, 342
Environmental Health Criteria Programme, 319
Epidermis
germinal basal, 185
keratinocytes, 185
keratinosomes, 186
keratohyalin, 186
langerhans, 186
malpighian layer, 185
melanocytes, 186
merkel cells, 187
odland bodies, 186
statum corneum, 185
stratum granulosum, 185
stratum spinosum, 185
tonofilaments, 186
Erythema, 123, 161, 323, 332
Erythemally weighted exitance, 32
Escherichia coli
SOS genes, 247
UV mutagenesis, 247
Excimer laser, 154, 155, 241
Exciton states, 114, 116
Exposure duration, 344
Exposure limits, 134
IRPA/INIRC guidelines, 341
IRPA/INIRC 1988 revision, 239
occupational, 263
Q-switched laser, 263
Extended-source limits, 345
Extinction coefficient, 94, 100
Extraterrestrial irradiance, 21

Eye
accomodation, 183
anatomy, 181
astigmatism, 184
hyperopia, 183
myopia, 183
optical physiology, 183
physiology, 181
presbyopia, 183
protection parameters
absorption, 290
filter damage threshold, 292
glass, 291
goggles, 291
lateral protection, 291
optical density, 288
plastic, 291
reflection, 290
spectacles, 291
spectral attenuation, 289
user acceptability, 292
visual transmittance, 291
wavelength, 287
refractive power, 183
vision mechanism, 184
Eye-safe lasers, 266
Eyeballs, 181
Eyelids, 181

Federal Laser Product Performance Standard, 331
Feynman diagrams, 112
Fibres of Mueller, 182
Film badges
diazo film, 129
polysulphone, 128
thermoluminescence, 128
Filters, 148
Flash lamps, 54
Flash tubes, 54
Fluorescent lamps
spectral emission, 38
Food and Drug Administration, 265
Fovea, 183
Free electron metal, 102
Frequency, 4
Fresnel formulas, 6

Galilean telescopes, 81, 82
Ganglion cells, 183
Gas discharge, 25
Gas discharge lamps,
neon lamp, 37
production of ozone, 37
Gas welding, 277
Gaussian beams, 134, 208
center beam, 82
cutoff parameter, 89
diameter, 82
diffraction limit, 90
focussing, 89
irradiance, 82
local irradiance, 82
Rayleigh range, 84
transformation, 85
waist, 84
Germicidal type lamps, 37
Global spectral irradiance, 19
Glow lamps, 54
Golay cell, 125
Grolux lamp
spectral emission, 43, 49

Hamiltonian function, 109
Hazard
assessment
control of outdoor, 263
corneal exposure, 265
laser output parameters, 259
nominal hazard zone, 266
classifications, 329
functions, 294
Health Council of the Netherlands, 329
HID, *see* High intensity discharge lamps, 48
High pressure mercury lamps,
basic construction, 56
High pressure sodium lamps, 52
basic construction, 58
spectral emissions, 46
Hot mirrors
spectral reflectance of aluminum, 274
spectral reflectance of gold, 274
transmittance of aluminum, 274
transmittance of gold, 274
Huygens principle, 12
Hyperbilirubinaemia, 43, 47

Incandescence, 25
Incandescent lamps
halogen lamps, 31
luminous efficacy, 30
quartz, 31
tungsten filaments, 30
Induced dipole momentum, 96
Industrial applications, 29
Infrared radiation
aqueous flare, 220
biological effects, 191, 229
biological parameters, 194
blister formation, 220
cataract, 221
chrio-retinal injury, 204
chromosomic aberrations, 213
clinical effects, 219
coherence, 194
corneal lesions, 205
damage mechanisms, 195, 219
definition of threshold, 195

effects on skin, 209, 229
epidemiological studies, 221, 222
erythema, 220
exposure conditions, 193
eye examinations, 224
filters, 269, 270
flash burns, 230
functional changes, 195
glass blowens cataract, 208, 221
glass works, 222
health hazards, 229
histological changes, 195
hyperemia, 220
industrial application, 269
limiting aperture, 193
mechanical effects, 196
miosis, 220
non thermal effects, 196
ocular effects, 219
ocular exposure, 197
opacification of the epithelium, 220
paralytic mydriasis, 220
penetration into tissues, 195
photochemical retinal hazards, 294
phototoxic damage, 197
protective screens, 269
radiometry, 223
reflectance, 229
response criteria, 194
retinal damage, 202
sources, 192, 269
spectral irradiance, 270
spectral transmittance, 271, 272
steel industry, 222
stromal inflammation, 221
synergistic effects, 231
thermal effects, 196, 230
thermal retinal hazards, 294
tissue models, 210
transmittance, 165
in the eye, 219
ultrashort pulses, 196
IR-A, 191
IR-B, 191
IR-C, 191
Q-switched exposure, 203
Threshold Limit Values, 293
Infrared Laser
exposure limits, 344
Integrated radiance, 121
Interference, 11
International Commission on Illumination, 191
International Radiation Protection Association, 309
International Commission on Radiological Protection, 301
International Electrotechnical Commission, 329
International Non-Ionizing Radiation Committee, 309, 319, 329, 3431
International Radiation Protection Association, 319, 329
International Radiation Protection Association, 341
Iris, 181
Irradiance, 121, 129, 260

Keratitis, 123
Kirchhoff's theory, 12
Kramers-Kronig relations, 95

Lacrimal ducts, 182, 182
Lacrimal gland, 182
Lacrimal passages
canaliculi, 182
lacrimal sac, 182
nasal cavity, 182
nasal duct, 182
Lambda bacteriophage, 247
Lambda prophage, 248
Larmor formula, 62
Laser beam
calculations
aperture diameter, 135
beam diameter, 135
effective, 135
beam divergence, 136
beam waist, 136
effective beam diameter, 135
Rayleigh range, 136
models
constant irradiance, 133
gaussian shape, 133
Laser diagnostics, 133
Laser eye protectors, 287
Laser radiation
detectors
photoelectromagnetic, 144
photoemissive, 142
radiometric, 141
semiconductor, 143
thermal, 142
thermoelectric, 144
infrared, 345
limits of exposure, 341
ultraviolet, 345
visible, 345
wavelengths, 331
Laser rangefinders, 263
Laser safety officer, 351
Laser safety standards
historical development, 329
rationale, 329
Laser target designators, 263
Laser welding, 277
Lasers
acoustic effects, 79

average power, 79
exposure by repetitive pulses, 78
peak, 79
photochemical effects, 79
plasma formation, 79
pulse period, 79
pulse width, 79
single pulse exposure, 78
temporal emission characteristics, 77
wavelength, 77
Latent image, 128
LED, *see* Light emitting diodes, 54
Lens, 182
Light emitting diodes
luminous intensities, 54
Lighting fluorescent lamps, 39, 40, 41
Limbus, 181
Limiting aperture, 260
Limiting membrane, 182
Linear arc lamps
luminance, 54
Lithotrypsy, 359
Lorentz-Drude model, 96
Low pressure sodium lamps, 42
spectral emissions, 46

Macula lutea, 183
Maximum permissible exposure, 288
Maxwell's equations, 4, 92
Medical lasers
bladder tumors, 364
carcinogenetic effect, 362
chromosomic aberrations, 362
clinical use, 362
cutting, 354
decrease of cell's growth, 362
dermatology, 355, 363
efficiency, 362
endoscopy, 354
gastro enterology, 355
gynecology, 355, 363
induced shock wave, 360, 361
interaction mechanisms, 357
laser-tissue interactions, 358
lithotrypsy, 356
long term effects, 362
microsurgery, 354
neurosurgery, 354
neurosurgery, 363
oncogenic factors, 362
otorhino laryngology, 354
pediatric surgery, 355
photo abrasion, 354
photo coagulation, 354
repair surgery, 355
stomatology, 354
therapeutic effects, 353
thermo coagulation, 354
thoracic surgery, 354
tonsillectomy, 363
tumoral treatment, 354
urology, 355, 364
vaporization, 354
vascular surgery, 355
vascular suture, 354
Medical surveillance, 347
acute eye damage, 352
end-of-job examination, 348
eye examination, 348
laser hazard evaluation, 349
medical history, 349
periodic examinations, 348
physical examination, 349
pre-placement examination, 348
responsibility, 351
risk groups, 349
Mercury vapour lamps, 48
ultra-violet radiation emission, 50
Metal halide lamps, 52
spectral emission, 56
Microvilli, 183
Midryasis, 182
Mie scattering, 10, 19
Model, 95
Momentum conservation, 113
Monochrometers
measurement pitfalls, 147
stray light, 148
Monomode optical fibers, 88
MPE, *see* Maximum permissible exposure
Mucosa conjunctiva, 181
Mutagenesis, 250, 252
genetoxic potency, 252
in mammalian cells, 250
mutagenesis, 250, 252
two-step model, 250
UV photoproducts, 252
Myosis, 182

Nadir, 171
National Institute forOccupational Safety and Health, 309
National Radiological Protection Board, 309
Natural sources
aerosol scattering, 19
altitude, 19
moon and planets, 22
ozone layer, 18
Rayleigh scattering, 19
stellar radiation, 22
sun, 18
UV spectral irradiance, 21
Near-infrared thermal hazards, 332
Neonatal jaundice, 43, 47
Nominal hazard zone, 262
Nominal ocular hazard distance,

260, 263
atmospheric attenuation, 138
reflections, 139
Non-coherent sources, 15, 80
Non-ionizing radiation, 332
Nuclear layer, 182, 182
NATO, 329

Occupational exposure limits, 343
Ophthalmoscopy, 351
Optic papilla, 183
Optical constants
quantum description, 109
Optical density
unaided viewing, 137
viewing aids, 137
Optical physics, 1
Ora serrata, 182
Organic dyes, 290
Ozone layer, 18, 170
depletion, 301

Paramedical lasers, 354
anti inflamatory action, 356
anti-cellulitis effects, 356
anti-oedematous action, 356
biostimulation, 356
effects on dental pulp, 358
effects on membrane potential, 358
lymphatic drainage, 356
microcirculation, 356
Periorbital cavities, 181
Permeability, 93
Permittivity, 93
Perturbation theory, 110
Phase velocity, 94
Photobiologically effective irradiance, 122
Photochemical reactors, 52
Photoelectric effect, 112, 116
Photographic developer, 128
Photokeratitis, 161, 332
Photokeratoconjunctivitis, 324
Photopic vision, 184
Photosensitizing agents, 321
Photosensitizing reaction, 321
Physiology
of the eye, 181
of the skin, 185
Pigment epithelium, 182
Pigmentation, 123
Planck's law, 16
Planck's constant, 5, 16
Plane of incidence, 6
Plasma arcs, 157, 158
Plasma frequency, 99, 104
Plexiformis, 182
Polarization density, 97
Polysulphone film, 153
Power, 259

Probit curve, 333
Projector systems
flash distance, 230
ocular skin hazards, 229
Protection standards, 299
American Conference of Governmental Industrial Hygienists, 309
Comite' International de l'Eclairage, 307
Dutch Health Council, 310
International Electrotechnical Commission, 306
International Non-Ionizing Radiation Committee, 309
International Radiation Protection Association, 309
National Institute for Occupational Safety and Health, 309
National Radiological Protection Board, 309
non-laser, 293
primary standards, 300
secondary standards, 300
skin tanning, 305
stochastic effects, 304
Swedish legislation, 306
ultraviolet radiation, 299, 303
USA standard, 306
Protective filters
electro-optic, 166
hot mirror type, 165
light transmittance, 165
luminous transmittance, 165
shade number, 165
IR transmittance, 165
UV transmittance, 165
Protective screens, 269
glasses and plastics, 270
heat absorbing filters, 272
heat reflecting filters, 273
welding filters, 271
Psoriasis UV-A (PUVA), 43
Pulse repetition frequency, 261
Pulse width, 261
Pulsed xenon arc lamps, 54
Pupil, 182
diameter, 134
Purkinjie shift, 184
Pyrimidine dimers, 249

Quantum detectors
photoconductive, 127
photodiode, 127
photoelectromagnetic, 127
photoemissive detectors, 126
photographic, 128
photomultipliers, 126
photovoltaic, 127
Quantum electrodynamics, 109

Radiance, 121
Radiant emittance, 121
Radiant energy, 259
Radiant exposure, 121, 260
Radiant intensity, 121
Radiant power, 259
Radiometer, 141
Radiometric measurements
  broadband optical sources, 121
Radiation protection
  justification, 301
  optimisation, 301
Rayleigh's law, 10
Reflectance, 177
  gold, 106
  silver, 106
Reflection, 5, 116
  coefficient, 6
Reflectivity, 101
  angle of incidence, 5
  angle of reflection, 5
Refraction, 8
  angle of incidence, 5
  angle of refraction, 8
Refractive index, 8, 94, 100
Repetitive laser exposures, 344
Repetitive pulses
  exposure limits, 335
Resonance fluorescence, 114
Retina, 182, 184
Retinal effects, 325
Retinal injury threshold
  dependence upon pulse duration, 337
Rhodopsin, 184
Robinson-Berger meter, 130
Rodes, 182
Ruby laser, 5

Scanning beams, 344
Scattering, 8, 10
  Raman, 113, 114, 115
  Rayleigh, 10, 19, 19, 113, 114
Schwinger's formula, 66
Scintillation, 264
Sclera, 181
Scotopic vision, 184
Secondary standards, 131
Skin
  action spectrum, 307
  anatomy, 185
  cancer, 324
  cutaneous appendages, 188
  cutaneous muscles, 188
  dermis, 187
  epidermis, 185
  fascia superficialis, 188
  physiology, 185
  subcutaneous, 188
Slit lamp microscopy, 350
Snell's law, 8
Solar constant, 18
Solid state lamps, 54
Soviet Ministry of Health, 329
  temporary sanitary norms, 330
Spatial coherence, 15
Spectral effectiveness curve, 321
Spectral irradiance, 121
Spectral lenses, 177
Spectral radiance, 17, 17
Spectroradiometers, 124
Spectroradiometry, 124
Standard light source, 23
Stefan-Boltzmann constant, 16
Stefan-Boltzmann law, 16
Stochastic processes, 333
Storage rings, 62
Stray light, 124
Stroma, 181
  ipoderm, 188
Sum rules, 107
Sunbeds
  spectral emission, 53
Suspensory ligament, 182
Synchrotron radiation, 61, 63, 65
  angular distribution, 63
  critical energy, 74
  critical wavelength, 66, 67, 74
  deflection parameter, 75
  degree of polarization, 69
  emission angle, 68
  energy loss, 74
  frequency cut-off, 67
  general properties, 66
  opening angle, 74
  spectral brilliance, 69
  total power radiated, 74
  undulators, 71
  wigglers, 71

Temporal coherence, 15
Thermal injury, 332
Thermal retinal hazard, 296
Thermoregulatory system, 189
Threshold Limit Values, 331
Transmission coefficient, 6
Transmittance, 8
Tritiated phosphor lamps, 58
Tungsten halogen lamps
  infrared radiation emission, 36
  spectral emission, 34, 35
  ultraviolet radiation emission, 33

Ultraviolet radiation
  actinic spectral region, 321
  action spectrum, 153, 172, 240
  adverse biological effects, 242
  ambient levels, 169
  angular distribution, 176

aphakic eyes, 241
carcinogenic effects, 332, 334
cataractogenesis, 237, 241, 325
delayed effects, 324
DNA lesions, 247
envelope action spectrum, 325
epidemiological studies, 169
erythema, 323
exposure limits, 311, 321
exposure of the eye, 237
genotoxic potency, 252
Hymalayan mountains, 170
IRPA/INIRC guidelines, 319, 322
1988 revision, 239
measurements, 171, 174
mutagenesis, 237, 247
mutational spectrum, 251
occupational exposure limits, 238
ocular exposure, 178
outdoor environment, 169
permissible exposure time, 321
photochemical damage mechanisms, 241
photokeratoconjunctivitis, 241, 324
premutational, 247
protective measures, 243, 323
quantitative determination, 147
reflectance factors, 175
retinal effects, 325
skin cancer, 169
solar spectrum, 172
spectral weighting function, 311
spectroradiometric measurements, 323
stochastic mechanisms, 241
thermal injury mechanisms, 241
UV-A, 321
PUVA (psoriasis UV-A), 43
solaria, 43
spectral irradiance, 44
sunbeds, 43
tanning purposes, 43
UV-B, 321
irradiation, 43
UV-C, 321
welding arcs, 159
United Nations Environmental Programme, 319, 341
Uveal tunic, 182
UV-sensitive film, 129

Visibility, 264
Visible infrared
photochemical effects, 294
photochemical retinal hazards, 294
thermal retinal hazards, 294
Threshold Limit Values, 293
Visual receptors, 183
Vitamin D, 189
Vitreous body, 182

Wave equation, 3
Wave vector, 94
Wave-particle duality, 4
Wavelength, 4
Welders' flash, 161
Welding arcs
arc current and voltage, 160
arc ignition, 161
cutting processes, 158
electrode, 160
exhaust/ventilation, 161
gas tungsten arc welding, 158
hazard evaluation, 158
industrial processes, 158
infrared radiation, 165
IR radiation, 160
light, 159, 161
manual metal arc, 157
measurement methods, 158
metal active gas, 157
metal inert gas, 157
shielded metal, 157
shielded metal arc welding, 157
shielding gas, 160
spraying processes, 158
tungsten inert gas, 158
typical data, 164
UV radiation, 159, 161
Welding filters
active filters, 283
blue light damage, 282
passive filters, 281
shade number, 282
spectral transmittance, 282
standard for passive filters, 281
Welding safety measures
activity coupled to welding, 281
contact lens wear, 283
damage from sparks, 278
eye damage, 278
eye protection, 281
infrared radiation, 279
photochemical damage, 278
photomechanical disruption, 278
radiation damage, 278
spectral radiance, 279
thermal damage, 278
toxic vapours, 281
ultraviolet radiation, 280
visual radiation, 280
Welding sparks, 277
Wien's displacement, 17
Wiggler wavelength shifter, 73
Wigglers, 71
deflection parameter, 71
World Health Organization, 319, 341
Zenith angle, 171
Zinn's zonule, 182

GPSR Compliance

The European Union's (EU) General Product Safety Regulation (GPSR) is a set of rules that requires consumer products to be safe and our obligations to ensure this.

If you have any concerns about our products, you can contact us on ProductSafety@springernature.com

In case Publisher is established outside the EU, the EU authorized representative is:

Springer Nature Customer Service Center GmbH
Europaplatz 3
69115 Heidelberg, Germany

Batch number: 10761599

Printed by Printforce, the Netherlands